The mechanics of earthquakes and faulting

CHRISTOPHER H. SCHOLZ

LAMONT–DOHERTY GEOLOGICAL OBSERVATORY
AND DEPARTMENT OF GEOLOGICAL SCIENCES,
COLUMBIA UNIVERSITY

CAMBRIDGE
UNIVERSITY PRESS

PUBLISHED BY THE PRESS SYNDICATE OF THE UNIVERSITY OF CAMBRIDGE
The Pitt Building, Trumpington Street, Cambridge, United Kingdom

CAMBRIDGE UNIVERSITY PRESS
The Edinburgh Building, Cambridge CB2 2RU, UK   http://www.cup.cam.ac.uk
40 West 20th Street, New York, NY 10011-4211, USA   http://www.cup.org
10 Stamford Road, Oakleigh, Melbourne 3166, Australia
Ruiz de Alarcón 13, 28014 Madrid, Spain

First published 1990
First paperback edition 1990
Reprinted 1992, 1994, 1997, 2000

Printed in the United States of America

Typeset in Times

A catalog record for this book is available from the British Library

Library of Congress Cataloging in Publication Data is available

ISBN 0 521 40760 5 paprback

...and then there would not be friction any more, and the sound would cease, and the dancers would stop...

Leonardo da Vinci
From a notebook dated September, 1508
MacCurdy, E. 1958. *The Notebooks of Leonardo da Vinci*, p. 282, New York: George Braziller.

Contents

5   The Seismic Cycle

6   Seismotectonics

Preface

It has now been more than thirty years since the publication of E. M. Anderson's *The Dynamics of Faulting* and C. F. Richter's *Elementary Seismology*. Several generations of earth scientists were raised on these texts. Although these books are still well worth reading today for their excellent descriptions of faults and earthquakes, the mechanical principles they espoused are now well understood by the undergraduate student at the second or third year. In the meantime a great deal has been learned about these subjects, and the two topics, faulting and earthquakes, described in those books have merged into one broader field, as earthquakes have been more clearly understood to be one manifestation of faulting. During this period of rapid progress there has not been a single book written that adequately fills the gap left by these two classics. As a result it has become increasingly difficult for the student or active researcher in this area to obtain an overall grasp of the subject that is both up-to-date and comprehensive and that is based firmly on fundamental mechanical principles. This book has been written to fill this need.

Not least among the difficulties facing the researcher in this field is the interdisciplinary nature of the subject. For historical reasons earthquakes are considered to be the province of the seismologist and the study of faults is that of the geologist. However, because earthquakes are a result of an instability in faulting that is so pervasive that on many faults most slip occurs during them, the interests of these two disciplines must necessarily become intertwined. Moreover, when considering the mechanics of these processes the rock mechanicist also becomes involved, because the natural phenomena are a consequence of the material properties of the rock and its surfaces.

It is a consequence of the way in which science is organized that the scientist is trained by discipline, not by topic, and so interdisciplinary subjects such as this one tend to be attacked in a piecemeal fashion from the vantage of the different specialties that find application in studying it. This is disadvantageous because progress is hindered by lack of communication between the different disciplines, misunderstandings can abound, and different, sometimes conflicting schools of thought can flourish in the relative isolation of separate fields. Workers in one field may be ignorant of relevant facts established in another, or, more likely, be unaware of the skein of evidence that weights the convictions of workers in another field. This leads not only to a neglect of some aspects in considering a question, but also to the quoting of results attributed to another field with greater confidence than workers in that field would themselves maintain. It is not enough to be aware, second-hand, of the contributions of another field – one must know the basis, within the internal structure of the evidence and tools of that field, upon which that result is maintained. Only then is one in a position to take the results of all the disciplines and place them, with their proper weight, in the correct position of the overall jigsaw puzzle. Because the literature on this topic has become both large and diverse, a guide is useful in this process, together with some unifying mechanical principles that allow the contours of the forest to be seen from between the trees.

Although I have dabbled, to one degree or another, in the various different disciplinary approaches to this problem and therefore have a rudimentary working knowledge of them, my own specialty is rock mechanics, and so this approach is the one most emphasized in this book. Faults are treated as shear cracks, the propagation of which may be understood through the application of fracture mechanics. The stability of this fault movement, which determines whether the faulting is seismic or aseismic, is determined by the frictional constitutive law of the fault surface, and so that is the second major theme applied throughout this treatment. The application of these principles to geology is not straightforward. One cannot actually do a laboratory experiment that duplicates natural conditions. Laboratory studies can only be used to establish physical processes and validate theories. To apply the results of this work to natural phenomena requires a conceptual jump, because of problems of scale and because both the nature of the materials and the physical conditions are not well known. In order to do this one must have constant recourse to geological and geophysical observations and, working backwards, through these physical principles determine the underlying cause of the behavior of faults. For this reason, much of this book is taken up in describing observations of natural cases.

Because rock mechanics is not taught universally in earth science curricula, the first two chapters present an account of brittle fracture and friction of rock, beginning from first principles. These chapters provide the basis for the later discussion of geological phenomena. The subsequent chapters assume a beginning graduate level understanding of the earth science disciplines involved. In these chapters the results of geology, seismology, and geodesy are presented, but the techniques employed by the various specialties are not described at any length. The emphasis is on providing an overall understanding of a scientific topic rather than teaching a specific craft. A goal was to describe each topic accurately, but at such a level that it could be understood by workers in other fields.

A book may be structured in many different ways. In this case, I found it difficult to choose between organizing the book around the physical mechanisms or around the natural phenomena in which they are manifested. The latter scheme would be more familiar to the earth scientist, the former to the mechanicist. Ultimately, I adopted a system arranged around mechanics, but which still retains many of the more familiar traditional associations. Because some mechanisms are important in a number of different phenomena, which might otherwise be considered quite distant, and some earthquakes provide example of several phenomena, there are often more than two connections to other topics. Therefore, it was not always possible to present the subject matter in a serial sequence. I consequently adopted a system of cross-referencing that allows the reader to traverse the book in alternative paths. I hope this system will be more helpful than confusing.

When I first entered graduate school twenty-five years ago, most of the material described in this book was not yet known. The first generation of understanding, outlined in Anderson's and Richter's books, has been augmented by a second generation of mechanics, much more thorough and quantitative than the preceding. This has been a most productive era, which this book celebrates. I owe my own development to associations with many people. My first mentor, W. F. Brace, set me on this path, and the way has been lit by many others since. I have also been a beneficiary of an enlightened system of scientific funding during this period, which has allowed me to pursue many interesting topics, often at no little expense. For this I particularly would like to thank the National Science Foundation, the U.S. Geological Survey, and NASA.

Many have helped in the preparation of this book. In particular I acknowledge the assistance of my editor, Peter-John Leone, Kazuko Nagao, who produced many of the illustrations, and those who have reviewed various parts of the manuscript: T.-F. Wong, W. Means, J. Logan, S. Das, P. Molnar, J. Boatwright, L. Sykes, D. Simpson, and C. Sammis. Particular thanks are due to T. C. Hanks, who offered many helpful comments on the text, and who, over the course of a twenty-year association, has not failed to point out my foibles. I dedicate the book to my wife, Yoshiko, who provided me with the stability in my personal life necessary for carrying out this task.

Acknowledgements

In addition to many colleagues who have allowed me to reproduce their graphical material herein, I would like to thank the following copyright holders who have graciously permitted the reproduction of material in this book.

ACADEMIC PRESS JAPAN, INC.

From Mogi, K. (1985) *Earthquake Prediction*: Figure 7.6, his Figure 4.11; Figure 7.7, his Figure 14.11; Figure 7.13, his Figure 13.16; Figure 7.15, his Figure 15.12.

AMERICAN ASSOCIATION FOR THE ADVANCEMENT OF SCIENCES

From *Science*: Figures 2.23 and 2.30, Raleigh et al. (1976) 191: 1230–7; Figures 2.11 and 2.33, Shimamoto (1986) 231: 711–4; Figures 7.21 and 7.27, Scholz et al. (1973) 181: 803–9; Figure 7.20, Scholz (1978) 201: 441–2.

AMERICAN GEOPHYSICAL UNION

From *Journal of Geophysical Research*: *Figure 2.2, Brown and Scholz* (1985) 90: 5531; *Figure 2.3b, Brown and Scholz* (1985) 90: 12575; *Figures 2.22 and 5.12, Tse and Rice* (1986) 91: 9452; *Figures 2.26 and 2.27, Scholz et al.* (1972) 77: 6392; *Figures 3.21 and 3.22, Scholz et al.* (1979) 84: 6770; *Figure 4.16, Scholz et al.* (1969) 74: 2049; *Figures 4.33 and 7.16, Das and Scholz* (1981) 86: 6039; *Figures 5.3 and 5.7, Fitch and Scholz* (1971) 76: 7260; *Figure 5.9, Scholz and Kato* (1978) 83: 783; *Figure 5.11, Li and Rice* (1987) 92: 11533; *Figure 5.27, Schwartz and Coppersmith* (1984) 89: 5681; *Figure 6.7, Scholz* (1980) 85: 6174; *Figure 7.28, Wesnousky et al.* (1983) 88: 9331.
From *Earthquake Prediction, an International Review*. M. Ewing Ser. 4 (1981). *Figure 4.28, Einarsson et al., p.* 141; *Figure 5.28, Sykes et al., p.* 1784; *Figures 7.2 and 7.3, Ohtake et al., p.* 53.
From *Earthquake Source Mechanics. AGU Geophys. Mono. 37* (1986): Figure 2.28, Ohynaka et al., p. 13; Figure 3.33, Sibson, p. 157; Figure 4.12, Shimazaki, p. 209; Figures 7.17 and 7.18, Dieterich, p. 37.
From *Geophysical Research Letters*: Figure 3.27, Power et al. (1987) 14: 29; Figure 3.34, Scholz (1986) 12: 717; Figure 4.8, Aki (1967) 72:

1217; Figure 4.30, Hudnut et al. (1989) 16: 199; Figures 5.13 and 5.26, Shimazaki and Nakata (1980) 7: 279.

From *Reviews of Geophysics and Space Physics*: *Figure* 4.23, *Whitcomb et al.* (1973) 11: 693.

From *Tectonics*: Figure 6.4, Yeats and Berryman (1987) 6: 363; Figure 6.6, Byrne et al. (1988) 7: 833.

AMERICAN ASSOCIATION OF PETROLEUM GEOLOGISTS

From *AAPG Bulletin*: Figure 3.9b, Barnett et al. (1987) 71: 925–37.

ANNUAL REVIEWS, INC.

From *Annual Reviews of Earth and Planetary Sciences*: *Figures* 1.16, 3.26, 6.13, 6.20, *and* 6.21, *Scholz* (1989) 17: 309–34; *Figure* 5.2, *Mavko* (1981) 9: 81–111; *Figure* 6.18, *Kanamori* (1986) 14: 293–322; *Figure* 6.25, *Simpson* (1986) 14: 21–42.

BIRKHÄUSER VERLAG

From *Pageoph*: Figure 2.7, Byerlee (1978) 116: 625; Figure 4.10, Hanks (1977) 115: 441; Figures 5.32 and 7.1, Scholz (1988) 126: 701; Figure 7.9, Sato (1988) 126: 465; Figure 7.10, Roeloffs (1988) 126: 177; Figure 7.11, Wakita (1988) 126: 267; Figures 7.22 and 7.23, Rudnicki (1988) 126: 531; Figures 7.24, 7.25, and 7.26, Stuart (1988) 126: 619.

ELSEVIER SCIENCE PUBLISHERS

From *Tectonophysics*: Figure 6.11, Uyeda (1982) 81: 133.

GEOLOGICAL SOCIETY OF AMERICA

From *GSA Bulletin*: *Figure* 3.31, *Tchalenka and Berberian* (1975) 86: 703; *Figure* 5.22, *Sieh and Jahns* (1984) 95: 883.

From *Geology*: Figures 2.13 and 3.11, Scholz (1987) 15: 493; Figure 3.32, Sibson (1987) 15: 701; Figure 3.17, Simpson (1984) 12: 8.

MACMILLAN MAGAZINES, LTD.

From *Nature*: Figure 3.12, Wesnousky (1988) 335: 340; Figure 2.31, Scholz (1988) 336: 761.

PERGAMON PRESS

From *Journal of Structural Geology*: *Figure* 3.6, *Cox and Scholz* (1988) 10: 413.

From *International Journal of Rock Mechanics and Mineral Science*: Figure 2.19, Scholz and Engelder (1976) 13: 149.

ROYAL ASTRONOMICAL SOCIETY

Figure 4.24, Berberian (1982) 68: 499.

ROYAL SOCIETY OF NEW ZEALAND

From *Royal Society of New Zealand Bulletin 24* (1986): Figure 5.1, Thatcher, p. 517; Figure 6.5, Bibby et al. p. 427.

SEISMOLOGICAL SOCIETY OF AMERICA

From *Bulletin of the Seismological Society of America*: Figure 4.4, *Andrews* (1985) 75: 1; *Figure 4.17, Eaton et al.* (1970) 60: 1151; *Figure 5.34, Nishenko and Buland* (1987) 77: 1382.

UTAH GEOLOGICAL AND MINERAL SURVEY

From *Utah Geological and Mineral Survey Special Studies 62* (1983): Figure 5.24, Schwartz et al., p. 45.

YORKSHIRE GEOLOGICAL SOCIETY

From *Proceedings of the Yorkshire Geological Society*: Figure 3.9a, Rippon (1985) 45: 147.

A listing is given of the most important symbols in alphabetical order, first in the Latin, then the Greek alphabets. The point of first appearance is given in brackets, which refers to an equation unless otherwise indicated. In some cases the same symbol is used for different meanings, and vice versa, as indicated, but the meaning is clear within the context used. Arbitrary constants and very common usages are not listed.

a	atomic spacing [1.1]; crack radius [4.24]
$a(\mathrm{H_2O})$	chemical activity of water [1.49]
$\mathbf{a}_i$	direct friction velocity parameter(s) [2.27]
$(\mathbf{a} - \mathbf{b})$	combined friction velocity parameter [2.28]
A	area [Sec. 2.1.2]
A_r	real area of contact [2.1]
$\mathscr{A}_s$	age of subducted slab [6.9]
$\mathbf{b}_i$	steady-state friction velocity parameter(s) [2.27]
B	exponent in the earthquake size distribution [4.31]
$\mathscr{B}$	Skempton's coefficient [6.11]
c	crack length [1.5]
C_0	uniaxial compressive strength [1.34]
d	contact diameter [2.18]
d_e	effective working distance of contact [2.19]
d_s	offset of jog (or step) [Fig. 3.26]
d_0	critical slip distance (slip-weakening model) [4.14]
D	total fault slip [2.21]; asthenospheric diffusivity [5.1]
E	Young's modulus [1.2]
$\underline{E}$	effective modulus [1.8]
E^*	activation energy [1.49]
E_s	seismic energy [3.5]
$f_{ij}(\theta)$	stress function [1.20]
$f_i(\theta)$	displacement function [1.20]
f_0	corner frequency [Sec. 4.3.1]
F	shear force [2.2]
$\mathscr{F}_i$	plate interface normal force [6.10]
g	acceleration of gravity [3.2]
G	energy release rate [1.21]
$\mathscr{G}$	shear modulus [6.11]

$\mathbf{G}_c$	critical energy release rate (fracture energy) [1.24]
h	hardness parameter [2.18]
H	lithospheric thickness [Fig. 5.10]
k, K	stiffness [5.3], [2.25]
K_n	stress intensity factor (mode) [1.20]
$\mathbf{K}_c$	critical stress intensity factor (fracture toughness) [1.24]
$\mathbf{K}_0$	stress-corrosion limit [Fig. 1.19]
K_s	static stress intensity factor [Sec. 4.5.2]
K_d	dynamic stress intensity factor [Sec. 4.5.2]
ℓ	slip zone length [2.30]
ℓ_c	critical slip zone length [2.31]
L	length of rupture [Sec. 4.3.2]
L_c	critical crack length [4.13]
$\mathscr{L}_i$	critical slip distance(s) [2.26]
m	mass [2.33]
M	magnitude [Sec. 4.3.1]
M_s	surface wave magnitude [4.26]
M_w	moment magnitude [Sec. 4.3.1]
M_w'	equivalent moment magnitude [6.9]
M_0, M_{0ij}	seismic moment [4.25]
$\dot{M}_0^s$	seismic moment release rate [6.5]
$\dot{M}_0^g$	geologic moment release rate [6.5]
n	stress-corrosion index [1.46]
N	normal force [2.1]
$\mathscr{N}$	number of contact junctions [2.19]
p	pressure [2.37]; pore pressure [1.43]; penetration hardness [2.1]
Δp_p	change in pore pressure [6.11]
q	heat flow [3.6]
Q	heat [3.5]
R	gas constant [1.49]
s	shear strength [2.2]
S	dynamic strength parameter [4.20]
t	time [4.21]
t_h	healing time [4.21]
t_r	rise time [2.36]
$\langle t \rangle$	mean fracture time [1.50]
T	temperature [1.49]; thickness of gouge layer [2.22]; earthquake recurrence time [Sec. 5.2.2]
T_{exp}	expected recurrence time [7.6]
T_{ave}	average recurrence time [Sec. 7.4.3]
T_0	uniaxial tensile strength [1.29]
$\mathbf{T}_1$	lower stability transition [Sec. 3.4.1]
$\mathbf{T}_2$	semibrittle–plastic transition [Sec. 3.4.1]

$\mathbf{T_3}$	schizosphere–plastosphere boundary [Sec. 3.4.1]
$\mathbf{T_4}$	upper stability transition [Sec. 3.4.1]
$u,\ u_i$	displacement [1.20]
u_p	afterslip [6.6]
$\underline{\Delta u_i}$	slip in earthquake (offset) [4.25]
$\overline{\Delta u}$	mean slip in earthquake [4.5]
U	total energy [1.6]
U_e	internal strain energy [1.6]
U_s	surface energy [1.6]
U_k	kinetic energy [4.1]
U_f	frictional work [4.1]
v	crack-tip velocity [1.46]
v	particle velocity [2.35]
v_{max}	maximum particle velocity [4.22]
v_0	asymptotic particle velocity [4.23]
v_{pl}	remote plate velocity [Sec. 5.2.2]
v_c	plate convergence rate [6.9]
V_P	P wave velocity [Sec. 7.2.2]
V_S	S wave velocity [Sec. 7.2.2]
v_r	rupture velocity [4.19]
V	volume of wear fragments [2.20]; slip velocity [2.26]; volume [4.3]
V_c	coseismic slip velocity [6.8]
W	work [1.6]; width of fault (or rupture) [Sec. 3.2.2]
W_{fr}	frictional work [Sec. 3.2.2]
W_f	work of faulting [3.5]
Z	thickness of schizosphere [Fig. 5.10]
β	shear wave velocity [4.16]
γ	specific surface energy [1.4]
Γ	Irwin's energy dissipation factor [1.27]
δ	joint closure [2.7]
δ_{ij}	Kronecker delta [1.43]
ε_{ij}	strain [Sec. 6.2.2]
$\Delta\varepsilon_v$	volume strain change [6.11]
η	seismic efficiency [4.8]; asthenospheric viscosity [Sec. 5.2.2]
θ	dilatant strain [2.24]
θ_s	angle of jog (or step) [Fig. 3.26]
κ	wear coefficient [2.22]
λ_c	critical aperture wavelength [2.39]
μ	friction coefficient [1.30]; coefficient of internal friction [1.31]; shear modulus [4.9]
μ_0	base friction coefficient [2.27]

μ^{ss}	steady-state friction coefficient [2.28]
μ_s	static coefficient of friction [2.26]
μ_d	dynamic coefficient of friction [2.26]
$\Delta\mu$	change in friction [2.29]
ν	Poisson's ratio [1.23]
ρ	density [3.2]; radius of curvature [2.10]
$\sigma,\ \sigma_{ij}$	stress [1.1]
σ_t	theoretical strength [1.1]
σ_f	Griffith strength [1.12]
σ_n	normal stress [1.30]
$\bar{\sigma}_{ij}$	effective stress [1.43]
σ_c	mean contact stress [2.15]
σ_1	initial stress [4.6]
σ_2	final stress [4.6]
σ_f	frictional stress [4.6]
σ_y	yield stress [4.20]
σ_D	intrinsic standard deviation [7.5]
$\Delta\sigma$	static stress drop [4.7]
$\Delta\sigma_d$	dynamic stress drop [Sec. 4.2.1]
τ	shear stress [1.30]; asthenospheric relaxation time [Sec. 5.2.2]
τ_0	cohesion [1.31]
τ_c	coseismic stress jump [6.7]
Φ	friction characteristic [5.3]
χ	seismic coupling coefficient [6.5]
ψ_i	friction state variable(s) [2.27]
ω_{ij}	rotation [Sec. 6.2.2]

Cover photograph: A photograph by G. K. Gilbert of the surface rupture produced by the 1906 San Francisco earthquake. (Photo courtesy of the U.S. Geological Survey.)

1   Brittle Fracture of Rock

Under the low temperature and pressure conditions of the Earth's upper lithosphere, silicate rock responds to large strains by brittle fracture. The mechanism of brittle behavior is by the propagation of cracks, which may occur on all scales. We begin by studying this form of deformation, which is fundamental to the topics that follow.

1.1   THEORETICAL CONCEPTS

1.1.1   *Historical*

Understanding the basic strength properties of rock has been a practical pursuit since ancient times, both because of the importance of mining and because rock was the principal building material. The crafting of stone tools required an intuitive grasp of crack propagation, and mining, quarrying, and sculpture are trades that require an intimate knowledge of the mechanical properties of rock. The layout and excavation of quarries, for example, is a centuries-old art that relies on the recognition and exploitation of preferred splitting directions in order to maximize efficiency and yield. One of the principal properties of brittle solids is that their strength in tension is much less than their strength in compression. This led, in architecture, to the development of fully compressional structures through the use of arches, domes, and flying buttresses.

Rock was one of the first materials for which strength was studied with scientific scrutiny because of its early importance as an engineering material. By the end of the nineteenth century the macroscopic phenomenology of rock fracture had been put on a scientific basis. Experimentation had been conducted over a variety of conditions up to moderate confining pressures. The Coulomb criterion and the Mohr circle analysis had been developed and applied to rock fracture with sufficient success that they remain the principal tools used to describe this process for many engineering and geological applications.

The modern theory of brittle fracture arose as a solution to a crisis in understanding the strength of materials, brought about by the atomic

1

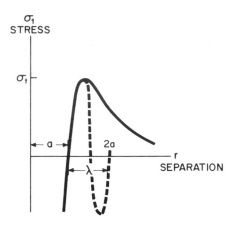

Fig. 1.1 Sketch of an anharmonic model of interatomic forces, showing the relationship between stress and atomic separation (solid curve) and a sinusoidal approximation (dashed curve).

theory of matter. In simplest terms, strength can be viewed as the maximum stress that a material can support under given conditions. Fracture (or flow) must involve the breaking of atomic bonds. An estimate of the *theoretical strength* of a solid is therefore the stress required to break the bonds across a lattice plane.

Consider a simple anharmonic model for the forces between atoms in a solid, as in Figure 1.1, in which an applied tension σ produces an increase in atomic separation r from an equilibrium spacing a (Orowan, 1949). Because we need only consider the prepeak region, we can approximate the stress–displacement relationship with a sinusoid,

$$\sigma = \sigma_t \sin\left(\frac{2\pi(r-a)}{\lambda}\right) \tag{1.1}$$

For small displacements, when $r \approx a$, then

$$\frac{d\sigma}{d(r-a)} = \frac{E}{a} = \frac{2\pi}{\lambda}\sigma_t \cos\left(\frac{2\pi(r-a)}{\lambda}\right) \tag{1.2}$$

but because $(r-a)/\lambda \ll 1$, the cosine is equal to 1, and

$$\sigma_t = \frac{E\lambda}{2\pi a} \tag{1.3}$$

where E is Young's modulus. When $r = 3a/2$, the atoms are midway between two equilibrium positions, so by symmetry, $\sigma = 0$ there and $a \approx \lambda$. The theoretical strength is thus about $E/2\pi$. The work done in separating the planes by $\lambda/2$ is the specific surface energy γ, the energy per unit area required to break the bonds, so

$$2\gamma = \int_0^{\lambda/2} \sigma_t \sin\left(\frac{2\pi(r-a)}{\lambda}\right) d(r-a)$$

$$= \frac{\lambda\sigma_t}{\pi} \tag{1.4}$$

which, with $\sigma_t \approx E/2\pi$, yields the estimate $\gamma \approx Ea/4\pi^2$.

The value of the theoretical strength from this estimate is 5–10 GPa, several orders of magnitude greater than the strength of real materials. This discrepancy was explained by the postulation and later recognition that all real materials contain defects. Two types of defects are important: cracks, which are surface defects; and dislocations, which are line defects. Both types of defects may propagate in response to an applied stress and produce yielding in the material. This will occur at applied stresses much lower than the theoretical strength, because both mechanisms require that the theoretical strength be achieved only locally within a *stress concentration* deriving from the defect. The two mechanisms result in grossly different macroscopic behavior. When cracks are the active defect, material failure occurs by its separation into parts, often catastrophically: this is brittle behavior. Plastic flow results from dislocation propagation, which produces permanent deformation without destruction of the lattice integrity.

These two processes tend to be mutually inhibiting, but not exclusive, so that the behavior of crystalline solids usually can be classed as brittle or ductile, although mixed behavior, known as semibrittle, may be more prevalent than commonly supposed. Because the lithosphere consists of two parts with markedly different rheological properties, one brittle and the other ductile, it is convenient to introduce two new terms to describe them. These are *schizosphere* (literally, the broken part) for the brittle region, and *plastosphere* (literally, the moldable part) for the ductile region. In this book we will assume, for the most part, that we are dealing with purely brittle processes, so that we will be concerned principally with the behavior of the schizosphere.

1.1.2 *Griffith theory*

All modern theories of strength recognize, either implicitly or explicitly, that real materials contain imperfections that, because of the stress concentrations they produce within the body, result in failure at much lower stresses than the theoretical strength. A simple example, Figure 1.2a, is a hole within a plate loaded with a uniform tensile stress σ_∞. It can be shown from elasticity theory that at the top and bottom of the

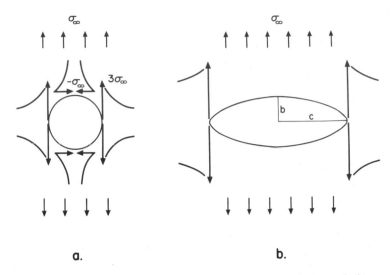

Fig. 1.2 Stress concentration around (a) a circular hole, and (b) an elliptical hole in a plate subjected to a uniform tension σ_∞.

hole a compressive stress of magnitude $-\sigma_\infty$ exists and that at its left and right edges there will be tensile stresses of magnitude $3\sigma_\infty$. These stress concentrations arise from the lack of load-bearing capacity of the hole and their magnitudes are determined solely by the geometry of the hole and not by its size. If the hole is elliptical, as in Figure 1.2b, with semiaxes b and c, with $c > b$, the stress concentration at the ends of the ellipse increases proportionally to c/b, according to the approximate formula

$$\sigma \approx \sigma_\infty(1 + 2c/b)$$

or   (1.5)

$$\sigma \approx \sigma_\infty\left(1 + 2(c/\rho)^{1/2}\right) \approx 2\sigma_\infty(c/\rho)^{1/2}$$

for $c \gg b$, where ρ is the radius of curvature at that point. It is clear that for a long narrow crack the theoretical strength can be attained at the crack tip when $\sigma_\infty \ll \sigma_t$. Because Equation 1.5 indicates that the stress concentration will increase as the crack lengthens, crack growth can lead to a dynamic instability.

Griffith (1920, 1924) posed this problem at a more fundamental level, in the form of an energy balance for crack propagation. The system he considered is shown in Figure 1.3a and consists of an elastic body that contains a crack of length $2c$, which is loaded by forces on its external boundary. If the crack extends an increment δc, work W will be done by the external forces and there will be a change in the internal strain energy U_e. There will also be an expenditure of energy in creating the

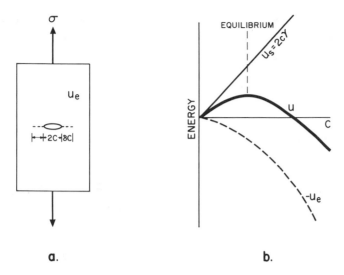

Fig. 1.3 Griffith's model for a crack propagating in a rod (a), and the energy partition for the process (b).

new surfaces U_s. Thus the total energy of the system, U, for a static crack, will be

$$U = (-W + U_e) + U_s \qquad (1.6)$$

The combined term in parentheses is referred to as the mechanical energy. It is clear that, if the cohesion between the incremental extension surfaces δc were removed, the crack would accelerate outward to a new lower energy configuration: Thus, mechanical energy must decrease with crack extension. The surface energy, however, will increase with crack extension, because work must be done against the cohesion forces in creating the new surface area. There are two competing influences; for the crack to extend there must be reduction of the total energy of the system, and hence at equilibrium there is a balance between them. The condition for equilibrium is

$$dU/dc = 0 \qquad (1.7)$$

Griffith analyzed the case of a rod under uniform tension. A rod of length y, modulus E, and unit cross section loaded under a uniform tension σ will have strain energy $U_e = y\sigma^2/2E$. If a crack of length $2c$ is introduced into the rod, it can be shown that the strain energy will increase an amount $\pi c^2 \sigma^2/E$, so that U_e becomes

$$U_e = \sigma^2(y + 2\pi c^2)/2E \qquad (1.8)$$

The rod becomes more compliant with the crack, with an effective modulus $\underline{E} = yE/(y + 2\pi c^2)$. The work done in introducing the crack is

$$W = \sigma y(\sigma/\underline{E} - \sigma/E) = 2\pi\sigma^2 c^2/E \qquad (1.9)$$

and the surface energy change is

$$U_s = 4c\gamma \qquad (1.10)$$

Substituting Equations 1.8–1.10 into Equation 1.6 gives,

$$U = -\pi c^2 \sigma^2/E + 4c\gamma \qquad (1.11)$$

and applying the condition for equilibrium (Eq. 1.7), we obtain an expression for the critical stress at which a suitably oriented crack will be at equilibrium,

$$\sigma_f = (2E\gamma/\pi c)^{1/2} \qquad (1.12)$$

The energies of the system are shown in Figure 1.3b, from which it can be seen that Equation 1.12 defines a position of unstable equilibrium: When this condition is met the crack will propagate without limit, causing macroscopic failure of the body.

Griffith experimentally tested his theory by measuring the breaking strength of glass rods that had been notched to various depths. He obtained an experimental result with the form of Equation 1.12 from which he was able to extract an estimate of γ. He obtained an independent estimate of γ by measuring the work necessary to pull the rods apart by necking at elevated temperatures. By extrapolating this result to room temperature, he obtained a value that was within reasonable agreement with that derived from the strength tests.

Griffith's result stems strictly from a consideration of thermodynamic equilibrium. Returning to our original argument, we may ask if the theoretical strength is reached at the crack tip when the Griffith condition is met: That is, is the stress actually high enough to break the bonds? This question was posed by Orowan (1949), who considered the stress at the tip of an atomically narrow crack, as described before. Combining Equations 1.3 and 1.4, we obtain

$$\sigma_t = (E\gamma/a)^{1/2} \qquad (1.13)$$

This stress will exist at the ends of a crack of length $2c$ when the macroscopic applied stress σ_f is (Eq. 1.5)

$$\sigma_t = 2\sigma_f(c/a)^{1/2} \qquad (1.14)$$

so that

$$\sigma_f = (E\gamma/4c)^{1/2} \qquad (1.15)$$

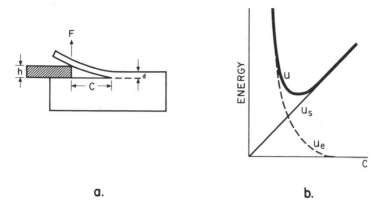

a.   b.

Fig. 1.4 The configuration of Obriemoff's mica cleaving experiment (a), and the energy partition for this process (b).

which is very close to Equation 1.12. The close correspondence of these two results demonstrates both necessary and sufficient conditions for crack propagation. Griffith's thermodynamic treatment shows the condition for which the crack is energetically favored to propagate, while Orowan's calculation shows the condition in which the crack-tip stresses are sufficient to break atomic bonds. For a typical value of $\gamma \approx Ea/30$ (Eq. 1.4), commonly observed values of strength of $E/500$ can be explained by the presence of cracks of length $c \approx 1 \ \mu$m. Prior to the advent of the electron microscope, the ubiquitous presence of such microscopic cracks was hypothetical, and this status was conferred upon them with the use of the term *Griffith crack*.

Griffith's formulation has an implicit instability as a consequence of the constant stress boundary condition. In contrast, the experiment of Obriemoff (1930) leads to a stable crack configuration. Obriemoff measured the cleavage strength of mica by driving a wedge into a mica book using the configuration shown in Figure 1.4a. In this experiment the boundary condition is one of constant displacement. Because the wedge can be considered to be rigid, the bending force F undergoes no displacement and the external work done on the system is simply

$$W = 0 \qquad (1.16)$$

From elementary beam theory, the strain energy in the bent flake is

$$U_e = Ed^3h^2/8c^3 \qquad (1.17)$$

and, using $U_s = 2c\gamma$ and the condition $dU/dc = 0$, we obtain the equilibrium crack length

$$c = \left(3Ed^3h^2/16\gamma\right)^{1/4} \qquad (1.18)$$

The energies involved in this system are shown in Figure 1.4b. It is clear that in this case the crack is in a state of stable equilibrium; it advances the same distance that the wedge is advanced. This example shows that the stability is controlled by the system response, rather than being a material property, a point that will be taken up in greater detail in the discussion of frictional instabilities in Section 2.3. In this case the loading system may be said to be infinitely stiff, and crack growth is controlled and stable. Griffith's experiment, on the other hand, had a system of zero stiffness and the crack was unstable. Most real systems, however, involve loading systems with finite stiffness so that the stability has to be evaluated by balancing the rate at which work is done by the loading system against the energy absorbed by crack propagation.

Obriemoff noticed that the cracks in his experiment did not achieve their equilibrium length instantly, but that on insertion of the wedge they jumped forward and then gradually crept to their final length. When he conducted the experiment in vacuum, however, he did not observe this transient effect. Furthermore, the surface energy that he measured in vacuum was about 10 times the surface energy measured in ambient atmosphere. He was thus the first to observe the important effect of the chemical environment on the weakening of brittle solids and the *subcritical crack growth* that results from this effect. This effect is very important in brittle processes in rock and will be discussed in more detail in Section 1.3.2.

1.1.3 *Fracture mechanics*

Linear elastic fracture mechanics is an approach that has its roots in the Griffith energy balance, but that lends itself more readily to the solution of general crack problems. It is a continuum mechanics approach in which the crack is idealized as a mathematically flat and narrow slit in a linear elastic medium. It consists of analyzing the stress field around the crack and then formulating a fracture criterion based on certain critical parameters of the stress field. The macroscopic strength is thus related to the intrinsic strength of the material through the relationship between the applied stresses and the crack-tip stresses. Because the crack is treated as residing in a continuum, the details of the deformation and fracturing processes at the crack tip are ignored.

The displacement field of cracks can be categorized into three modes (Fig. 1.5). Mode I is the tensile, or opening, mode in which the crack wall displacements are normal to the crack. There are two shear modes:

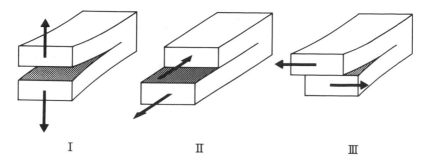

Fig. 1.5 The three crack propagation modes.

in-plane shear, Mode II, in which the displacements are in the plane of the crack and normal to the crack edge; and antiplane shear, Mode III, in which the displacements are in the plane of the crack and parallel to the edge. The latter are analogous to edge and screw dislocations, respectively.

If the crack is assumed to be planar and perfectly sharp, with no cohesion between the crack walls, then the near-field approximations to the crack-tip stress and displacement fields may be reduced to the simple analytic expressions:

$$\sigma_{ij} = K_n (2\pi r)^{-1/2} f_{ij}(\theta) \qquad (1.19)$$

and

$$u_i = (K_n/2E)(r/2\pi)^{1/2} f_i(\theta) \qquad (1.20)$$

where r is the distance from the crack tip and θ is the angle measured from the crack plane, as shown in Figure 1.6. K_n is called the *stress intensity factor* and depends on mode, that is, K_I, K_{II}, and K_{III}, refer to the three corresponding crack modes. The functions $f_{ij}(\theta)$ and $f_i(\theta)$ can be found in standard references (e.g., Lawn and Wilshaw, 1975), and are illustrated in Figure 1.6. The stress intensity factors depend on the geometry and magnitudes of the applied loads and determine the intensity of the crack-tip stress field. They also can be found tabulated, for common geometries, in standard references (e.g., Tada, Paris, and Irwin, 1973). The other terms describe only the distribution of the fields.

In order to relate this to the Griffith energy balance it is convenient to define an *energy release rate*, or *crack extension force*,

$$\mathbf{G} = -d(-W + U_e)/dc \qquad (1.21)$$

which can be related to K by (Lawn and Wilshaw, 1975, page 56)

$$\mathbf{G} = K^2/E \qquad (1.22)$$

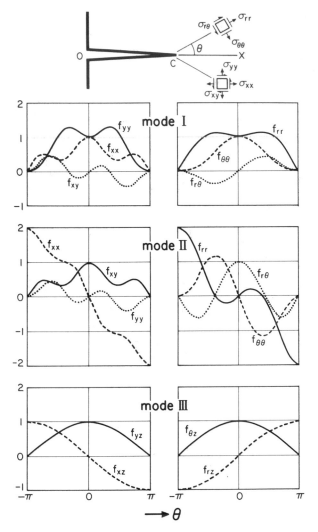

Fig. 1.6 The stress functions near the tips of the three modes of cracks, using both Cartesian and cylindrical coordinates, as shown in the geometrical key. (After Lawn and Wilshaw, 1975.)

for plane stress, or

$$\mathbf{G} = K^2(1 - \nu^2)/E \tag{1.23}$$

for plane strain (ν is Poisson's ratio). In Mode III, the right-hand sides of the corresponding expressions must be multiplied by $(1 + \nu)$ for plane stress and divided by $(1 - \nu)$ for plane strain, respectively. From Equations 1.6 and 1.7, it is clear that the condition for crack propagation will

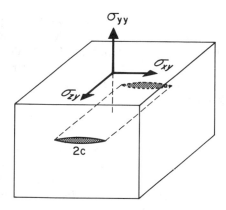

Fig. 1.7 Geometry of a crack in a uniform stress field.

be met when

$$G_c = K_c^2/E = 2\gamma \tag{1.24}$$

for plane stress, with a corresponding expression for plane strain. Thus K_c, the *critical stress intensity factor*, and G_c are material properties that, because they can be related to the applied stresses through a stress analysis, provide powerful and general failure criteria. K_c is also sometimes called the *fracture toughness*, and G_c the *fracture energy*.

A simple and useful case is when uniform stresses σ_{ij} are applied remote from the crack, as in Figure 1.7. In this case the stress intensity factors are given by

$$K_I = \sigma_{yy}(\pi c)^{1/2}$$

$$K_{II} = \sigma_{xy}(\pi c)^{1/2} \tag{1.25}$$

$$K_{III} = \sigma_{zy}(\pi c)^{1/2}$$

and, using Equation 1.22, the corresponding crack extension forces, for plane stress, are

$$G_I = (\sigma_{yy})^2 \pi c/E$$

$$G_{II} = (\sigma_{xy})^2 \pi c/E \tag{1.26}$$

$$G_{III} = (\sigma_{zy})^2 \pi c(1+\nu)/E$$

In plane strain, E is replaced by $E/(1-\nu^2)$ for Modes I and II.

Equation 1.25 may be compared with the approximate expression for the stress concentration at the tip of an elliptical crack, Equation 1.5. However, inspection of Equation 1.19 indicates that there is a stress singularity at the crack tip. This results from the assumption of perfect sharpness of the slit. This is nonphysical, both because it internally violates the assumption of linear elasticity, which implies small strains, and because no real material can support an infinite stress. There must be a region of nonlinear deformation near the crack tip that relaxes this singularity. This can be ignored in the fracture mechanics approach, because it can be shown that the strain energy in the nonlinear zone is bounded, and because the small nonlinear zone does not significantly distort the stress field at greater distances from the crack. It is, of course, of paramount importance for studies concerned with the detailed mechanics of crack advancement. Various models, such as that of Dugdale (1960) and Barenblatt (1962) have been advanced to describe this nonlinear zone, but it suffices here to state that linear elastic fracture mechanics is not applicable at that scale or if there is large-scale yielding.

Within the nonlinear zone distributed cracking, plastic flow, and other dissipative processes may occur that contribute to the crack extension force. To account for these additional contributions we can rewrite Equation 1.24 as

$$\mathbf{G_c} = 2\Gamma \tag{1.27}$$

where Γ is a lumped parameter that includes all dissipation within the crack-tip region. This failure criterion is associated with the work of Irwin (1958). The fact that we do not usually know the specific processes that contribute to Γ is not normally of practical significance because $\mathbf{G}$ still can be evaluated if mechanical measurements can be made suitably outside the nonlinear zone [because integration around the crack tip is path-independent (Rice, 1968)].

A more serious problem, for geological applications, lies in the fracture mechanics assumption that the crack is cohesionless behind the crack tip. In shear motion on a fault, friction will exist over all the fault, and work done against this friction will become a significant term in an energy balance describing this process. As will be discussed in more detail in Section 4.2.1, it is not possible to evaluate this frictional work term and so solve for the energy partition in earthquakes. In terms of the present context, this means that, for the shear modes, Equations 1.24 and 1.27 are reduced to the status of local stress fracture criteria as opposed to global criteria tied to an energy balance.

1.1.4   *Macroscopic fracture criteria*

The theory of fracture discussed above specifies the conditions under which an individual crack will propagate in an elastic medium. We will show in Section 1.2, however, that only in one special case, that of tensile fracture of a homogeneous elastic material, do these theories also predict the macroscopic strength. In describing the strength of rock under general stress conditions, we are forced to use criteria which are empirical or semiempirical. Such fracture criteria had been well established by the end of the nineteenth century and hence pre-date the theoretical framework that has been described so far.

In formulating a fracture criterion we seek a relationship between the principal stresses $\sigma_1 > \sigma_2 > \sigma_3$ (compression is positive) that defines a limiting failure envelope of the form

$$\sigma_1 = f(\sigma_2, \sigma_3) \tag{1.28}$$

with some few parameters with which we can characterize the material.

One such criterion, which experiment shows is generally adequate, is that tensile failure will occur, with parting on a plane normal to the least principal stress, when that stress is tensile and exceeds some value T_0, the tensile strength. Thus,

$$\sigma_3 = -T_0 \tag{1.29}$$

Shear failure under compressive stress states is commonly described with the Coulomb criterion (often called the Navier–Coulomb, and sometimes the Coulomb–Mohr criterion). This evolved from the simple frictional criterion for the strength of cohesionless soils,

$$\tau = \mu \sigma_n \tag{1.30}$$

by the addition of a "cohesion" term τ_0. Thus

$$\tau = \tau_0 + \mu \sigma_n \tag{1.31}$$

where τ and σ_n are the shear and normal stresses resolved on any plane within the material. The parameter μ is called the *coefficient of internal friction* and is often written $\tan \phi$, ϕ being called the *angle of internal friction*. This criterion is shown in Figure 1.8, together with a Mohr circle from which the relationships between the failure planes and stresses can be deduced readily. From the Mohr circle it can be seen that failure will occur on two *conjugate* planes oriented at acute angles

$$\theta = \pi/4 - \phi/2 \tag{1.32}$$

on either side of the σ_1 direction and will have opposite senses of shear. From the geometry of Figure 1.8 one also can derive an expression of Equation 1.31 in principal axes, which, after some trigonometric manipulation, is found to be

$$\sigma_1 \left[(\mu^2 + 1)^{1/2} - \mu \right] - \sigma_3 \left[(\mu^2 + 1)^{1/2} + \mu \right] = 2\tau_0 \tag{1.33}$$

which is a straight line in the σ_1, σ_3 plane with intercept at the uniaxial

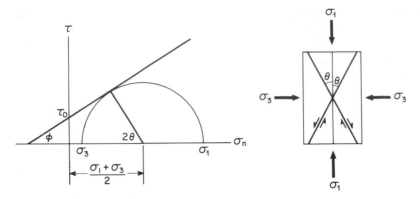

Fig. 1.8 Illustration of the Coulomb fracture criterion by means of a Mohr diagram. The relationships between the parameters at failure may be worked out from the geometry of the figure. On the right are shown the angular relationship between fracture planes and the principal stresses.

compressive strength,

$$C_0 = 2\tau_0 \left[\left(\mu^2 + 1 \right)^{1/2} + \mu \right] \qquad (1.34)$$

This criterion is defined only for compressive stresses. To form a complete criterion, we can specify this and combine Equation 1.33 with the tensile strength criterion, Equation 1.29 (Jaeger and Cook, 1976, page 89),

$$\sigma_1 \left[\left(\mu^2 + 1 \right)^{1/2} - \mu \right] - \sigma_3 \left[\left(\mu^2 + 1 \right)^{1/2} + \mu \right] = 2\tau_0,$$

$$\text{when } \sigma_1 > C_0 \left[1 - C_0 T_0 / 4\tau_0^2 \right] \quad \text{and} \qquad (1.35)$$

$$\sigma_3 = -T_0 \quad \text{when } \sigma_1 < C_0 \left[1 - C_0 T_0 / 4\tau_0^2 \right]$$

This criterion is strictly two dimensional: There is no predicted effect of the intermediate principal stress σ_2 on the strength.

The simple criterion for cohesionless soils, Equation 1.30, can be understood in terms of a microscopic failure process. The parameter μ is the friction coefficient between adjacent grains, which, in principle, can be determined independently of the criterion. Also, ϕ has a physical meaning: It is the steepest angle of repose that the material can support. In contrast, the coefficient of internal friction in the Coulomb criterion cannot be identified with any real friction coefficient, because the failure surface is not yet present at the ultimate stress. For the same reason, one cannot simply interpret the cohesion term as a pressure-independent strength that can be added simultaneously to this friction term. The Coulomb criterion thus may be viewed as strictly empirical.

Griffith (1924) developed a two-dimensional fracture criterion in terms of his theory of crack propagation. The underlying assumption of this criterion is that macroscopic failure can be identified with the initiation of cracking from the longest, most critically oriented Griffith crack. He analyzed the stresses around an elliptical crack in a biaxial stress field and found the most critical orientations that yielded the greatest tensile stress concentrations. He compared these results with that for a crack in uniaxial tension by normalizing them to the uniaxial tensile strength. The resulting criterion is

$$(\sigma_1 - \sigma_3)^2 - 8T_0(\sigma_1 + \sigma_3) = 0 \quad \text{if } \sigma_1 > -3\sigma_3$$

and $\hspace{11cm}$ (1.36)

$$\sigma_3 = -T_0 \quad \text{if } \sigma_1 < -3\sigma_3$$

The corresponding Mohr envelope is a parabola,

$$\tau^2 = 4T_0(\sigma_n + T_0) \tag{1.37}$$

(see Jaeger and Cook, 1976, pages 94–9). For the tensile fracture portion of this failure envelope, the most critically oriented crack is normal to σ_3. For the shear portion, it is inclined at an angle θ from the σ_1 direction given by,

$$\cos 2\theta = \tfrac{1}{2}(\sigma_1 - \sigma_3)/(\sigma_1 + \sigma_3)$$

This criterion is based on a microscopic failure mechanism. It has the attractive feature of combining tensile and shear failure in a single criterion. It predicts that $C_0 = 8T_0$, which, though smaller than generally observed, is of the correct order. Like the Coulomb criterion, it does not predict a σ_2 effect.

McClintock and Walsh (1962) pointed out that, under compressive stress states, cracks would be expected to close at some normal stress σ_c and thereafter crack sliding would be resisted by friction. They reformulated the Griffith criterion to admit this assumption and obtained the *modified Griffith criterion*

$$\left[(1 - \mu^2)^{1/2} - \mu\right](\sigma_1 - \sigma_3) = 4T_0(1 + \sigma_c/T_0)^{1/2} + 2\mu(\sigma_3 - \sigma_c)$$

$$\tag{1.38}$$

which has the corresponding Mohr envelope

$$\tau = 2T_0(1 + \sigma_c/T_0)^{1/2} + 2\mu(\sigma_n - \sigma_c) \tag{1.39}$$

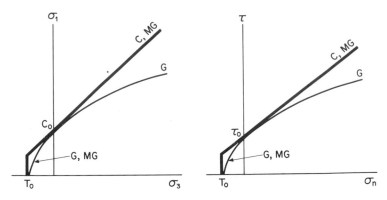

Fig. 1.9 Comparison between the forms of the Coulomb criterion (C) and the Griffith and modified Griffith (G and MG) criteria in (σ_1, σ_3) and (τ, σ_n) space.

This criterion, like the Coulomb criterion, predicts a linear relationship between the stresses. If we assume further that σ_c is negligibly small, we obtain the simplified forms,

$$\left[(1 + \mu^2)^{1/2} - \mu\right](\sigma_1 - \sigma_3) = 4T_0 + 2\mu\sigma_3 \qquad (1.40)$$

and

$$\tau = 2T_0 + \mu\sigma_n \qquad (1.41)$$

which are identical with the Coulomb criterion, with $\tau_0 = 2T_0$, and μ now identified with the friction acting across the walls of preexisting cracks. This led Brace (1960) to suggest that this formed the physical basis for the Coulomb criterion.

These several criteria are compared in Figure 1.9 in (σ_1, σ_3) and (τ, σ_n) coordinates. They all, to an extent, account for the first-order strength properties of rock and cannot be distinguished on the basis of experimental data. Whereas the Coulomb criterion is strictly empirical, the generalized forms of the Griffith criterion are attempts to predict macroscopic failure based on a correct description of the micromechanics. However, as we show in the next section, none adequately describe the complexity of the process, and under compressive stress states the micromechanics assumed in the Griffith formulations is incorrect.

1.2   EXPERIMENTAL STUDIES OF ROCK STRENGTH

When discussing the strength of rock in terms of both theoretical concepts and geological applications, it is important to keep in mind the vast range of scales over which we must consider this phenomenon.

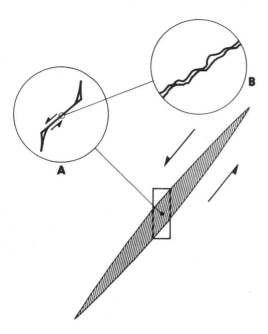

Fig. 1.10 An illustration of the way fracture may be viewed on different scales. The shaded form may represent a fault, and the rectangular shape a laboratory-size specimen. Inset A shows a microcrack within the specimen and inset B a detail of the surface contact on the microcrack.

Thus, beyond consideration of the scaling of strength, which will be discussed in Section 1.2.4, there are different regimes of scale for which it is convenient to conceptualize the process of fracture in different ways, which basically embrace different levels of approximation or complexity. This idea is illustrated in Figure 1.10. The large-scale feature in the figure is meant to represent a fault. The small rectangle in the center represents a laboratory-scale rock specimen undergoing brittle fracture, and Inset A shows microscopic cracking occurring in the rock. When considering the propagation of the fault as a whole, it may be natural and useful to discuss it in terms of fracture mechanics. This neglects the details of the process on the scale of the laboratory specimen, at which scale the deformation is bound to be far too complex to apply fracture mechanics and another approach must be used. On a still finer scale, on the level of an individual microcrack, as in Inset A (Fig. 1.10), fracture mechanics again may be a useful technique. Inset B shows a detail of the shear crack in Inset A, to show that what may be treated as friction on the scale of Inset A may involve fracture of asperities on the scale of Inset B. Thus, these levels of complexity may be considered to be nested within

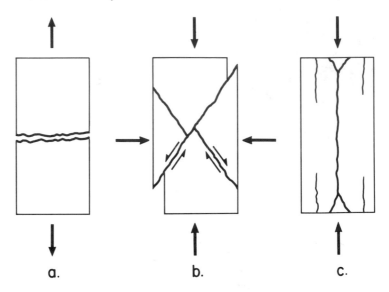

Fig. 1.11 Three modes of fracture observed in laboratory experiments: (a) tensile fracture; (b) faulting in a compression test; (c) splitting observed in a compression test at low confining pressure.

one another, and the approach used depends on the scale and degree of approximation demanded.

1.2.1   *Macroscopic strength*

The strength of rock is commonly measured in the laboratory by means of the uniaxial and triaxial compression test. Tensile strength can be determined by one of several direct and indirect means. Descriptions of the various experimental procedures used can be found in standard texts (e.g., Jaeger and Cook, 1976; Paterson, 1978). The two principal modes of failure observed are sketched in Figure 1.11a, b, namely, tensile failure involving parting on a surface approximately normal to the σ_3 axis and, in uniaxial and triaxial compression, shear on a surface inclined at an acute angle to σ_1. Also, splitting parallel to σ_1 often is observed in uniaxial compression tests (Fig. 1.11c), although this behavior is suppressed by quite small confining pressures, and it is not clear if it is an effect of end conditions or an intrinsic fracture mode.

The effect of pressure on the strength of rock is quite dramatic and is illustrated in Figure 1.12. Although the pressure effect is often represented as being linear, the curvature shown in the figure is more typical of careful studies. Experimental scatter in such strength data can be as low as 1–2% if adequate care is taken in both sample preparation and experimental technique (Mogi, 1966).

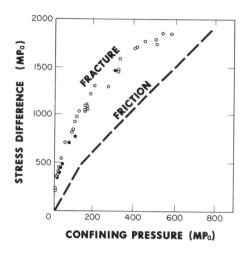

Fig. 1.12 The strength of Westerly granite as a function of confining pressure. Also shown, for reference, is the frictional strength for sliding on an optimally oriented plane. Data sources are: open circles, Brace et al. (1966) and Byerlee (1967); closed circles, Hadley (1975); friction, Byerlee (1978). Stress difference is $\sigma_1 - \sigma_3$.

A complete stress–strain curve for a compressive strength test is shown in Figure 1.13. This was obtained with a stiff loading system so that the instability usually associated with brittle fracture did not terminate the test. Notice that the ultimate strength occurs only after a period of pronounced nonlinear deformation and that the loss of load-bearing capacity after peak stress is not abrupt but occurs gradually. These features will be discussed later, when we consider the nature of instabilities in Section 2.3 and earthquake precursory phenomena in Section 7.3.

The nature of the prefailure deformation may be understood by examining all strain components. In Figure 1.14 axial, lateral, and volumetric strain are plotted versus stress for a typical uniaxial compression test. These stress–strain curves can be divided into four characteristic regions, as shown (Brace, Paulding, and Scholz, 1966). On initial loading (Stage I) the stress–axial-strain curve is concave upwards and the rock undergoes more volume compaction than would be expected from solid elasticity. This behavior is caused by the closing of preexisting cracks, primarily those oriented at high angles to the applied stress. This stage is not observed in triaxial compression tests, because the confining pressure closes the cracks before the deviatoric stress is applied. After these cracks largely have closed, the rock deforms in a nearly linear elastic manner, according to its intrinsic elastic constants (Stage II). At a stress commonly found to be at about half the fracture stress, the rock is observed to dilate relative to what would be expected from linear

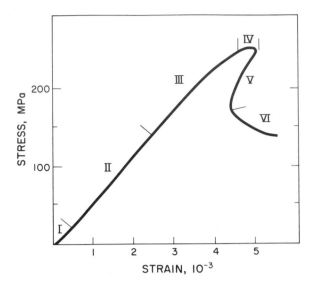

Fig. 1.13 A complete stress–strain curve for a brittle rock subjected to compression. (After Wawersik and Brace, 1971.)

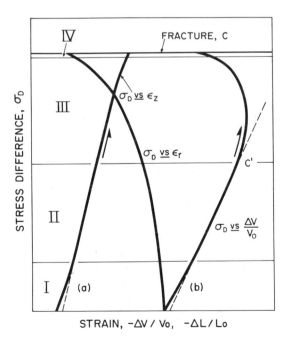

Fig. 1.14 Curves of stress versus axial strain (ε_z), lateral strain (ε_r), and volumetric strain ($\Delta V / V$) for a brittle rock in a compressive test to failure.

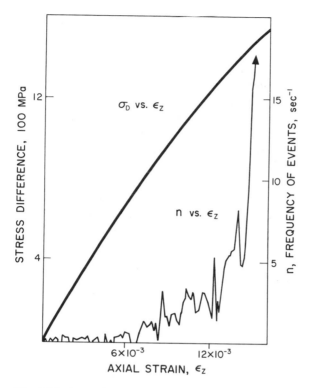

Fig. 1.15 Acoustic emission observed during the brittle fracture of rock in triaxial compression. (After Scholz, 1968a.)

elasticity (Stages III and IV). This is accompanied by a reduction in the axial modulus, but is primarily due to nonelastic lateral expansion of the rock, as may be seen in the record of lateral strain. This rheological property, in which volume dilation occurs as a result of application of a deviatoric stress, is called *dilatancy*. Prior to the work of Brace et al. (1966), this was primarily known only as a property of granular materials. Brace et al. (1966) interpreted it as being due to the development of pervasive microcracking within the rock, with a concomitant increase in void space.

The acoustic emission (AE) that occurs during compressive failure experiments confirms this interpretation of the strain data (Scholz, 1968a). As is shown in Figure 1.15, AE begins at the onset of dilatancy and this activity accelerates in proportion to the rate of dilatancy throughout Stage III. Stage IV is distinguished from Stage III because there is often an observed localization of both deformation and acoustic emission in this stage (Scholz, 1968b; Sondergeld and Esty, 1982; Lockner and Byerlee, 1977; Soga et al., 1978). Stage IV thus appears to involve the coalescence of microcracks leading to the formation of the macroscopic fracture.

Strain data such as in Figure 1.14 indicate that most of the microcracks responsible for dilatancy are nearly parallel to the maximum principal stress direction. Scanning electron microscope (SEM) examination of specimens following their stressing to various stages prior to failure have confirmed the preponderance of axial cracks among those that are stress induced (Tapponnier and Brace, 1976; Kranz, 1979).

The few studies that have investigated the radial variation in dilatancy show that it is probably rarely axially symmetric, even under an axially symmetric stress field. Hadley (1975) and Scholz and Koczynski (1979) found in triaxial compression tests that the sample cross section normal to the maximum compression deformed into an ellipse, with the axes of the ellipse corresponding with preexisting crack microstructure directions in the rock. The orthorhombic symmetry of dilatancy becomes very pronounced under true triaxial conditions, when $\sigma_1 \neq \sigma_2 \neq \sigma_3$. Mogi (1977) showed that in true triaxial tests, as σ_2 is increased from σ_3 to σ_1, the dilatant strain parallel to σ_2 is progressively suppressed until eventually all dilatancy occurs by expansion in the σ_3 direction. The dilatant cracks form preferentially in the σ_1, σ_2 plane, with the degree of anisotropy being controlled by the stress ratios. He also observed a marked effect of σ_2 on strength. As σ_2 was increased above σ_3 and dilatancy was suppressed in the σ_2 direction, the strength rose, but at a lesser rate than if σ_3 and σ_2 had been increased simultaneously (as in the pressure effect, Fig. 1.12). The macroscopic fracture formed at an acute angle to σ_1, and parallel to σ_2, in accordance with the Coulomb criterion. The minimum lateral dilatancy direction was thus in the fracture plane, and this was true also of the previously cited cases of dilatancy anisotropy when $\sigma_1 = \sigma_2$, that is, it appears that the fracture orientation is controlled by the dominant orientation of microcracks, which is usually, but not always, controlled by the stresses. This is particularly clear in studies of the strength of highly anisotropic rocks (e.g., Donath, 1961).

The peak stress (Fig. 1.13), usually referred to as the failure stress, is followed by Stage V, in which there is a loss of load-bearing capacity that is first gradual, and then usually very abrupt. In this stage, the strain localization, which began in Stage IV, becomes much more pronounced, with intense microcracking in a progressively more brecciated zone that forms the incipient macroscopic shear fracture (Wawersik and Brace, 1971; Wong, 1982). There is often a region in which the unloading curve has a positive slope, which in most testing machines would result in the instability so characteristic of brittle fracture. Stage V is followed by a stress plateau, Stage VI, in which the stress is determined by residual friction on the shear zone. In Figure 1.12 it can be seen that brittle fracture involves a substantial stress drop, from the strength to the friction level.

1.2.2   *Fracture energies*

Studies have been made of the propagation of single fractures in rocks and minerals, which yield the fracture mechanics parameters $\mathbf{K}_c$ and $\mathbf{G}_c$. Cracks are propagated in these experiments from starter notches in specimens loaded either in the double cantilever beam configuration (similar to Fig. 1.4a) or with the double torsion method (described by Atkinson, 1984), both of which result in fracturing in Mode I.

Experiments of this type with glasses or single crystals result in the growth of a single crack, and measured values of $\mathbf{G}_c$ can be equated with the specific surface energy of the material, using Equation 1.24. However, in polycrystalline materials like rock, deformation near the crack tip occurs as complex microcracking distributed over a region known as the brittle process zone (Friedman, Handin, and Alani, 1972; Evans, Heuer, and Porter, 1977). As the crack advances, at some distance behind the tip of the process zone the microcracks link up to form a macroscopic fracture, which still can support tensile stress because of geometrical interlocking and friction (Swanson, 1987). There is thus a cohesion zone, typically many grain diameters long, between the tip of the process zone and the traction-free region of the crack. As a result, fracture energies measured on rocks typically increase with "crack length" until a steady state is reached when the entire cohesion zone has been produced (Peck et al., 1985). Thus measures of $\mathbf{G}_c$ for such materials must be interpreted with the Irwin criterion (Eq. 1.27), where Γ is a lumped parameter that includes *all* energy dissipation in the process zone.

The energy expended in fracturing an intact rock sample in compression is even more difficult to assess. The pervasive microcracking that produces dilatancy in Stage III requires energy dissipation in the form of surface energy and the kinetic energy of acoustic emission. Energy is dissipated in forming the macroscopic shear fracture in Stages IV and V, and any further deformation requires that work be done against friction in Stage VI. Wong (1982) evaluated these terms for fracture of a granite in triaxial compression. He estimated the density of stress-induced cracks produced during dilatancy with a SEM survey. By multiplying this by Mode I fracture energies of single crystals and rocks, he obtained an estimate of the total surface energy involved in microcracking which compared satisfactorily with the work done against the inelastic component of the strains. He estimated $\mathbf{G}_c$ for the formation of the shear fracture using a method suggested by Rice (1980), in which the work

Table 1.1. *Fracture mechanics parameters for some geological materials*

Test	K_c MPa m$^{1/2}$	G_c J m^{-2}
Mode I		
single crystals		
quartz (1011)	0.28	1.0
orthoclase (001)	1.30	15.5
calcite (10$\underline{1}$1)	0.19	0.27
rocks		
Westerly granite	1.74	56.0
Black gabbro	2.88	82
Solnhofen limestone (normal to bedding)	1.01	19.7
Mode III		
Westerly granite (no end load)	2.4	100
Solnhofen limestone (normal to bedding, no end load)	1.3	35
Triaxial compression		
Westerly granite		10^4
Earthquakes		10^6–10^7

References: Mode I, Atkinson (1984); Mode III, Cox and Scholz (1988a); triaxial compression, Wong (1982); earthquakes, Li (1987).

done in shear during the stress breakdown (Stage V) is calculated. When estimated this way, his value of G_c was $\sim 10^4$ J m^{-2}, which is one to two orders of magnitude higher than an estimate based on surface energy created in Mode I during the formation of the brecciated zone. He concluded that, either G_{IIc} is intrinsically much larger than G_{Ic}, or there is considerable frictional dissipation during the shear fracture formation. Cox and Scholz (1988a) found that a frictional component was present in their measured values of G_c for shear cracks growing under an applied normal load.

Typical ranges of measured values of fracture energy are given in Table 1.1. We can take the single-crystal data as representative values of the intrinsic specific surface energies of the materials, that is, 1–10 J m^{-2}. Measured values of G_{Ic} for rocks are typically about one order of magnitude higher. This is because much greater surface energy is required per unit crack length to form the process zone, and because deformation of this complexly cracked zone involves some frictional work as well. The values of G_c estimated from triaxial compression tests, as described above, are much higher: in the range of 10^4 J m^{-2}. In this case the high value of G_c is due in part because of the greater degree of brecciation involved in creating the shear zone, but probably mostly due to frictional work, because there is a large compressive normal stress present across the shear zone. The positive dependence of G_c on normal stress, found in several studies (Wong, 1986; Cox and Scholz, 1988a) is most likely due to this latter effect. Measurements of shear fracture energy made without a normal stress yield values of G_{IIIc} larger than that for Mode I fractures but much less than the value obtained from triaxial compression tests (Cox and Scholz, 1988a). In this latter case, it is important to point out that although the starter notch was in pure Mode III loading, the fracture propagated as a process zone dominated by Mode I microcracks. As will be discussed in the next section and again in Section 3.2, true shear cracks really do not propagate in their plane as a single crack; at some scale they are arrays of Mode I cracks. A shear rupture will propagate only along a weakness plane such as a preexisting fault.

1.2.3 *Discussion of fracture criteria in the light of experimental results*

The observation that the fracture toughness of rock is much greater than that of the single crystals that comprize it is consistent with the typical observation that single crystals are stronger than rocks composed of the same mineral. Thus the uniaxial compressive strength of single-crystal quartz is around 2,000 MPa, whereas the strength of quartzites is more typically in the range 200–300 MPa. This riddle has a simple answer: The single crystal measured is usually of gem quality, and hence contains only submicroscopic flaws, whereas flaws in the quartzite will be at least as large as the grain diameter. According to Equation 1.25, the stress intensity factor is proportional to the crack length, so the strength is correspondingly smaller for the quartzite. This example is illustrative of the distinction that usually must be made between microscopic and macroscopic fracture criteria.

On the other hand, the apparent success of the macroscopic criteria (Sec. 1.1.4) in qualitatively predicting the gross features of rock strength requires some discussion, considering the obvious failure of the microscopic criterion upon which they are based, according to the results of Section 1.2.1. The Griffith-type criteria are "weakest link" theories that are based on the assumption that macroscopic failure occurs when the most critical Griffith crack propagates. This is clearly violated by the observation that dilatancy begins at about half the failure stress and that microcracking occurs pervasively throughout the specimen prior to failure. This problem has been discussed at some length by Paterson (1978, pages 64–6): Here we focus on two important points.

As discussed in Section 1.1.4, the most critically oriented Griffith crack in a compressive stress field is a shear crack oriented at an acute angle to σ_1. An analysis of the stress field about such a crack shows that the maximum crack extension force does not lie in the plane of the crack but is such as to deflect the crack into an orientation parallel to σ_1 (Nemat-Nasser and Horii, 1982; see also Lawn and Wilshaw, 1975, pages 68–72). The result (depicted in Inset A of Fig. 1.10) is that tensile cracks propagate from the tips of the shear crack (Brace and Bombalakis, 1963). These cracks become stable after propagating a short distance and further loading is required to cause further propagation. As the stress is increased, similar cracking will occur from other nucleation sites, leading to an array of axial cracks that produce dilatancy, primarily by lateral expansion (Fig. 1.14), but which do not individually cause failure of the specimen. The shear crack discussed as the initiator of such a microcrack should be taken as only an example; in a rock there will be many more complex sites of initial stress concentration. Scholz, Boitnott, and Nemat-Nasser (1986) found that Mode I cracks could grow stably for long distances from an initial flaw in a direction parallel to σ_1 in an all-around compressive stress field, and Nemat-Nasser and Horii (1982) suggested that this could be the cause of splitting in uniaxial compressive tests (Fig. 1.11c).

The most basic problem, which was anticipated in Figure 1.10, is that although rock, both in the laboratory and in nature, is macroscopically observed to fail in compression by the formation of shear fractures (faults) inclined at an acute angle to the σ_1 direction, it has also been found that it is not possible for a shear crack in an isotropic elastic medium to grow in its own plane. Instead, the propagation of a shear crack inclined to σ_1 occurs by the generation of Mode I cracks parallel to σ_1, as shown in three dimensions in Figure 1.16. The upper and lower tips of this crack, which are in a Mode II configuration, grow by propagation of single Mode I cracks in the axial direction, as in the experiments of Brace and Bombalakis (1963). The lateral edges, which are in Mode III, generate an array of Mode I axial cracks (Cox and Scholz, 1988b). Whereas this behavior is expected from a stress analysis, it poses major problems for the formation and growth of faults (Sec. 3.2). Part of the answer to this problem was found by Cox and Scholz (1988b), who found that although the initial cracks that propagated from a Mode III crack were as shown in Figure 1.16, further shearing resulted in the array of Mode I cracks being broken through by cracks parallel to the shear plane, which formed a shear process zone, the first stage in forming a fault.

The other major factor that complicates brittle fracture of rock, emphasized by Mogi (1962) and Scholz (1968a) is the importance of the heterogeneity of rock and its role in promoting the development of a strongly heterogeneous stress field. Thus, stresses locally within the rock

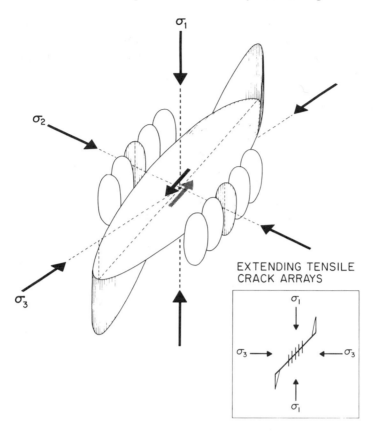

Fig. 1.16 Schematic diagram showing the propagation of tensile cracks from the edges of a shear crack in a brittle material. The patterns at the Mode II and Mode III edges are quite different.

can depart markedly from the applied stresses so that cracking can be initiated at points of high local stress and inhibited from propagating into adjacent regions of lower stress. Similarly, fracture toughness will be highly variable on the grain scale, due to the presence of grain boundaries, cleavage, and so forth. This heterogeneity alone would prohibit macroscopic failure from occurring by the propagation of a single crack. In the case of the fracture of a polycrystalline material in tension, where the considerations discussed in the previous two paragraphs do not apply, the effects of heterogeneity are manifest in the formation of a process zone (Swanson, 1984) and the occurrence of abundant AE prior to failure, just as in the case of failure in compression (Mogi, 1962).

Griffith-type theories strictly predict only local crack initiation, so we may ask why rock fracture, in obeying the Coulomb criterion, qualitatively mimics their predictions? A rationalization is that the macroscopic fracture forms from a coalescence of microcracks, which themselves initiate and develop further locally according to the Griffith theory. Therefore the whole process scales in the same way as the Griffith theory predicts. In essence, the principle espoused is that the property of the whole (the rock) is a function of the property of the parts (the cracks). The functional relationship between the two has not been subject to rigorous analysis because of the complexity of the process and of the microstructure of the rock. One suggestion, for example, is that failure occurs at a critical dilatancy (Scholz, 1968a; Brady, 1969), which implies the self-similarity necessary for such a relationship, although the critical dilatancy has been found to vary under widely different conditions (Kranz and Scholz, 1977). Recent attempts to discuss this have focussed on self-similarity in a stochastic sense (Allegre, Le Mouel, and Provost, 1982).

If we accept that, under given conditions, failure occurs when the density of microcracks reaches some critical value, then fracture becomes a three-dimensional process, as opposed to the behavior of an individual crack, which is two-dimensional. This view of fracture has an advantage inasmuch as it allows us to understand intuitively the σ_2 effect on strength, since the effect of σ_2 is to inhibit microcracking on planes at high angle to it and thus reduce the total microcrack population at any given value of σ_1.

1.2.4   *Effect of scale on strength*

In any application of laboratory studies to geological processes, we must be concerned with scaling to dimensions and times much larger than those accessible in the laboratory. The problem of time effects on brittle fracture will be deferred to Section 1.3.2; here we will briefly discuss the effect of size on strength. This topic is well covered in Jaeger and Cook (1976, pages 184–5), and Paterson (1978, pages 33–5). To summarize the important empirical result, it is commonly found that if d is a characteristic dimension of a test specimen, the compressive strength is found to decrease with d according to

$$\sigma_c = md^{-\zeta} \tag{1.42}$$

where ζ is some constant less than 1 and m is a normalization factor. A typical example is shown in Figure 1.17, where it is found that a value of $\zeta = \frac{1}{2}$ fits the data over the range 0.05 m $< d <$ 1.0 m. This behavior usually has been attributed to a dependence of flaw size on sample size. Even though, as we have pointed out, rock fracture is not well described by catastrophic failure of a weakest link, we can expect that crack initiation and ultimate strength will be controlled by the longest preexist-

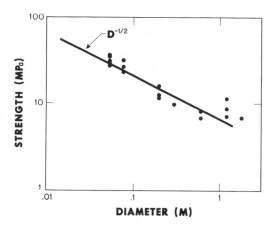

Fig. 1.17 An illustration of the relationship between specimen size and strength. The data are for quartz diorite from Pratt et al. (1972).

ing cracks. If the intrinsic strength of the rock is given by a fracture toughness K_c, which is scale independent, and the mean length of longest cracks increases proportionally with d, Equation 1.42 is predicted from Equation 1.25, with $\zeta = \frac{1}{2}$. This value agrees very well with the data in Figure 1.17. In laboratory specimens, which usually are selected to be free of macroscopic cracks, strength is often found to be dependent on grain size, with a form similar to Equation 1.42 and $\zeta = \frac{1}{2}$ (Brace, 1961). This scaling, known as Petch's law, can also be understood from the above argument if the length of the longest cracks is proportional to the grain diameter.

The data shown in Figure 1.17 indicate a plateau at scales greater than 1 m. Since that study was explicitly of "unjointed" rock, this may indicate a sampling bias in which macroscopic flaws larger than a certain size were avoided. Otherwise, the lower limit on rock strength determined by scale is the frictional strength, in the case where the rock is completely broken into rubble. As will be discussed in Section 2.1, friction itself has no size dependence, so that the lower limit on strength should be given by laboratory friction values, shown in Figure 1.12 for comparison with strength. In reality, this lower limit rarely is reached, since the earth is not pervasively broken, and faults act as planes of weakness even when not ideally oriented in the stress field (Sec. 3.1.1).

1.3 PORE FLUID EFFECTS ON FRACTURE

The brittle part of the earth can be assumed to be almost universally permeated with fluids. This produces two very important but unrelated effects on rock strength and fracturing that have a strong influence on geological processes. The first is a purely physical effect, owing to the pressure the fluid transmits to the pore space, which in fracture and friction is akin to certain types of lubrication. The second is a physicochemical effect, which results in the strength of rock being time dependent.

1.3.1   *Laws of effective stress*

If we describe the behavior of a crack using linear elastic fracture mechanics (Section 1.1.3) for the case in which the crack contains a fluid at pressure p, we can superimpose linearly this pressure with the applied stresses σ_{ij} and find that the stress intensity factor depends only on the difference

$$\bar{\sigma}_{ij} = \sigma_{ij} - p\delta_{ij} \qquad (1.43)$$

where δ_{ij} is the Kronecker delta. The tensor $\bar{\sigma}_{ij}$ is called the *effective stress*. Most of the physical properties of porous solids obey a *law of effective stress*, which means that they change in response to changes in the effective stress, as opposed to solely the externally applied stress. The nature of the effective stress law depends on the property under consideration, P, and generally takes the form

$$P(\bar{\sigma}_{ij}) = P(\sigma_{ij} - \alpha p\delta_{ij}) \qquad (1.44)$$

(Nur and Byerlee, 1971; Robin, 1973). In the case of strength, we are concerned with the behavior of cracks, which are generally in frictional contact. In that case (see Eq. 2.16), $\alpha = (1 - A_r/A)$, where A_r is the real, and A the nominal, area of contact. Because in most cases $A_r \ll A$, so that $\alpha \approx 1$, the simple effective stress law (Eq. 1.43) can be used. Because this effective stress law governs the behavior of cracks, we can expect it to govern the macroscopic strength as well, so rewriting the Coulomb criterion, we should expect the strength of rock to follow

$$\tau = \tau_0 + \mu(\sigma_n - p) \qquad (1.45)$$

which agrees very well with experiment.

In this treatment we have assumed that the applied stresses and the local pore pressure are independent of one another, whereas in real cases they may be coupled through deformation of the medium. One interesting example of coupling is caused by dilatancy, which produces an increase in void space and therefore a reduction in the pore pressure. If the rate of pore volume increase via dilatancy is faster than the rate fluids can flow into the dilating region, which is governed by the permeability of the rock, then the pore pressure will drop and *dilatancy*

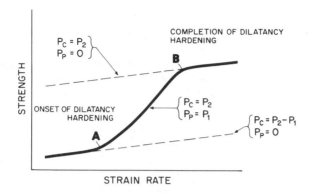

STRAIN RATE

Fig. 1.18 An illustration of dilatancy hardening for a compact crystalline rock. P_c is confining pressure and P_p is pore pressure. (After Brace and Martin, 1968.)

hardening will occur (Frank, 1965). This effect was demonstrated by Brace and Martin (1968), and is illustrated in Figure 1.18. The strength of rock was measured at various strain rates under two conditions: one with a confining pressure $p_c = p_2$ and pore pressure $p = p_1$; the other with $p_c = p_2 - p_1$, saturated, with $p = 0$. At low strain rates, the pore pressure remained constant and the strength under the two conditions was the same. Above a critical strain rate, however, the strength measured under the first condition became larger than under the second, due to dilatancy hardening, and at sufficiently high strain rates dilatancy hardening became complete and the strength approached the strength for $p_c = p_2$, $p = 0$. This effect plays an important role in dilatancy theories of earthquake precursors (Sec. 7.3).

1.3.2 *Environmental effects on strength*

It was mentioned in Section 1.1.2 that Obriemoff noticed an effect of the environment on the cleavage strength of mica. This turns out to be a physicochemical effect, in which certain chemically active species in the environment react with the solid at a rate that is enhanced by the tensile stress field at crack tips and at the same time, by virtue of the reaction, locally reduce the fracture toughness of the solid (Orowan, 1944). This process thus leads to the strength of the solid being inherently time dependent as long as it is subject to this environment. Water is the principal active species in the case of silicates. Water is ubiquitously present in the lithosphere of the earth, and displacement rates are very low in geological processes, so this reaction must control the strength of rock in the earth. If the rate of geological loading were uniformly low, these processes would not be noticed, as some equilibrium state would prevail. It is only when rates change suddenly, such as at the time of earthquakes, that their effects become noticeable (Secs. 4.4 and 4.5), and they result in a crustal response that is viscoelastic (Sec. 6.5.2). These environmental effects on rock strength are important not only in brittle fracture, but in friction, discussed in more detail in Section 2.3.

The presence of water also enhances plasticity in silicates (Griggs and Blacic, 1965). In that case, called hydrolytic weakening, the reaction takes place at the core of a dislocation rather than at a crack tip. A special issue of the *Journal of Geophysical Research* (vol. 89, no. B6, 1984) is devoted to these various effects; excellent reviews of brittle fracture effects by Atkinson (1984), Freiman (1984), Dunning et al. (1984), and Swanson (1984) are contained therein.

If a crack is driven through rock or glass, a marked departure from the prediction of linear elastic fracture mechanics is observed. The crack is found to extend, subcritically, at stress intensity factors much below K_{Ic}, with a well-defined relationship between K and v, the crack-tip velocity. A typical plot of K versus v exhibits several regions of different behavior (Fig. 1.19). The most dominant of these is Region II, in which the K–v relation is usually determined experimentally as

$$v = AK_I^n \tag{1.46}$$

where n is called the stress-corrosion index. For theoretical reasons the expression

$$v = v_0 \exp(bK_I) \tag{1.47}$$

is often preferred. In this region the rate of crack-tip advance is controlled by the rate of a stress-enhanced chemical reaction at the crack tip. Although such reactions are many and complex, the principle can be illustrated with the simple example of hydration of the Si–O bond, given by

$$[Si—O—Si] + H_2O \leftrightarrow \chi \rightarrow 2[SiOH] \tag{1.48}$$

(Freiman, 1984), where χ is an activated complex. This type of reaction has two important features: First, it produces a large extensional strain, so it is driven thermodynamically by the high tensile stress at the crack tip and thus occurs preferentially there; second, it replaces a covalent Si–O bond with a much weaker hydrogen bond that can then be broken readily under the existing stress field. An expression can be written for the rate of this reaction, from which we can rewrite Equation 1.47 (Freiman, 1984) as

$$v = v_0 a(H_2O)\exp[(-E^* + bK_I)/RT] \tag{1.49}$$

where E^* is an activation energy and the stress dependence (related to the activation volume) is given by b. We notice that crack velocity

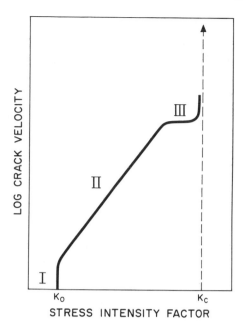

Fig. 1.19 Schematic diagram showing a typical relationship between crack-tip velocity and stress intensity factor for subcritical crack growth in a brittle solid. K_0 is the stress-corrosion limit, region II is where the rate-controlling process is the chemical reaction, and in region III the rate is controlled by diffusion of the reactive species to the crack tip. At K_c the crack propagates dynamically.

depends on temperature and the chemical activity of water, $a(H_2O)$. These are the typically observed effects and many of the observed parameters in Equations 1.47–1.49 are tabulated in Atkinson (1984).

At higher crack velocities, in Region III (Fig. 1.19), the rate-limiting process is the diffusion of the chemically active species to the crack tip. Though well defined in glasses, it is less so for rock, which may result from the more complex diffusion path afforded by the process zone (Swanson, 1984). A stress-corrosion limit $\mathbf{K}_0$ is included in Figure 1.19, although it has been very hard to demonstrate experimentally because of the very low crack velocities involved (typically $< 10^{-9}$ m s^{-1}). Atkinson (1984) suggests that it may be as low as $0.2\mathbf{K}_c$. Various mechanisms, including crack blunting (Freiman, 1984) and healing (Atkinson, 1984) have been discussed to support its existence, which is predicted from thermodynamics (Cook, 1986). This parameter determines the long-term strength of the material.

As a result of these environmental effects, rock and many other brittle materials have strengths which are inherently time dependent. If a glass

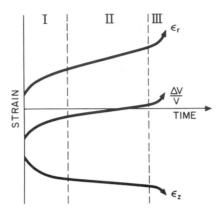

Fig. 1.20 The stages of brittle creep. (ϵ_z, axial strain; ϵ_r, radial strain; $\Delta V/V$, volume strain.) Stages I, II, and III are primary, secondary, and tertiary creep. Brittle fracture occurs at the end of the tertiary stage.

rod is subjected to a uniform tension well below its instantaneous breaking strength, it will fail spontaneously after a characteristic time $\langle t \rangle$. This behavior, called *static fatigue*, has been found for many materials, for example, for fused silica glass in tension (Procter et al., 1967) and for quartz in compression (Scholz, 1972). For quartz in compression the mean time to fracture is found experimentally to be described by

$$\langle t \rangle = f\left[a(\mathrm{H_2O})\right]\exp\left[(E - n\sigma)/RT\right] \qquad (1.50)$$

(Scholz, 1972). The similarity to Equation 1.49 is obvious, and Weiderhorn and Bolz (1970) showed that this observed static fatigue law could be predicted by integrating Equation 1.46 or Equation 1.49.

Similarly, when rock is subjected to a fixed load less than its instantaneous fracture strength, it undergoes the same internal processes, but because of its greater heterogeneity, the cracking is more pervasive and leads to macroscopic strains that define a typical creep curve (Figure 1.20). What distinguishes this type of creep is that, because its micromechanism involves cracking, the strains are dilatant and are accompanied by acoustic emission. Thus, although the rheology of the rock (the macroscopic stress–strain–time behavior) has become viscoelastic, the underlying mechanism is a brittle one. Scholz (1968c) and Cruden (1970) developed models to explain this creep based on subcritical crack growth and local static fatigue within the rock. Other than the time dependence of the deformation, the overall mechanism of dilatancy and its role in macroscopic fracture for "creep loading" does not seem markedly different from that in triaxial compression. In particular, the tertiary creep stage, where there is an acceleration of strain rate prior to failure, may be similar to Stage IV in compressive failure (Sec. 1.2.1), in which a considerable localization of deformation occurs prior to rupture (Yanigadani et al., 1985). The static fatigue law for rock has the same form as that described above for single crystals and glass (Kranz, 1980).

Therefore, the strength of rock cannot be described by a single size-invariant parameter, K_c, because subcritical crack growth can occur at much lower stress intensity factors, and such growth ultimately can lead to critical fracture. This behavior leads to some interesting earthquake phenomena, to be taken up in Section 4.5.2. In terms of the scaling of rock strength from laboratory to natural conditions (Sec. 1.2.4), the time dependence of strength described here leads to the conclusion that the fracture strength of rock at geological strain rates should be considerably less than measured in the laboratory. There will be a lower limit on this weakening, due to the stress-corrosion limit K_0. In any case, for the same reason given in Section 1.2.4, the lower limit on crustal strength is given by friction, which, because of competing effects, has a more complex time dependence than strength (Sec. 2.3).

1.4 THE BRITTLE–PLASTIC TRANSITION

Purely brittle behavior in crystalline materials gives way, at sufficiently elevated pressures and temperatures, to crystalline plasticity. There is usually a broad transition between these distinct regimes, in which the deformation is semibrittle, involving on the microscale a mixture of brittle and plastic processes, and the rheology is macroscopically ductile. Thus, we might anticipate that brittle faulting at shallow depths will give way, over a depth range that depends largely on the lithology and thermal environment, to plastic shearing at greater depth. This transition between the schizosphere and plastosphere marks a profound change in the mechanical properties of the earth and limits the depth to which earthquakes and other brittle phenomena normally can occur. Here we give an account of this transition without recapitulating the fundamentals of crystalline plasticity, which may be found elsewhere (e.g., Nicolas and Poirier, 1976; Poirier, 1985). In what follows, we shall distinguish *ductility*, which is used to refer to a rheology that allows large strain and is characterized by macroscopically homogeneous deformation, regardless of the micromechanisms involved, from *plasticity*, which is taken to imply that the underlying mechanism is some form of dislocation motion, perhaps diffusion aided. The term brittle–ductile transition is therefore avoided, because it has produced confusion between the mode and mechanism of deformation, which has resulted in a misunderstanding of this transition in the mechanical behavior of the earth, especially as regards faulting (Rutter, 1986). We use instead the term *brittle–plastic transition* to denote the entire transition from purely brittle to purely plastic behavior, which therefore encompasses the intervening semibrittle field. It is bounded on one side by the brittle–semibrittle transition and on the other by the semibrittle–plastic transition.

1.4.1 *General principles*

The brittle–plastic transition for rock is only qualitatively understood. The transition occurs under conditions where dislocations can propagate as easily as cracks. Thus we may regard it in terms of a competition between cracking and dislocation motion. Since cracking involves both a volume increase and frictional work, it will be inhibited by increased pressure. Dislocation glide, however, involves no volume change or friction, so it is insensitive to pressure. On the other hand, the mobility of dislocations, because they are lattice defects, is enhanced by thermal activation so dislocation motion is favored by increased temperature. Surface defects like cracks are not so affected, and therefore, if we ignore environmental effects, ease of stress-induced cracking is temperature insensitive. Hence plastic flow is favored by both high temperature and pressure and the deformation of any crystalline solid will be marked by a transition between brittle and plastic fields.

This change in deformation mode is not abrupt, but involves a gradual transition through a semibrittle field (Carter and Kirby, 1978; Kirby, 1983) as shown schematically in Figure 1.21. There are a number of reasons for this gradual transition, one of which is that the onset of

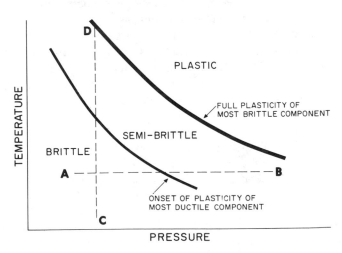

Fig. 1.21 Diagram that schematically shows the brittle–plastic transi-toin in *P–T* space. A–B and C–D are experimental paths described in the text.

plasticity in polycrystals is itself intrinsically gradual. The onset of the brittle–plastic transition usually occurs at low temperatures where diffusion is unimportant and the principal mode of plastic flow is dislocation glide (though diffusion-assisted dislocation motion will become increasingly important farther into the semibrittle field). With the exception of cleavage, which occurs on crystallographic planes of particularly low specific surface energy, cracking is only weakly controlled by crystal structure. In contrast, dislocation glide requires a bond to be broken and another to be made following an incremental vector shear offset **b** that connects two equivalent lattice sites. The Burgers vector **b** must therefore be in a rational crystallographic direction on a rational crystallographic plane and cannot contain any unstable transitional positions. Dislocation glide is consequently highly restricted by crystal structure and only a limited number of slip systems are possible. This is particularly true for silicates, which have complex structures dominated by the strong and directionally restricted covalent bond, and which also have low symmetry, so that the number of possible slip directions in any slip system is not greatly multiplied by symmetry. The critical resolved shear stress required to produce glide in one slip system is independent of that for another, so that under any given set of conditions there will always be a slip system which is "easy," as opposed to others that are "hard" and only activated at much higher stresses. For fully plastic flow of a polycrystal, each grain must be capable of straining arbitrarily, which requires the operation of five independent slip systems, a condition known as the von Mises–Taylor criterion. The brittle–plastic transition can therefore begin when the easiest slip system becomes active but cannot become complete until five independent slip systems are active or, as is more often the case with silicates, when the temperature becomes high enough that diffusion-controlled processes like climb and recovery partially obviate the von Mises–Taylor criterion.

There are important interactions between brittle and plastic processes within the transition. Plastic flow will tend to be concentrated at crack tips because of the high stresses there and will have the effect of both inhibiting and stabilizing the crack, both because crack advance will then involve the work of plastic flow, which will increase **G**, and because the plastic flow will blunt the crack, reducing the stress concentration there. On the other hand, plastic flow can also induce cracking. When an insufficient number of slip systems are active, the crystallographic mismatch at grain boundaries will often cause dislocation pile-ups that produce stress concentrations which may nucleate microcracks (the Zener–Stroh mechanism).

The term *cataclastic flow* refers to a type of distributed deformation characterized by ductile stress–strain behavior, but which to some degree involves internal cracking and frictional sliding. In the case of granular materials or high-porosity rocks, which crush and become granular under load, cataclastic flow may be entirely brittle–frictional. Cataclastic flow is also typical of the semibrittle behavior within the brittle–plastic transition, where it is a mixture of brittle–frictional and plastic deformation. It can be distinguished easily from fully plastic flow because it produces dilatancy and has a strongly pressure-dependent strength.

1.4.2   *The transition induced by pressure*

Silicates are not known to become ductile at room temperature even at very high confining pressure. The effect of pressure in inducing the brittle–ductile transition along the path A–B in Figure 1.21 is therefore more conveniently demonstrated by experimenting with an inherently more ductile mineral, such as calcite, which can deform plastically by mechanical twinning at room temperature and pressure.

Stress–strain curves for marble deformed at several confining pressures are shown in Figure 1.22 (Scholz, 1968a; see also Edmond and Paterson, 1972). The rock fails by crushing in uniaxial compression, but becomes ductile when deformed at a confining pressure as low as 25 MPa. Tests at higher pressures show that this behavior, while ductile, is highly pressure dependent. Both the yield strength and the strain hardening index, defined as the slope of the postyield stress–strain curve, increase strongly with pressure up to 300 MPa, above which there is no further change. The volume strain is also plotted versus axial strain in the figure. At the lower pressures, axial strain in the postyield region is accompanied by a steady rate of dilatancy, indicating that microcracking is occurring within the rock. The effect of pressure is to suppress progressively the dilatancy until above 300 MPa, where the postyield deformation occurs with no volume change. The ratio between dilatancy and axial strain mimics the effect of pressure on the axial stress–strain curve, being greatest at low pressure and gradually decreasing, with both vanishing at 300 MPa. The rock deforms cataclastically at lesser pressures; there is a gradual reduction of the contribution of brittle processes to the deformation as the pressure is raised until, at 300 MPa, the brittle–plastic transition is complete and the rock becomes fully plastic.

1.4.3   *The transition induced by temperature*

The principles discussed in Section 1.4.1 are aptly demonstrated by a study of the brittle–plastic transition of polycrystalline MgO by Paterson and Weaver (1970). Studies of the deformation of single crystals of MgO show that at low temperatures only the easy $\{110\}\langle110\rangle$ slip system is active. The $\{100\}\langle110\rangle$ slip system only becomes active at high temperatures, where it shows a yield strength that is strongly temperature

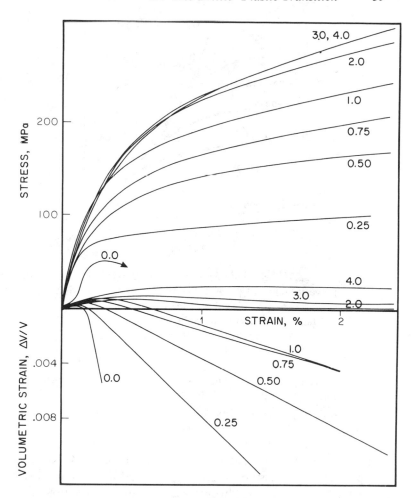

Fig. 1.22 The deformation of marble at a series of confining pressures passing through the brittle–plastic transition, following path A–B in Figure 1.21. At the top are shown stress–strain curves, showing a strong pressure dependence at low pressures that vanishes at high pressures. The bottom part of the figure is of volumetric strain plotted versus axial strain, showing the progressive suppression of dilatancy by confining pressure. The transition to full plasticity is complete at about 300 MPa. (After Scholz, 1968a.)(Pressures labeled in kb.)

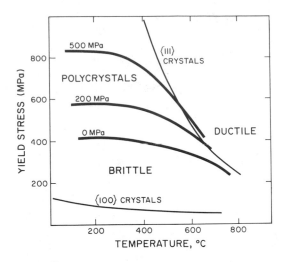

Fig. 1.23 The relationship between the brittle–plastic transition in polycrystalline MgO and the single crystal behavior. The easy glide system is activated in the {100} crystals, the hard system in the {111} crystals. The polycrystals only become plastic when both systems are active. (After Paterson and Weaver, 1970.)

dependent (Figure 1.23). The $\{110\}\langle110\rangle$ system provides only two independent slip systems, so that the von Mises–Taylor criterion for polycrystalline plastic flow is not satisfied until the $\{100\}\langle110\rangle$ system becomes active. As a result, polycrystalline MgO is brittle at room temperature and pressure.

Paterson and Weaver deformed polycrystalline MgO, at a variety of temperatures, at pressures of 0, 200, and 500 MPa. The yield stresses are shown in Figure 1.23. At low temperatures the deformation is cataclastic and there is a marked effect of pressure on the yield strength. Above 300°C the yield strengths decrease with temperature and the effect of pressure on the strength also begins to decrease. At 650°C the polycrystalline strength curves have intersected the yield strength curve for activation of the $\{100\}\langle110\rangle$ slip system. At that point, the strengths at 200 and 500 MPa have converged and coincide with that for activation of the $\{100\}\langle110\rangle$ system, which they follow at higher temperatures. At this point, there is no longer any pressure dependence of the strength and the behavior is fully plastic. The transition to full plasticity thus does not occur until the von Mises–Taylor criterion is satisfied.

The situation becomes more complicated for rocks of more geologic significance, because they are often composed of several mineral species that contrast markedly in their ductility. This is the case with quartzo-feldspathic rocks, which have been studied in the region of their brittle–plastic transition by Tullis and Yund (1977 and 1980). Their study followed a path of increasing temperature, like C–D in Figure 1.21. In these rocks, quartz undergoes the brittle–plastic transition at temperatures several hundred degrees lower than feldspar, which because of its complex structure and very low symmetry is highly resistant to plastic flow. At the strain rate used, 10^{-6} s^{-1}, the granite they studied was brittle at the lowest temperatures and pressures. At higher temperatures and pressures the rock became cataclastic, and they observed the following changes in microstructure in both the quartz and the feldspar as temperature was increased: abundant microcracks; abundant microcracks with some dislocations inhomogeneously distributed in planar zones; higher but still inhomogeneous densities of tangled dislocations with fewer microcracks; a cellular substructure of dislocations with rare microcracks; a lower uniform density of straighter dislocations with subgrains and no microcracks; and a uniform density of straight dislocations with recrystallized grains. In "dry" granite the transition between dominant microcracking and dominant dislocation motion occurred at approximately 300–400°C in the quartz and 550–650°C in the feldspar. The transition from cataclastic flow to dislocation flow in feldspar is described in more detail by Tullis and Yund (1987). Over all conditions the deformation was inhomogeneous because of the polyphase nature of the rock.

Tullis and Yund (1980) found that the addition of 0.2 wt% water to granite deformed under these same conditions produced a pronounced weakening and enhanced ductility. This was not due to a pore-pressure effect but was the combined effect of stress corrosion enhancing microcracking and hydrolytic weakening inducing dislocation mobility. Of these two effects, hydrolytic weakening dominated, so that the temperature for the aforementioned stages in the brittle–ductile transition was reduced by about 150–200°C for both quartz and feldspar. Water content is therefore an extremely important variable in determining the flow behavior of silicates, and one that is only beginning to be explored. It has been found that the brittle–plastic transition can be induced at high temperature simply by the addition of water to quartzites (Mainprice and Paterson, 1984), and clinopyroxenite (Boland and Tullis, 1986).

1.4.4 *Extrapolation to geological conditions*

We may summarize the transition by showing, in Figure 1.24, data for Solnhofen limestone at pressures and temperatures simulating depth within the earth (Rutter, 1986). Two transitions are indicated in the figure, one between brittle and semibrittle (cataclastic) behavior, and a deeper one between semibrittle and plastic deformation. The latter transition corresponds with the limit of dilational strain and with the

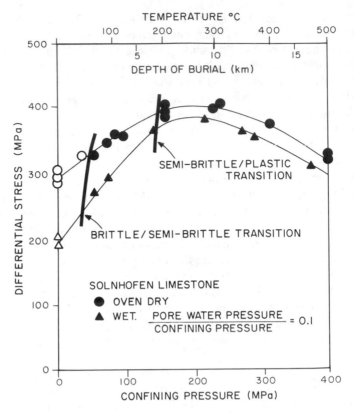

Fig. 1.24 Stages in the brittle–plastic transition for solnhofen limestone deformed at 10^{-5} s^{-1}. The rock has been deformed at a series of pressures and temperatures consistent with different depths of burial. A broad semibrittle field is seen between the cessation of brittle faulting and that of dilatancy, which occurs at the peak strength. (After Rutter, 1986.)

peak in strength. The figure is drawn for a strain rate of 10^{-5} s^{-1}. Because plasticity is strain-rate sensitive, at lower strain rates the peak strength will be less and the transitions at shallower depths.

It is difficult at present to extrapolate these results to geological conditions (Paterson, 1987). None of the experiments conducted to date on silicate rocks in the brittle–plastic transition region can be said to be at equilibrium, either with respect to the microstructure or the water content. The latter is a particular problem, both because of the sluggishness of the diffusion of water into the lattice and because it is not clear what the equilibrium water content should be under natural conditions. The solubility of water in silicates increases with pressure, producing an inverse pressure effect on strength (Tullis and Yund, 1980; Mainprice and Paterson, 1984). The experiments of Tullis and Yund (1980), conducted at 1.0–1.5 GPa, which are much higher pressures than would occur in nature at the same temperature, are therefore likely to be more ductile than at the lower pressures of the natural environment. In addition to these problems, there are a host of other deformation mechanisms that involve transport of aqueous fluids that have not been considered here (Etheridge et al.,1984).

These considerations cast considerable doubt on the practice of extrapolating high-temperature flow laws through the brittle–plastic transition to intersect the friction law in modeling the strength of the lithosphere, a topic that will be discussed in more detail in Section 3.4.1. This intersection is commonly associated with the depth limitation of seismic faulting, which is, strictly speaking, an oversimplification: The depth limitation is controlled by the lower stick-slip to stable sliding friction transition (Sec. 2.3.3), which is related to, but distinct from, the brittle–plastic transition for bulk deformation.

For the present we must rely on geological observations to estimate the conditions under which the brittle–plastic transition occurs in nature. Since both quartz and feldspar have only limited slip systems, fully plastic flow in those materials cannot occur by glide alone, but must be associated with recovery and recrystallization. For quartz this appears at about 300°C (Voll, 1976; Kerrich, Beckinsdale, and Durham, 1977) and, for feldspar, at about 450–500°C (White, 1975; Voll, 1976). Thus quartz begins to flow at about the beginning of greenschist facies metamorphism, but feldspar does not begin to flow until well into the amphibolite grade. Between these two states quartzo-feldspathic rocks behave as composite materials, with the quartz flowing and the feldspar responding in a rigid, brittle manner, forming *porphyroclasts*, a common textural feature of mylonites (Sec. 3.3.2).

2   Rock Friction

Once a fault has been formed its further motion is controlled by friction, which is a contact property rather than a bulk property. In the schizosphere the micromechanics of friction involve brittle fracture, but frictional behavior is fundamentally different from bulk brittle fracture. Here we examine this property in some detail and, in particular, discuss the stability of friction, which determines if fault motion is seismic or aseismic.

2.1   THEORETICAL CONCEPTS

2.1.1   *Historical*

Friction is the resistance to motion that occurs when a body is slid tangentially to a surface on which it contacts another body. It plays an important role in a great variety of processes. It is always present in machines in which there are moving parts, and a significant part of people's energy consumption is used in overcoming friction. In engineering, therefore, considerable effort has been made to reduce friction. In other situations it is important to improve traction by making things less slippery. Because friction plays a role in everyday life, its basic properties are common knowledge and have been since ancient times. The wheel was an important invention primarily because it substitutes rolling friction for the much higher sliding friction. Simple lubrication was also employed by the ancients to reduce friction. In spite of our familiarity with friction, its basic nature remained obscure until recent times.

The first systematic understanding of friction was obtained in the middle of the fifteenth century by Leonardo da Vinci. By careful experimentation, he discovered the two main laws of friction and further observed that friction is less for smoother surfaces. Leonardo's discoveries remained hidden in his codices and were rediscovered 200 years later by Amontons. In his paper of 1699, Amontons described the two main

44

laws of friction:

> Amontons's first law: The frictional force is independent of the size of the surfaces in contact.
>
> Amontons's second law: Friction is proportional to the normal load

In addition, he observed the general rule that the friction force is about one-third the normal load, regardless of surface type or material. Rock friction, as we shall see, is about twice that.

From Amontons's time through the time of Coulomb, nearly 100 years later, a mechanism of friction that could explain these two laws was sought vigorously. During this period the importance of surface roughness in friction was recognized. The subject was dominated by explanations for friction involving various types of interactions between protrusions on surfaces, called *asperities*. Amontons and his contemporaries thought the cause of friction was the interactions of asperities, which acted as either rigid or elastic springs. In the first case, friction was viewed as being due to gravitational work done in asperities riding up over one another. In the second case, asperities were seen as being elastically bent over during sliding; the amount of deflection and hence friction increased with the load. For a given load, the deflection of each asperity would be inversely proportional to the number of asperities, so that the summed resistance would be the same regardless of the number of asperities, in accordance with the first law. Although the shearing through or fracture of asperities was also thought to occur sometimes, its role in friction was considered minor. In that case, friction would depend on the number of asperities sheared, and hence, it was thought, would depend on the size of the surfaces in contact.

The difference between static and kinetic friction was also recognized at that time, and various explanations for it were proposed. Coulomb noticed, particularly for wooden surfaces, that initial friction increased with the time the surfaces were left in stationary contact. To explain this he imagined that the surfaces were covered with protuberances like bristles on a brush. When brought together the bristles interlocked, and this process became more pronounced the longer the surfaces were in contact. Coulomb used this mechanism to explain the general observation that static friction is higher than kinetic friction.

The principal weaknesses of these early theories of friction were that they failed to account for the energy dissipation characteristic of friction and for frictional wear. Both of these point to asperity shearing as an important mechanism, but to establish that required two developments: a model for asperity shearing that is still compatible with Amontons's first law, and microscopes that would allow examination of the surface damage produced during friction.

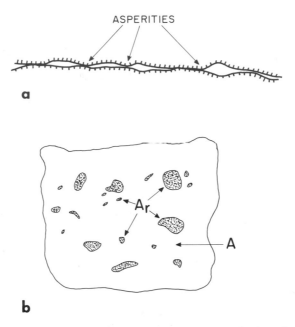

Fig. 2.1 Schematic diagram, in section and plan view, of contacting surfaces. The stippled areas in the plan view represent the areas of asperity contact, which together comprize the real contact area A_r.

2.1.2   *The adhesion theory of friction*

The modern concept of friction is generally attributed to Bowden and Tabor (1950, 1964), who, in a large body of work summarized in their treatises, investigated many different frictional phenomena for a wide range of materials. Central to their work was their adhesion theory for friction of metals, and many of their experimental studies were aimed at testing various aspects of this theory. Its concept is at once simple and subtle. They envisioned all real surfaces as having topography, as in Figure 2.1, so that when they are brought together they only touch at a few points, or asperities. The sum of all such contact areas is the real area of contact A_r, which is generally much smaller than the apparent, or geometric, area A of the contacting surfaces. It is only A_r, not A, that is responsible for friction. In their basic model, they assumed that yielding occurs at the contacting asperities until the contacting area is just sufficient to support the normal load N, thus

$$N = pA_r \tag{2.1}$$

where p is the penetration hardness, a measure of the strength of the material. They then supposed that, due to the very high compressive stress at the contact points, adhesion occurred there, welding the surfaces together at "junctions." In order to accommodate slip, these junctions would have to be sheared through, so that the friction force F is the sum of the shear strength of the junctions,

$$F = sA_r \tag{2.2}$$

where s is the shear strength of the material. Combining the two equations we can describe friction with a single parameter, the *coefficient of friction* μ,

$$\mu \equiv F/N = s/p \tag{2.3}$$

These results are elegantly simple. The crucial step occurs in Equation 2.1, in which we see that the real area of contact is controlled by the deformation of asperities in response to the normal load. This idea, which was not contained in any of the classical theories of friction, explains Amontons's first law. Because any constitutive law governing the shear interaction of asperities (Eq. 2.2) is bound to predict a shear force proportional to A_r regardless of the exact mechanism assumed, Equation 2.1 implicitly satisfies the second law as well, as long as the equation itself is linear in N.

The friction coefficient as given by Equation 2.3 is the ratio of two different measures of strength of the same material, being that of the softer material if two different materials are in contact. Therefore, to first order μ should be independent of material, temperature, and sliding velocity, because s and p, though both depend strongly on those parameters, differ between themselves by only a geometric constant. These predictions are, to first order, all borne out by experiment (cf. Rabinowicz, 1965). Furthermore, dramatic and important exceptions to Equation 2.3, as for example, in the case of various types of lubrication, usually can be found to result from specific mechanisms that allow Equations 2.1 or 2.2 to be violated.

Although the adhesion theory of friction conceptualizes the physical essence of the frictional interaction, in most cases it does not predict the correct value for μ. If the materials are ideally plastic, which is close to true for the ductile metals, and have a yield strength in compression σ_y, then $p = 3\sigma_y$ and $s = \sigma_y/2$. In this case then, Equation 2.3 predicts $\mu = \frac{1}{6}$, whereas the unlubricated friction of metals is usually two or three times that value. This is because overcoming junction adhesion is usually not the only work done in friction. Asperities often plough through the adjacent surface, or interlock, requiring additional deformation that needs to be specified with additional terms in Equation 2.2. Furthermore, reflection shows that A_r, as defined in Equation 2.1, is a minimum value, and that it may increase with shearing. In normally brittle materials like rock, it may be argued that the basic assumption behind Equation 2.1, that the surfaces yield plastically, is incorrect: Perhaps their contacts are elastic. Moreover, in that case asperities may fail by brittle fracture rather than plastic yielding.

The adhesion theory of friction therefore can be used only as a conceptual framework. In studying the friction of any class of materials over any given range of conditions, one is often concerned with interface deformation mechanisms that are specific to those conditions and materials. One therefore needs to ignore the overall triumph of the adhesion theory in explaining the general constancy of μ and to study in detail the effects on friction of variables such as material type, temperature, sliding velocity, and surface roughness. These may provide clues to the deformation mechanisms involved and may also be important in specific applications. It is especially important for geological applications, where, in order to scale from laboratory to geological conditions, we must understand the micromechanisms involved in the process.

In so doing, it is important to recognize that a constitutive law that describes μ, as in Equation 2.3, is actually a combination of two constitutive laws that describe different processes: the contact of the surfaces, as in Equation 2.1, and the shearing of contacting surfaces, as in Equation 2.2. There may be different mechanisms involved in these two processes, and their interactions may be complex.

2.1.3   *Contact of surfaces*

The view that the contact of surfaces occurs entirely by the mechanism of plastic yielding of asperities, as expressed by Equation 2.1, seems counterintuitive in many cases. Certainly for the harder materials, such as the silicates, we might expect contact to be largely elastic. If this were true, the contact area for an asperity would obey Hertz's solution for contact of an elastic sphere on an elastic substrate, and

$$A_r = k_1 N^{2/3} \tag{2.4}$$

where k_1 combines the elastic constants and a geometrical factor. Combining this with Equation 2.2, we predict the friction law

$$\mu = s k_1 N^{-1/3} \tag{2.5}$$

A formula of the same iorm was derived by Villaggio (1979). Bowden and Tabor (1964) found that friction of a diamond stylus sliding on a diamond surface obeys Equation 2.5. In general, however, friction on extended surfaces of the harder materials obeys a linear friction law, similar to Equation 2.3.

A solution to this paradox was found by Archard (1957). He studied the contact of surfaces with a model in which the surface was comprized of a large number of spherically tipped elastic indenters, each of which was covered by smaller spheres, and so on. Although each sphere obeyed Hertz's equation, he obtained an asymptotic solution in the limit of a

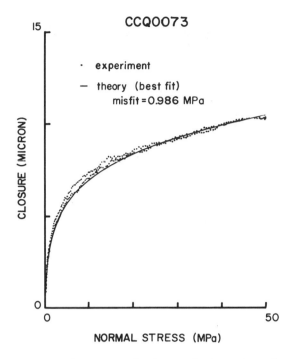

CCQ0073

Fig. 2.2 The closure δ of two elastic surfaces in contact under the action of a normal load. The theoretical fit is obtained from an independent analysis of the surface topography, using a modified version of the Greenwood–Williamson theory for contact of elastic surfaces. (From Brown and Scholz, 1985a.)

large number of hierarchies of sphere sizes that produced a linear law,

$$A_r = k_2 N \tag{2.6}$$

where k_2 contains elastic constants and the geometrical properties of the sphere distribution. Equation 2.6, when combined with a constitutive law of the form of Equation 2.2, predicts a linear friction law.

Greenwood and Williamson (1966) developed a more realistic model for contact of a rough surface with a flat surface in which the rough surface was described with a random distribution of asperity heights. This result was generalized to the case of two rough surfaces in contact by Brown and Scholz (1985a). The closure δ of the surfaces under the action of a normal stress σ_n obeys a constitutive equation approximately given by

$$\delta = B + D \log \sigma_n \tag{2.7}$$

where B and D are constants that scale with the elastic constants and are otherwise determined by the topography of the surfaces. An experimental result verifying Equation 2.7 is shown in Figure 2.2. The first time

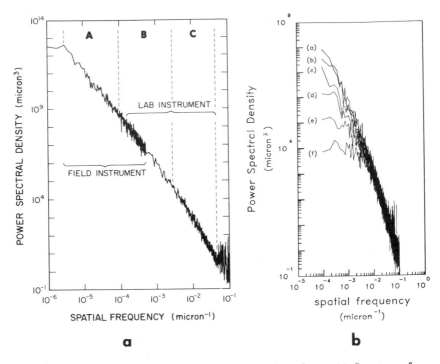

Fig. 2.3 Power spectra of the topography of rock surfaces: (a) Spectrum from 1 m–10 μm of a natural joint surface; (b) spectra (1 cm to 10 μm) of ground surfaces of various roughnesses. (After Brown and Scholz, 1985b.) In (a) the fractal dimension, proportional to the slope of the curve, is found to be different over different scale ranges, denoted A, B, and C. In (b) the corner in the spectra is a result of grinding and occurs approximately at the dimension of the grit size of the grinding wheel.

such surfaces are loaded there is a certain amount of permanent closure, due to brittle fracture or plastic flow of asperity tips. After several cycles the closure becomes completely recoverable, so that the contact is indeed elastic. In more ductile rocks, such as marble, the permanent closure becomes more pronounced and flattening of asperities by plastic flow can be observed (Brown and Scholz, 1986). Greenwood and Williamson (1966) evaluated the contact of spherical tips and found that plastic flow is expected when the closure exceeds a critical value $\delta_p \approx k_3 \rho (p/E')^2$, where p is hardness, ρ is the radius of curvature of the tips, E' is an elastic constant, and k_3 is a geometrical factor. Thus under sufficiently high loads, asperities of small ρ may flow plastically even for normally brittle materials of high p, although brittle fracture of the asperities often may intercede. This elastic contact model also predicts that A_r varies nearly linearly with load, as in Equation 2.6, where k_2 depends on the elastic constants and the topography of the surfaces.

Because the constitutive laws that govern the elastic contact of surfaces under the action of normal stress, Equations 2.6 and 2.7, are largely controlled by topography of the surfaces, it is important to understand the nature of surface topography. Brown and Scholz (1985b) made a general study of the topography of rock surfaces over the spatial band 10^{-5} m to 1 m. They found, as is summarized in Figure 2.3a, that natural rock surfaces resembled fractional Brownian surfaces over this bandwidth, with a fractal dimension $\mathbf{D}$ that was a function of spatial wavelength. A fractal surface has a power spectrum that falls off as $\omega^{-\xi}$, where ξ is between 2 and 3, ω is spatial frequency, and $(5 - \xi)/2 = \mathbf{D}$ (see Mandelbrot, 1983). Differences between surfaces could thus be characterized by differences in $\mathbf{D}$ and a roughness κ_0 defined at an arbitrary wavelength. For such a surface, then, roughness cannot be specified by rms asperity height, or similar measure frequently used in engineering practice, without specifying the size of the surface. In contrast, surfaces that have been prepared artificially with grinding machines exhibit a "corner frequency" in their spectra, which generally occurs at a spatial wavelength close to that of the dimension of the grinding compound used (Fig. 2.3b). Thus, such ground surfaces differ fundamentally from natural surfaces in their scaling properties. Since friction is a contact property, it is likely sensitive to the topography of the surfaces. Results to date, though suggestive, have been ambiguous on what the relationship might be. This point is particularly important regarding scaling, as measurements of fault topography (discussed in Section 3.5.1) show that faults also have this general fractal topography over a broad scale range.

2.1.4 *Other frictional interactions*

The real area of contact A_r is thus predicted to increase linearly with the load regardless of whether asperity deformation is elastic or plastic. As noted in Section 2.1.2, this alone can explain Amontons's laws, provided that the shear failure law for sliding is of the form of Equation 2.2, but the value of μ in different cases depends specifically on that equation. It was also noted that the shearing of adhered junctions, in the basic Bowden and Tabor model, is seldom sufficient to account for the observed friction. There are usually additional processes that come into play that increase the friction (Jaeger and Cook, 1976). The most common effects are shown in Figure 2.4. In the first of these, a hard asperity penetrates into a softer surface, which it then ploughs through during sliding. If p is the hardness of the softer material, a spherically tipped asperity with radius of curvature ρ will penetrate until the radius of contact r is given by

$$N = \pi r^2 p \tag{2.8}$$

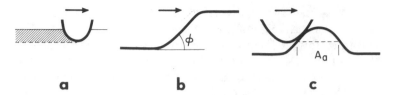

Fig. 2.4 Several types of asperity interactions: (a) ploughing; (b) riding up; (c) interlocking.

During sliding the asperity will plough a groove of cross-sectional area

$$A_{\mathrm{g}} = 2r^3/3\rho \tag{2.9}$$

The force necessary to plough this groove will be of the order pA_{g}, and that necessary to shear the junctions at the surface of the asperity will be of the order $\pi r^2 s$. Adding these and using Equation 2.8, we obtain

$$F = sN/p + 2N^{3/2}/3\pi^{3/2}p^{1/2}\rho \tag{2.10}$$

The first term is called the shearing term and is the same as in Equation 2.3. The addition of the ploughing term thus increases the friction and also makes it nonlinear.

If the surfaces are initially mated at some wavelength and have irregular topography with long-wavelength asperities, as the natural surfaces in Figure 2.3a, then when sliding commences the asperities may ride up on one another, so that sliding occurs at a small angle ϕ to the direction of the applied force F and the mean plane of the surfaces. If the friction coefficient is μ, then

$$F = \left[\mu + \phi(1 + \mu^2)\right]N \tag{2.11}$$

for small ϕ, so that there is an apparent increase in friction. This effect will also produce "joint dilatancy" during sliding, since a component of the motion will be normal to the mean sliding surface.

With natural surfaces with topography as in Figure 2.3a, asperities will occur on all scales and asperity interlocking, such as shown in Figure 2.4c, is likely to be common. If the normal load is sufficiently large relative to shear strength, so that riding up is suppressed, sliding will occur by shearing through the interlocked asperities. The force required for this is

$$F = sA_{\mathrm{a}} \tag{2.12}$$

where s is the shear strength of the asperity and A_{a} is its area. Since $A_{\mathrm{a}} > A_{\mathrm{r}}$, the contact area, this will result in a larger friction than predicted by Equation 2.2. This mechanism also predicts a large amount of wear, which is typical of rock friction.

The effects of ploughing and asperity interlocking in the case of friction of metals have been evaluated by Challen and Oxley (1979) and Suh and Sin (1981). They analyzed these effects by assuming plastic flow as the deformation mechanism and concluded that their contribution to friction was, in common cases, each of similar magnitude as junction shearing. The friction of silicate rock, on the other hand, must certainly involve a considerable amount of brittle fracture of asperities, as was first pointed out by Byerlee (1967a, b). The strongest evidence for this is that frictional sliding of rock produces abundant wear particles. Under conditions in which silicate rocks are macroscopically brittle, these particles are angular and have the appearance of being formed by brittle fracture. Byerlee (1967b) presented a rudimentary theory of rock friction based on a brittle fracture micromechanism, which explains some of the more salient features of rock friction. It is, however, highly idealized and is far from a realistic description of the actual process. Yamada et al. (1978) formulated the problem of contact of elastic surfaces in shear, obtaining the result of Equation 2.6, and also produced some experimental confirmation. Their contact model was expanded by Yoshioka and Scholz (1989, unpublished), who showed experimentally that it predicts the correct shear stiffness of the contact prior to any slip taking place. This treatment of friction, however, does not go beyond the assumption of Equation 2.2, so it remains highly simplified. All of the descriptions of frictional processes discussed in this section, however, are similarly crude, and can only serve as a qualitative guide in interpreting friction data. That there is no constitutive law for friction that is quantitatively built upon a micromechanical framework is not surprising, since not only is the shear contact problem itself complex, depending in detail on the topography of the contacting surfaces, but the topography itself constantly evolves due to wear during sliding.

2.2   EXPERIMENTAL OBSERVATIONS OF FRICTION

Rock friction studies are usually made in regard to two main applications: the stability of engineering structures in rock, and the mechanics of earthquakes. In the first case, experiments are usually conducted at relatively low loads, and the concern is with friction of joints, so that the starting surfaces are often mated joint surfaces with no initial wear material. The interest is primarily in determining the initial friction and peak friction at which major sliding commences. Studies aimed at understanding earthquakes, on the other hand, are concerned with friction of faults, which have unmated topography and abundant wear material, and the goal is to understand "steady-state" friction, achieved after initial frictional yielding. These studies usually start with ground

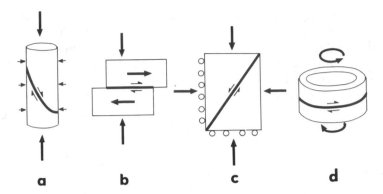

Fig. 2.5 Common types of experimental configurations used in friction studies: (a) triaxial compression; (b) direct shear; (c) biaxial loading; (d) rotary.

surfaces, sometimes covered with artificial gouge, and are typically conducted at higher loads.

There are a variety of different experimental configurations used in rock friction studies (Fig. 2.5), each with its own advantages and disadvantages. The triaxial test is best suited for high-pressure and high-temperature studies. However, its configuration is not compatible with slip, so that only limited amounts of slip can be achieved and dynamic measurements during stick slip are of doubtful validity because of the possibility of sample misalignment occurring during rapid slip. The other three types of experiments are limited to lower loads by the uniaxial failure strength of the rock. The direct shear test, shown in one configuration in Figure 2.5b, is simple and suitable for large specimens, but contains a moment so that the normal stress is not constant over the surface. A version of this, the "sandwich" configuration of Dieterich (1972), also contains a moment that results in nonuniform normal stress, and, in addition, has two sliding surfaces. The biaxial experiment, Figure 2.5c, produces uniform stresses, and moderate displacements are possible, whereas the rotary shear apparatus, Figure 2.5d, is particularly suited to producing large displacements.

2.2.1 *General observations*

Typical frictional behavior of rock is shown schematically in Figure 2.6 as shear stress versus displacement diagrams, at constant normal stress. Because the load point is usually at some distance from the sliding surface, displacement includes both sliding and the elastic distortion of the sample. Upon first loading of surfaces that have not previously slipped, sliding often begins gradually, beginning at a relatively low

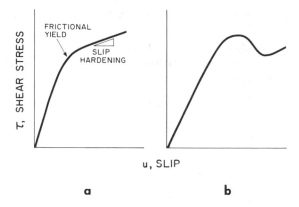

Fig. 2.6 Two common types of stress–displacement curves observed for frictional sliding of rock: (a) frictional yielding followed by slip hardening; (b) sliding following an upper yield point.

stress, the *initial friction* level. Frictional resistance continues to rise rapidly, and as stress is increased steadily, the rate of slip accelerates until a frictional *yield* point is reached. The inflection change at yielding is variably pronounced and marks the point at which the principal sliding begins. The frictional strength may increase further as *slip hardening* continues at a diminished rate (Fig. 2.6a), or it may fall from a peak to a residual value and then increase with slip hardening (Fig. 2.6b). At some large slip, friction may eventually reach a constant, steady-state level, but this is seldom actually observed in experiments.

Frictional strength for a variety of rocks and surface types is summarized in Figure 2.7a and 2.7b for low and high loads, respectively, by showing maximum friction, which is either friction measured at the peak (as in Figure 2.6b) or at some point after yielding (as in the case of the type of friction curve shown in Figure 2.6a).

At low loads, data for maximum friction taken for a given surface will generally follow a linear friction law quite closely, but great variations can be observed among surfaces and rock types, with values of μ ranging between 0.2 and 2. This variation is primarily due to the effect of surface roughness, which is dominant at low loads, particularly for joint friction.

At high loads, in contrast, there is very little variation in maximum friction for a wide variety of rock types, gouge types, and surface roughnesses (Fig. 2.7b). Some of the clay minerals are exceptional in having unusually low friction. This difference with the low-load data is because the effect of roughness diminishes with load and partly because these experiments were primarily conducted with "faults" (meaning surfaces with uncorrelated topography). Those in Figure 2.7a mainly were with "joints," in which the surfaces were "mated." These two cases differ significantly in their dependence on roughness, as will be discussed in Section 2.2.2.

Friction at high loads is nonlinear. Thus if friction is defined in the usual way, $\mu = \tau/\sigma_n$, one would find that μ decreases with σ_n. Alternatively, Jaeger and Cook (1976, page 56) suggested the law

$$\tau = \mu_0 \sigma_n^m \qquad (2.13)$$

where μ_0 and m are constants. Byerlee (1978) fit the data with two straight lines,

$$\tau = 50 + 0.6\sigma_n \qquad (2.14a)$$

(in MPa) for $\sigma_n > 200$ MPa, and

$$\tau = 0.85\sigma_n \qquad (2.14b)$$

at lower normal stress.

This friction law is, with very few exceptions, independent of lithology. It holds over a very wide range of hardness and ductility, from carbonates to silicates. To first order it is also independent of sliding velocity and roughness, and, for silicates, to temperatures of 400°C (Stesky et al., 1974). It has become known as *Byerlee's law*. Because of its universality one can use it to estimate the strength of natural faults.

Paradoxically, this universal law has been found to describe a strength property that, being dependent in detail on a great many variables, is one of the least reproducible strength properties known. Furthermore, the dynamic instability that often occurs during frictional sliding of rock (which is known on the laboratory scale as stick slip and on the geological scale as earthquakes) depends on friction properties that are of only second order. These two aspects of friction require explanation, which is only partially forthcoming at present.

2.2.2 *Effects of other variables on friction*

Many variables affect rock friction, some in such sensitive ways that it is often difficult to duplicate conditions well enough to obtain sufficiently reproducible results so as to recognize an effect and isolate its cause. This is particularly true of effects that control the stability of sliding, which are mainly connected with the velocity and time effects in friction (discussed in Section 2.3). Here we restrict the discussion to effects that cause variations in the "base" frictional strength.

Roughness The topography of natural fracture or fault surfaces is typically like that shown in Figure 2.3a. In contrast, many friction studies are made with ground surfaces, as Figure 2.3b, which have a

MAXIMUM   FRICTION

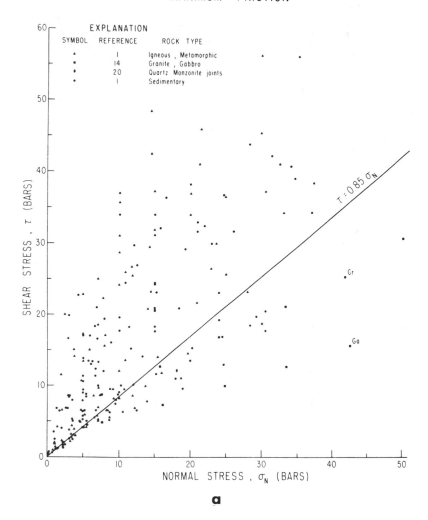

Fig. 2.7 Frictional strength for a wide variety of rocks plotted as a function of normal load: (a) at low loads; (b) at high loads. (From Byerlee, 1978.)

corner frequency in their spectrum and hence a stable roughness value at long period. Furthermore, initial friction of joints will differ from that of faults, because the former have mated walls and should experience the greatest topographic interference at slip onset, whereas fault surfaces will be unmated and less dominated by roughness.

The friction data taken at low stress, Figure 2.7a, are primarily from natural mated joints. The large scatter in those results is largely due to variations in roughness. At low loads, sliding of initially mated surfaces

MAXIMUM FRICTION

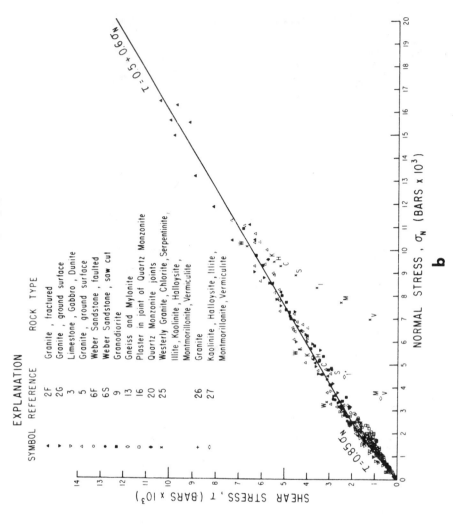

EXPLANATION

SYMBOL	REFERENCE	ROCK TYPE
◄	2F	Granite , fractured
▶	2G	Granite , ground surface
▷	3	Limestone , Gabbro , Dunite
△	5	Granite , ground surface
●	6F	Weber Sandstone , faulted
■	6S	Weber Sandstone , saw cut
□	9	Granodiorite
◇	13	Gneiss and Mylonite
◆	16	Plaster in joint of Quartz Monzonite
▲	20	Quartz Monzonite joints
×	25	Westerly Granite , Chlorite , Serpentinite, Illite , Kaolinite , Halloysite , Montmorillonite , Vermiculite
+	26	Granite
◇	27	Kaolinite , Halloysite , Illite , Montmorillonite , Vermiculite

$\tau = 0.5 + 0.6\sigma_N$

$\tau = 0.85\sigma_N$

SHEAR STRESS , τ (BARS x 10^3)

NORMAL STRESS , σ_N (BARS x 10^3)

b

Fig. 2.7 (cont.)

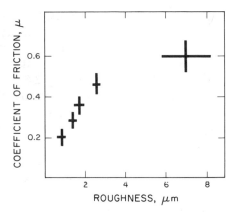

Fig. 2.8 The dependence of friction on surface roughness for very smooth surfaces at very low loads. (After Byerlee, 1967b.)

up to the maximum friction level is accompanied by riding up of asperities and joint dilatancy (Sundaram, Goodman, and Wang, 1976). Friction will be controlled by the effect given in Equation 2.11, and hence be strongly influenced by roughness. Barton and Choubey (1977) developed an empirical joint friction law to account for this effect. Their model, which includes an empirical topographic parameter, the *joint roughness coefficient*, satisfactorily explains this type of data. Swan and Zongqi (1985) have put this type of formulation on a better theoretical basis by adapting the elastic shear contact theory of Yamada et al. (1978) to the case of mated surfaces. They checked this with topographic and friction data from an actual joint and obtained good agreement between theory and experiment in predicting initial friction. Their theory predicts linear friction and only a nominal effect of scale on friction beyond a certain dimension.

As might be expected intuitively, friction of unmated ground surfaces becomes small for very smooth surfaces at low loads. This behavior, observed by Byerlee (1967b), is shown in Figure 2.8. He found, independent of the hardness of the material, that friction had a value of $\mu \approx 0.15$ for mirror-smooth surfaces and then approached a limiting value of $\mu = 0.4$–0.6 as roughness increased. This result agrees with his theory of friction, in which frictional sliding occurs by brittle fracture of asperities that are in contact only at their tips, and which predicts a value of μ between 0.1 and 0.15. He concluded that this condition is approached for very smooth surfaces and that for rougher surfaces friction increases because of the increasing effect of asperity interlocking.

At high loads a direct effect of roughness on frictional strength is difficult to discern for either mated or unmated surfaces. In the case of the former, asperity riding-up is suppressed by high loads; asperities shear through, and friction of mated surfaces does not differ greatly from unmated ones (Byerlee, 1967a). Accordingly, joint dilatancy becomes negligible (Scholz, Molnar, and Johnson, 1972). On the other hand, strong slip hardening is usually observed at high loads (Fig. 2.6), particularly during the first loading. Because wear is greatest at the highest loads, the roughness of the surfaces will change rapidly after sliding commences. It is not surprising that little correlation is found between postyield friction and initial roughness. It is often found, however, that relatively smooth surfaces exhibit friction curves of the type shown in Figure 2.6a, whereas very rough or mated surfaces are more likely to obey friction curves of the type shown in Figure 2.6b, with a peak stress at yield followed by a drop to a lower residual friction (e.g., Ohnaka, 1975). Because wear tends to roughen smooth surfaces and smooth rough ones, this suggests that the early part of the friction behavior is controlled by roughness, but after some amount of sliding the surfaces become more similar in both friction and roughness.

It is likely that the slip hardening itself is a result of the changing of surface topography with slip and wear. In general, slip hardening is almost universally observed and must involve an increase with sliding in the intimacy of contact of the surfaces. This is probably only partly due to changes in roughness. Wear products (gouge) also result from sliding, and these can play an important role in slip hardening, particularly for the softer rocks that tend to be plucked by adhesion rather than fragmented by abrasion (Ohnaka, 1975).

Hardness   At low and intermediate stresses, a mild effect of hardness on frictional strength can be observed sometimes (Byerlee, 1967b; Ohnaka, 1975). This effect becomes negligible at high loads (Fig. 2.7b), for the reasons given in Section 2.1.2.

While not greatly influencing the frictional strength, hardness plays an important role in determining the mechanism of friction. Logan and Teufel (1986) measured the real area of contact during sliding for several cases: sandstone sliding on sandstone (s/s); limestone sliding on limestone (l/l); and sandstone sliding on limestone (s/l). Their results, shown in Figure 2.9, are that A_r increases linearly with N in all cases, which is anticipated from either plastic or elastic contact theory (Eqs. 2.1 and 2.6). The growth of A_r was accomplished in the sandstone/sandstone case by a rapid increase in the number of contact spots with normal stress, which would be expected from elastic theory (Yamada et al., 1978), whereas in the other cases the number of spots did not increase greatly, but the spots grew in size, more like the observations of Brown and Scholz (1986) of plastic blunting of asperities in the contact of marble. In all cases N/A_r, the mean stress on the contacts, was

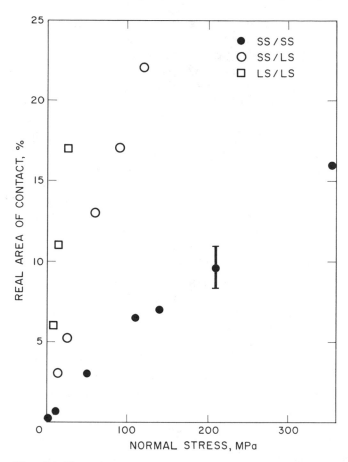

Fig. 2.9 The relationship between fractional real contact area and applied normal stress during frictional sliding in three cases: ss, sandstone on sandstone; ls, limestone on limestone; ss/ls, sandstone on limestone. (Data from Logan and Teufel, 1986.)

constant and independent of load. The approximate values of N/A_r for the three cases were: sandstone/sandstone, 2,200 MPa; limestone/limestone, 200 MPa; sandstone/limestone, 600 MPa. These are about the same as the uniaxial compressive strength of quartz, of calcite, and the penetration hardness of calcite, respectively. Thus friction can be interpreted as being controlled, in these three cases, by: s/s, the brittle fracture of quartz asperities; l/l, flattening and shearing of calcite asperities; and s/l, ploughing of quartz asperities through calcite. The real area of contact is such that it just balances differences in the shear resistance of asperities in each case so that there is little difference in overall friction. Stesky and Hannan (1987), in static contact experiments, obtained the same result as Logan and Teufel for quartz and calcite, but found that a more ductile material (alabaster) showed a nonlinear increase in A_r above a critical load. In friction, this may be a clue to the nature of a stability transition induced by plasticity, discussed in Section 2.3.3. Hardness and ductility also have strong influences on the wear processes (Engelder and Scholz, 1976; see also Sec. 2.2.3).

Temperature and ductility   It is observed that temperature has little or no effect on frictional strength of either metals or rock over a very broad range (Rabinowicz, 1965; Stesky, 1978). Because the effect of temperature is largely to change the hardness and ductility of the materials, which, as just discussed, have scant effect on friction, this observation is consistent with the concepts presented so far. If, however, the temperature is high enough so that the shear-zone materials become sufficiently ductile to weld the surfaces, so that $A_r \approx A$, then slip will occur by ductile shearing within this material, the constitutive behavior will resemble a flow law rather than a friction law, and the strength will become independent of normal stress.

The frictional strength of two rocks at high loads is shown as a function of temperature in Figure 2.10a, where they are compared with their fracture strengths under similar conditions (Stesky et al., 1974). The fracture strength of Westerly granite decreases rapidly with temperature, the effect being greater at higher pressure, so the strengths at 400 and 500 MPa confining pressures become equal at about 500°C. The frictional strength, in contrast, is independent of temperature until it becomes equal to the fracture strength at 500°C, above which it falls in concert with the fracture strength. This behavior is somewhat clearer for the data on San Marcos gabbro. The lack of an effect of pressure on the strength of the granite above 500°C suggests, from the arguments of Section 1.4, that the rock is flowing plastically above this temperature (compare with Fig. 1.23). Because the frictional strength is the same as the fracture strength above that temperature, this suggests that welding has occurred. This was confirmed by examination of the specimens (Stesky, 1978).

The situation is not quite so simple, however. The frictional strength of the gabbro, for which the temperature sensitivity was more pronounced than the granite, is shown in Figure 2.10b. Although above 400°C the frictional strength falls and the normal stress dependence also decreases, there is still a positive dependence of shear strength on normal stress at 700°C, that is, the essential frictional character of the constitutive behavior has been retained. The deformation involved during sliding was studied in more detail for the granite (Stesky, 1978). Above 500°C plastic flow in the quartz could be identified microscopically, and the activation energy for the process changed from about 30 kcal/mole to

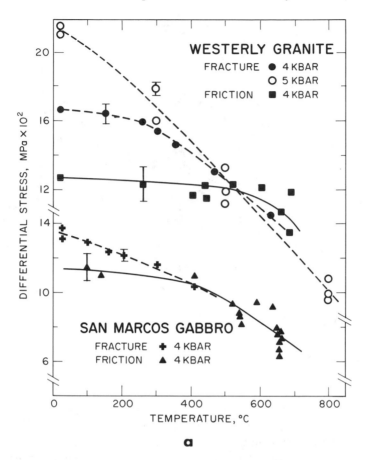

Fig. 2.10 In (a) the intact strength (at two confining pressures) and the frictional strength are shown for two rocks as a function of temperature. In (b) the frictional strength of San Marcos gabbro is shown versus normal stress at a variety of temperatures. (After Stesky et al., 1974.)

about 85 kcal/mole. The lower value is typical of stress-corrosion assisted crack growth (Scholz and Martin, 1971), whereas the higher is similar to that found for the thermal activation of plastic flow processes in granite (Goetze, 1971). The microscopic study of Stesky indicated that the feldspars remained brittle under all conditions and that brittle fracture was still prevalent in the gouge at the highest temperature studied, 700°C. That, together with the retention of the normal stress dependence of strength, indicates that the brittle–plastic transition was not complete in these experiments and the behavior of the gouge was still cataclastic. In the case of the gabbro, which contains no quartz, semibrit-

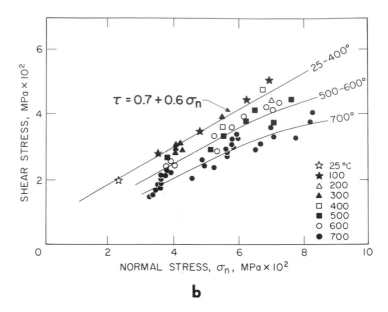

Fig. 2.10 *(cont.)*

tle behavior occurred throughout the experimental series and welding did not occur.

The completion of this process was demonstrated by Shimamoto (1986), who deformed halite gouge sandwiched between sandstone surfaces at room temperature. The halite underwent a brittle–plastic transition with increasing normal stress in much the same way as did the marble discussed in Section 1.4.2. The frictional strength of the halite gouge at three sliding rates is shown in Figure 2.11. At low normal loads, strength increased rapidly with σ_n, following a typical friction law with only a weak rate effect, partly negative. At intermediate normal loads the behavior is still frictional, with a strength that increases rapidly with normal load, but the rate effect becomes positive. At higher normal stress the shear strength dependence on σ_n falls off, at the lowest rate first, until above $\sigma_n \approx 200$ MPa there was no longer any normal stress dependence and there was a strong positive rate dependence. At 250 MPa normal stress the halite was fully plastic and the shearing behavior was obeying the flow law for halite (Shimamoto and Logan, 1986).

It appears, from this discussion, that shear-zone behavior should remain essentially frictional, with strength increasing with normal stress, up to the point where the brittle–plastic transition for the shear-zone material is complete, although there may be some thermal softening before that (see Sec. 3.4.1). For silicate rocks, no effect of temperature on frictional strength has been observed below 400°C. However, what

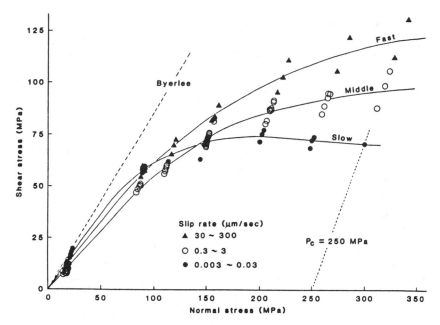

Fig. 2.11 Strength of sandstone surfaces sliding on a sandwich of halite at three sliding velocities, plotted versus normal stress. (From Shimamoto, 1986.)

remains to be discussed (Sec. 2.3.3) is the sign change in the velocity effect on friction induced by ductility, which controls the stability of faulting and hence the depth to which earthquakes can occur.

Pore fluids Just as with brittle fracture (Sec. 1.3), there are two distinct effects of pore fluids on friction: a purely physical effect due to fluid pressure, and an environmental effect due to chemomechanical weakening of the solid by the fluid, of which water is the most important agent.

The first leads to an effective stress law for friction. If two surfaces are in contact, subject to an applied normal stress σ_n, and a fluid pressure p exists within the noncontacting interstice between the surfaces, then

$$\sigma_n A = p(A - A_r) + \sigma_c A_r \qquad (2.15)$$

where σ_c is the mean stress on the asperity junctions. From Equations 2.1 or 2.6, $\sigma_c A_r = N$, and since $N/A = \bar{\sigma}_n$, the effective normal stress that controls contact and friction is then

$$\bar{\sigma}_n = \sigma_n - \left(1 - \frac{A_r}{A}\right) p \qquad (2.16)$$

Equation 2.16 defines the effective stress law for friction. Because in most cases $A_r/A \ll 1$, Equation 2.16 can often be approximated quite well with the simple effective stress law (Eq. 1.45). For some properties that are very sensitive to joint aperture, like permeability, the difference between Equations 2.16 and 1.45 may be significant (Kranz et al., 1979).

Byerlee (1967a) measured the friction of a gabbro in which water was present at various pressures. He observed the friction law

$$\tau = 10 + 0.6(\sigma_n - p) \qquad (2.17)$$

This, when compared with Equation 2.14a, demonstrates the two effects of water. The second term reflects the effective stress law, whereas the first term shows that there is an intrinsic reduction of friction due to the presence of water.

Dieterich and Conrad (1984) studied the effect of humidity on rock friction by conducting experiments in an atmosphere of dry argon and comparing them to identical experiments done at normal laboratory humidity. They found that the base friction coefficient μ increased from 0.55–0.65 under humid conditions to 0.85–1.0 under dry conditions, in agreement with Byerlee's results. They also found that several of the time and velocity effects on friction, which are important in determining the stability in slip, were suppressed under dry conditions. This confirmed earlier predictions regarding the mechanism of these effects, the discussion of which will be deferred until Section 2.3.

There are probably several underlying causes for the effect of humidity on friction. Because surfaces have unbalanced charge, they are normally covered with a layer of molecules, adsorbed from the environment, that reduces adhesion between surfaces in contact. Because Dieterich and Conrad cleaned and baked their specimens in a dry atmosphere before testing, they probably drove off much of this adsorbed layer. This resulted in higher friction. They observed friction to drop to normal levels once water vapor was admitted into their experimental chamber. Related to this is the effect of water on reducing the surface energy of silicates: the effect first observed by Obriemoff (Sec. 1.1.2; see also Parks, 1984). Finally, because rock friction often occurs by brittle fracture of asperities, the intrinsic reduction of brittle strength by water, which is related to the same processes, as discussed in Section 1.3.2, also plays a role.

2.2.3   *Wear and gouge behavior*

Frictional sliding is always accompanied by damage and erosion of the surfaces, a process that is known as wear. There are several important mechanisms involved, particularly adhesive wear and abrasive wear (Rabinowicz, 1965). Adhesive wear occurs when junction adhesion is so strong that junctions shear off part of the adjoining asperity, rather than at the junction itself, resulting in transfer of material from one surface to the other. When there is a hardness contrast between opposing materials, the harder material may plough through the softer and gouge material out, a process known as abrasive wear. Because rocks often contain minerals of differing hardness, this type of wear is often more important than adhesive wear. Furthermore, under the conditions in which the rock-forming minerals are brittle, the wear process may be dominated by brittle fracture, forming loose wear particles with angular forms. In the rock friction literature, such loose wear detritus is called gouge, adopting the geological term for the same material found in faults. Under more ductile conditions, adhesive wear may become more important, and may contribute to the formation of mylonites in ductile faulting (Sec. 3.3).

Several types of wear mechanism are illustrated in Figure 2.12. A groove made by the sliding of a hard rider on a softer surface is shown in Figure 2.12a. This type of wear track can be found on surfaces of materials that are macroscopically brittle, such as the fused silica shown, indicating that such materials may flow plastically under the high hydrostatic compression that exists beneath the point of contact. Often, however, a trail of partial ring cracks can also be seen (Fig. 2.12b), formed by the high tensile stresses that exist behind the trailing edge of a contacting asperity (Lawn, 1967). In rock friction at high loads, feather fractures that penetrate into the surfaces at acute angles to the sliding direction are another characteristic form. Continued sliding may cause such cracks to propagate well into the substrate and eventually form large fragments that are plucked loose. The carrot-shaped grooves shown in Figure 2.12c and d seem to be characteristic of stick-slip motion (Engelder, 1974). The sharp ends of the grooves always point in the direction of motion of the surface in which they lie, and the length of the grooves is usually equal to or less than the slip in an individual stick-slip event. The interpretation given by Engelder is that brittle fracture of the contacting asperities occurs at the onset of stick slip. Asperities then dig in progressively during slip, and finally come to rest in the blunt end of the grooves.

A theory of wear was developed by Archard (1953), which, although quantitatively describing the main properties of the process, is not specific about the mechanism, and hence finds general application. We begin by assuming that Equation 2.1 determines the real area of contact, where, following the results of Logan and Teufel (1986), we use an unspecified hardness parameter h in place of p. Then assume that the contacts are circular, of diameter d, so that there will be n contacts, given by

$$n = \frac{4N}{\pi h d^2} \tag{2.18}$$

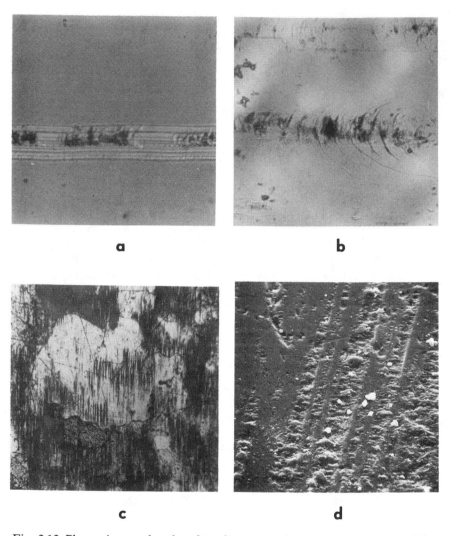

a   b

c   d

Fig. 2.12 Photomicrographs of surface damage produced during frictional sliding. (a) Groove made by sliding a topaz stylus over fused silica at a normal load
$N = 191$ g. Sliding direction, left to right; field of view, 0.02 cm. (b) Same as (a)
except with a normal load of 653 g. Notice partial ring cracks. (c) Carrot-shaped
grooves produced on granite surface during stick-slip motion. Grain near center
is quartz; field of view, 0.2 cm; sliding direction, bottom to top. (d) SEM image
showing close-up from (c). Field of view, 0.01 cm; sliding direction, parallel to
wear tracks. (Photographs by Terry Engelder.)

and that each contact junction exists for an effective working distance d_e. We assume that $d_e = \alpha d$, where α is a constant with a value near unity, for which there is some experimental justification (Rabinowicz, 1965). Then each junction must be replenished $1/d_e$ times per unit of travel, so that the number of junctions per unit of travel is

$$\mathcal{N} = \frac{n}{d_e} = \frac{4N}{\alpha\pi h d^3} \qquad (2.19)$$

The probability that any junction is sheared off during sliding is k, and on the assumption that the fragment formed by shearing is a hemisphere of diameter d, the wear rate (defined as the volume V of wear material formed per unit slip D) is given by

$$\frac{\partial V}{\partial D} = \frac{k\pi d^3}{12}\mathcal{N} = \frac{kN}{3h\alpha} \qquad (2.20)$$

so that the volume of wear fragments formed in sliding through a distance D is

$$V = \frac{kND}{3h\alpha} \qquad (2.21)$$

which, neglecting the porosity change, produces a gouge zone of thickness

$$T = \frac{\kappa\sigma D}{3h} \qquad (2.22)$$

where σ is the normal stress and $\kappa = k/\alpha$ is a dimensionless parameter called the *wear coefficient*. If σ is constant during sliding, then there will be a constant thickness ratio

$$\frac{T}{D} = \frac{\kappa\sigma}{3h}. \qquad (2.23)$$

Different assumptions can be made concerning the mechanism of wear, which results in different geometrical factors in the equations (Rabinowicz, 1965), but here such details can be lumped into κ.

This model can be checked and κ evaluated using some experimental results (Yoshioka, 1986). The data from these experiments are shown in Figure 2.13, where thickness ratio has been plotted against normal stress. Although there is considerable scatter, the results are in rough agreement with the linear prediction of Equation 2.23. The data have been fit

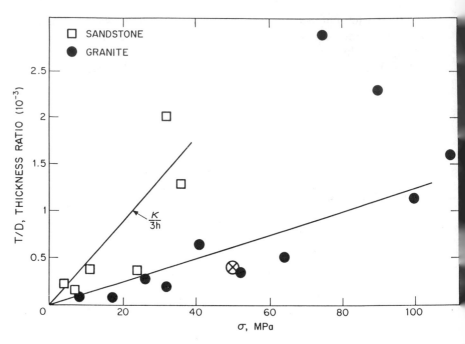

Fig. 2.13 The ratio of gouge thickness to slip (T/D) vs. normal stress for two different rock types. The open crossed symbol is the average for a sample that slipped 30 cm at several normal stresses. (After Scholz, 1987; data from Yoshioka, 1986.)

accordingly with straight lines, the slopes of which are $\kappa/3h$. The measurements of Logan and Teufel (1986) provide a value of $h = 2$ GPa during frictional sliding of sandstone. Using that and the slopes of the lines fit to the data in Figure 2.13, we obtain $\kappa = 0.3$ for the sandstone and 0.075 for the granite. These values of κ are fairly typical for abrasive wear such as occurs in grinding (Rabinowicz, 1965). The wear rate for the sandstone is always 3 or 4 times that of the granite. This seems inconsistent with the model, because both rocks are silica rich and we have used the same value of h in the calculation. The model, however, assumes a solid substrate. With rock we might expect that some grain plucking occurs during wear, so that we should also consider differences in the grain boundary strength of the two rocks (Hundley-Goff and Moody, 1980). Some idea of this can be obtained from the uniaxial strengths of the rocks, about three times greater for the granite than for the sandstone, which seems to explain the difference in wear rates. Although the contact area is controlled by the single-crystal strength, wear rate is controlled by the bulk-rock strength (Kessler, 1933). This analysis can be applied also to gouge production in faulting (Sec. 3.2.2),

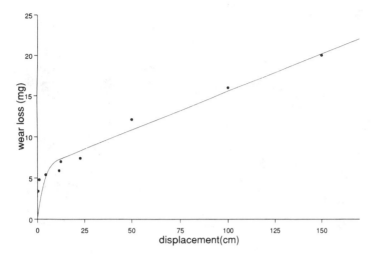

Fig. 2.14 A complete wear curve for Westerly granite, σ_n = 3 MPa. The initial exponential part is the running-in wear, which eventually evolves to steady-state wear at a constant rate. The curve fitting the data is $\partial V/\partial D = \kappa_1[V_0 - V(D)] + \kappa_2 A_r$, where V_0 is the initial excess roughness, A_r the real area of contact, and κ_1 and κ_2 are two wear coefficients.

where it is found that the wear coefficients for natural faults are higher than in this laboratory study.

The wear law described above is for the case of steady-state wear. A complete wear curve (Fig. 2.14) also contains an early "running-in" phase, in which high initial wear rates decay exponentially with sliding until a steady-state rate is finally achieved. The usual explanation for running-in wear is that the starting surfaces have greater roughness than that which would be present in equilibrium with the sliding conditions, and so have an initially high wear rate that is proportional to this excess roughness (Queener, Smith, and Mitchell, 1965). Later, we will see evidence that faults exhibit such running-in wear (Sec. 3.2.2).

As wear continues, the surfaces at some point may become completely separated by gouge, and the frictional properties may then become more a property of the gouge than of the surfaces. This is the case with the halite gouge in the experiments of Shimamoto (1986; see also Sec. 2.2.2), in which the frictional behavior follows the deformation of,the halite through its brittle–plastic transition and at the highest pressures obeys a flow law for halite (Fig. 2.11).

If the gouge is composed of brittle fragments, on the other hand, it shears as a granular material, with cataclastic flow dominated by grain comminution, and by dilatancy during shear (Sammis et al., 1986; Marone, Raleigh, and Scholz, 1990).   In experiments, it is found that

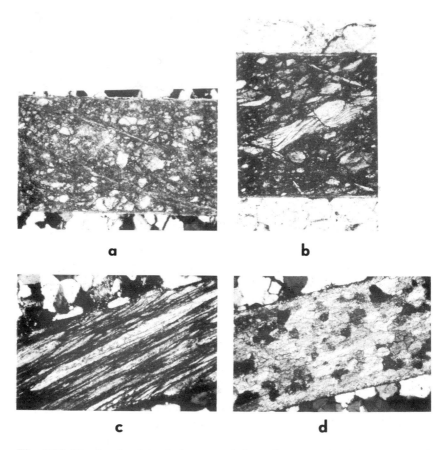

Fig. 2.15 Simulated calcite fault gouge, deformed as a sandwich between sandstone in triaxial friction experiments at 20 MPa confining pressure. All shears are right lateral; see Fig. 3.14 for fabric nomenclature. (a) $T = 25°C$; notice the development of Reidel (R_1) shears. (b) $T = 250°C$; R_1 still present, but twinning and translation gliding in calcite now contributing to the deformation with elongation in the S plane. (c) $T = 650°C$; shearing primarily by translation glide with homogeneous deformation in the shear zone. Notice the rotation of the S plane as compared to (b). (d) 900°C; deformation now accommodated by syntectonic recrystallization with a resulting annealed microstructure. (Photographs by John Logan.)

comminution produces a fractal, or power-law, size distribution of grains, also observed in natural fault gouges (Sammis et al., 1986). Once shear, or grain size reduction, has reached a critical level, further deformation becomes localized in Reidel shears at an acute angle to the shear zone, or in boundary shears, within which further grain size reduction occurs. The dilatancy that accompanies the shear of these granular materials has a strong stabilizing effect on their slip, which will be discussed further in Section 2.3.3.

Fault gouge found in near-surface exposures of active faults sometimes contains an abundance of clay minerals. Clays like montmorillonite and illite that have structural layers of molecular water may act as solid lubricants coating the sliding surfaces and thereby substantially reduce the frictional strength as well as stabilize sliding (Summers and Byerlee, 1977; Byerlee, 1978; see also Fig. 2.7b). This has led to the suggestion that the widespread occurrence of clay gouges in active fault zones may result in faults being far weaker than predicted by Byerlee's law (Wang, Mao, and Wu, 1980; see also Sec. 3.4.3). Friction studies with clay gouge from the San Andreas fault zone, show, however, that those lubricating properties are lost at high pressures and temperatures, due to dehydration of the clays (Logan, Higgs, and Friedman, 1981).

Photographs of a variety of experimentally deformed gouges are shown in Figure 2.15. These range from purely brittle cataclastic materials to those flowing plastically, with a range of mixed cataclastic behavior in between. They exhibit a number of features common to rocks found in faults and shear zones: mylonitic foliation and lineation; concentrated zones of shearing and grain size reduction; grain rotation and flattening; secondary (Reidel) shears; and syntectonic recrystallization (see Sec. 3.3.2).

2.3   STICK SLIP AND STABLE SLIDING

2.3.1   *Introduction*

If there is any variation of frictional resistance during sliding, a dynamic instability can occur, resulting in very sudden slip with an associated stress drop. This often occurs repetitively – the instability is followed by a period of no motion during which the stress is recharged, followed by another instability. This common frictional behavior is called *regular stick slip*. The conditions for an instability are illustrated in Figure 2.16a, where we consider a simple frictional slider loaded through a spring with stiffness K. The spring stiffness may represent either the stiffness of the loading machine used in the laboratory or the elastic properties of the medium surrounding a fault. Suppose that the frictional resistive force F of the slider is like that shown in Figure 2.16b in which it has a maximum followed by a decrease with continued slip. During the latter stage the spring unloads following a line of slope $-K$. If at some point B, F falls off faster with u than K, an instability will occur because there will be a force imbalance that will produce an acceleration of the slider. Beyond point C, F becomes greater than the force in the spring and the slider decelerates, coming to rest at point D, where (in the absence of

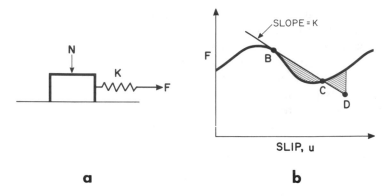

Fig. 2.16 Schematic diagram illustrating the origin of frictional instability. (a) A block slider model; (b) a force–displacement diagram showing a hypothetical case in which the frictional resistance force falls with displacement at a rate faster than the system can respond.

other dissipation) the area between the curves between B and C is just equal to that between C and D. The condition for instability,

$$\left| \frac{\partial F}{\partial u} \right| > K \tag{2.24}$$

is therefore both a property of the friction of the slider and the elastic properties of the environment that is loading it (recall the discussion of the role of boundary conditions on the stability of cracks, Section 1.1.2).

Regular stick slip is commonly observed in the frictional sliding of rock, which led Brace and Byerlee (1966) to propose it as the mechanism of earthquakes. Because earthquakes are recurring slip instabilities on preexisting faults, they are, by definition, a stick-slip phenomenon. The significance of the Brace and Byerlee paper is in the emphasis on stability, not strength, in explaining the mechanism of earthquakes. Their paper therefore initiated the modern era of earthquake mechanism studies.

In early exploratory work, friction experiments were conducted with a wide variety of rock types, gouges, and surface roughnesses over a broad range of experimental conditions in an effort to map out the conditions that favored stick slip over stable sliding. A variety of behavior was observed, which made it impossible to establish with certainty clear, universal fields of stability that could be reproduced from laboratory to laboratory and used to predict the behavior of any given rock. Part of this was because the inherent effect of the testing machine, shown in Equation 2.24, was not at first recognized. Because different laboratories used very different types of testing apparatuses, the results regarding the stability of sliding were not directly comparable. More fundamental, though, came the realization that stick slip, being a consequence of a rate of change of friction, was a second-order frictional phenomenon. The offsetting nature of surface contact and shear strengths leads to a frictional strength that is nearly independent of material and conditions, as in Byerlee's law, but the stability of the slip is controlled by a complex interplay of a variety of micromechanisms, with subtle differences being decisive in determining the behavior.

Nonetheless, some general conclusions could be made, as summarized by Brace (1972) and Paterson (1978). Stick slip is particularly favored in low-porosity siliceous rocks, especially those containing quartz. It is inhibited by the presence of soft, ductile minerals like calcite, serpentine, and clay, sometimes even in small amounts. Stick slip is favored by smooth surfaces with small amounts of gouge; slip tends to be stable with thick gouge and for very rough surfaces (which produce large amounts of gouge, so this may be the same effect). For a given rock and surface type, an increase of normal stress brings about a transition from stable sliding to stick slip and an increase in temperature brings about a transition in the reverse sense. In Section 2.3.3 these stability transitions will be discussed in more detail, because they are critical to understanding the limiting conditions for unstable faulting in the earth.

The frictional behavior in which strength falls with slip, as sketched in Figure 2.16b is known as *slip weakening*. This type of behavior may sometimes be observed, particularly on initial loading, as in the case shown in Figure 2.6b, and may result in stick slip. Early models of the stick-slip instability in rock friction concentrated on this type of behavior (Byerlee, 1970). However, slip weakening does not intrinsically provide a mechanism for the frictional strength to regain its prior level and hence will not lead to regular stick slip that oscillates around a mean friction level: A behavior that is often observed in experiments and which is the type of behavior that is required to explain earthquakes.

Regular stick slip is often observed in materials in which friction has a negative dependence on sliding velocity, which leads to a kinetic friction that is lower than the static friction (Ishlinski and Kraghelsky, 1944; Rabinowicz, 1965). This type of behavior, called *velocity weakening*, can lead to an instability for the same reason as does slip weakening. Because the weakening usually evolves over some finite displacement, it may result in a friction–slip relation that looks very much like that shown in Figure 2.16. However, it provides a mechanism by which the frictional strength is regained after the instability. Because it is not position dependent, the instability does not depend on the fault being in a local position of high friction at the outset.

In the next section, we will show that velocity weakening is the critical characteristic that leads to regular stick slip in rock under most conditions. This behavior occurs when the strength parameters that govern normal and shear contact in Equations 2.1 and 2.2 are themselves time dependent, because of environmental effects of strength of the same type as discussed in Sections 1.3.2 and 2.2.2.

2.3.2   *Rate effects on friction*

The modern understanding of regular stick slip was developed by Rabinowicz (1951, 1958), from whom the following discussion is derived. The idea from Coulomb's time is that a static friction coefficient μ_s must be exceeded for slip to commence, during which slip is resisted by a dynamic friction μ_d. If $\mu_s > \mu_d$, stick slip will occur. However, experiment shows that, in the case of systems in which stick slip occurs, if the surfaces are held in stationary contact for a time t, then μ_s increases approximately as $\log t$. Furthermore, if the surfaces are slid at constant velocity V, then μ_d is observed to decrease with $\log V$. Either of these phenomena could result in stick slip.

Rabinowicz also made the crucial observation that there was a critical slip distance in order for friction to change from one value to another. If a block in stationary contact on an inclined plane is impacted by a ball, it must slip a critical distance before friction breaks down from the static to the dynamic value. Similarly, if under steady sliding the slip velocity is suddenly changed, the friction level will evolve gradually to a new value over this same critical slip distance. In any friction experiment fluctuations occur in the friction force, and these are only correlated at distances shorter than the critical slip distance, which he found to be approximately equal to the mean diameter of the contact junctions (Rabinowicz, 1956).

From this he deduced that the contact had a memory of its previous state that faded over the critical slip distance. This concept allowed him to unify the rate effects on μ_s and μ_d, described above. If the surface contact state only resolves differences after sliding a critical distance $\mathscr{L}$, then an experiment in which the surfaces are held in stationary contact for time t, then slid $\mathscr{L}$, would be identical to an experiment in which sliding occurred at a constant velocity $V = \mathscr{L}/t$. Thus if one were to do "holding" experiments and measure the change in static friction with time in stationary contact, and observe the common result

$$\mu = \mu_0 + A \log\left(\frac{t}{\mathscr{L}}\right) \tag{2.25a}$$

where μ_0 is some base friction and A is a constant, it would imply that friction under steady sliding should depend on velocity $V = \mathscr{L}/t$ accord-

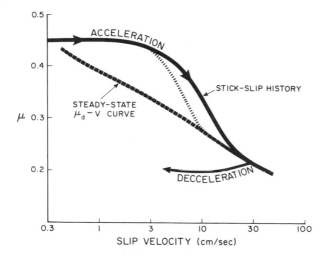

Fig. 2.17 During a stick-slip event (solid curve) the friction does not drop immediately along the steady-state μ–v curve (dashed curve) but retains a memory of its initial high value until it has slipped a critical distance. A simple model with a critical slip distance of 10 μm is shown by the dotted curve. Similar behavior occurs during deceleration, resulting in hysteresis. [Material, steel on steel; after Rabinowicz (1958).]

ing to

$$\mu = \mu_0 - A \log\left(\frac{\mathscr{L}}{t}\right) \qquad (2.25b)$$

which is the observed result. Thus the distinction made between static and dynamic friction is only an operational one. Notice that these empirical relations cannot hold over all times or velocities. There is experimental evidence that the time effect given in Equation 2.25a diminishes at short times and that there is a cutoff in Equation 2.25b at high velocities (Dieterich, 1978). At very low velocities there is evidence that friction has a positive velocity dependence, and hence in this case there is no true static coefficient and no true stationary state, the system always being in motion at very low rates even during the "stick" stage of the cycle (Rabinowicz, 1958).

To illustrate this further, consider the behavior during a stick-skip instability, in which friction and velocity are measured throughout (Fig. 2.17). If only the velocity dependence of friction were important, friction would follow the μ_d–V curve, as measured during steady slip experiments. Instead, it follows the looping trajectory shown. In the accelerating portion of the cycle it has a higher friction than the steady-state curve because, for slip less than $\mathscr{L}$, it has a memory of its higher friction

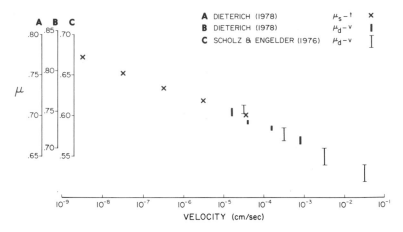

Fig. 2.18 Comparison of the μ_s–t and μ_d–v relations by assuming that μ_s after holding at t is equivalent to μ_d at a steady $v = \mathscr{L}/t$. Data are from Dieterich (1978) and Scholz and Engelder (1976): (A) μ_s–t, quartz sandstone, $\sigma_n = 18.7$ MPa; (B) μ_d–v, Westerly granite, 1.96 MPa;(C) μ_d–v, Westerly granite, 20 MPa. $\mathscr{L} = 5\ \mu$m (Dieterich, 1978). Friction scales are offset because base friction is not reproducible.

during stationary contact. Similarly, during the deceleration phase, it has a lower friction because it has a memory of sliding at a higher velocity.

These rate effects have been well documented for rock friction; the effect of stationary contact time (Eq. 2.25a) having been first described by Dieterich (1972) and the negative velocity dependence (Eq. 2.25b) by Scholz et al. (1972). The equivalence of the two, using the above analysis, is shown in Figure 2.18.

The type of micromechanism responsible for such behavior is illustrated in Figure 2.19. A hard indenter was pressed with constant load against a crystal face, and it is observed that the area of the indent increases with time, according to

$$A_r = A_{r_0} + B \log(t) \tag{2.26a}$$

We may substitute Equation 2.26a into Equation 2.1, and recognizing N as constant, solve for p. This then can be substituted into Equation 2.3, resulting in

$$\mu = \frac{sA_{r_0}}{N} + \frac{sB}{N} \log(t) \tag{2.26b}$$

which predicts the result given in Equation 2.25a.

The time dependence of contact area of the type shown in Figure 2.19 has also been demonstrated to occur during frictional sliding of rock (Teufel and Logan, 1978). This time dependence of contact area has been shown to occur for a wide variety of materials loaded in an aqueous

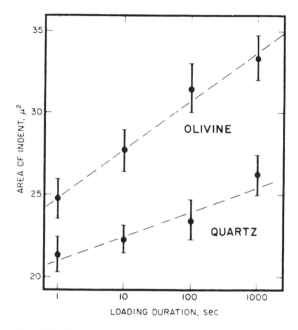

Fig. 2.19 Real contact area versus loading time measured in microindentation experiments. (After Scholz and Engelder, 1976.)

environment, but can be eliminated when tests are done under water-free conditions (Westbrook and Jorgensen, 1968). This explains why Dieterich and Conrad (1984) observed the rate effects on rock friction to disappear in experiments done in dry argon. These rate effects on friction, and therefore stick slip, are thus a consequence of the environmental effects on rock strength discussed in Sections 1.3 and 2.2.2.

Dieterich (1979a, 1981) and Johnson (1981) subsequently found that, on application of a sudden change in sliding velocity, there is a direct positive velocity effect on friction that decays exponentially to the steady-state value with a critical displacement $\mathscr{L}$. This suggests that the shear strength s in Equation 2.26b is also time dependent and exhibits a memory property. The various effects previously described can be combined in the framework of an empirical constitutive law in terms of sliding velocity V and state variables ψ_i,

$$\tau = \sigma_n \left[\mu_0 + \mathbf{b}_1 \psi_1 + \cdots + \mathbf{b}_i \psi_i + \mathbf{a} \ln(V/V^*) \right] \quad (2.27a)$$

$$d\psi_1/dt = -\left[(V/\mathscr{L}_1)(\psi_1 + \ln(V/V^*)) \right]$$

$$\vdots \quad\quad\quad\quad\quad\quad\quad\quad\quad (2.27b)$$

$$d\psi_i/dt = -\left[(V/\mathscr{L}_i)(\psi_i + \ln(V/V^*)) \right]$$

where the $\mathbf{b}_i$ and $\mathscr{L}_i$ are empirically determined constants and critical

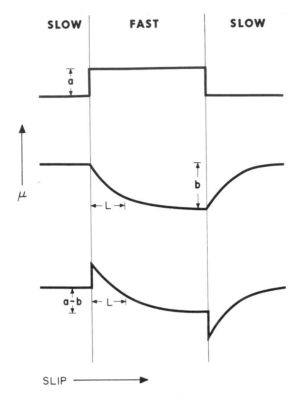

Fig. 2.20 Illustration of the rate effects on friction using a single-state-variable constitutive law: from top to bottom are shown the direct effect; the evolving effect; the combined effect.

slip distances, **a** is an empirical constant, and V^* is an arbitrary reference velocity. Several forms of this constitutive law have been described in the literature. Equation 2.27 is the compact form of Rice and Gu (1983), which closely follows that of Ruina (1983). An equivalent form has been given by Dieterich (1981, 1986), which employs only a single state variable and includes cutoffs in the time effects at high slip velocity. The basic effects can be discussed using a single state variable and critical slip distance, although there are both experimental and theoretical justifications for the more general form of Equation 2.27 (Tullis and Weeks, 1986; Horowitz and Ruina, 1985).

The last term in Equation 2.27a describes the direct velocity effect, which is positive. The state variables ψ_i represent negative velocity effects that evolve over critical distances $\mathscr{L}_i$ in the manner given by Equation 2.27b. With a single state variable formulation, the response to a factor of e step increase in sliding velocity is shown in Figure 2.20. There is first an immediate increase in friction, with magnitude determined by **a**, followed by a fall, of magnitude given by **b**, over the characteristic distance $\mathscr{L}$. After friction reaches steady state at the new velocity, $d\psi/dt = 0$, and thus the state variable at steady state is $\psi^{ss} = -\ln(V/V^*)$. Using ψ^{ss} in Equation 2.27a shows that

$$\frac{\partial \mu^{ss}}{\partial [\ln(V)]} = \mathbf{a} - \mathbf{b} \tag{2.28a}$$

where μ^{ss} is friction at steady state. We also find the definitions

$$\left. \frac{\partial \mu}{\partial [\ln(V)]} \right|_{\psi} = \mathbf{a} \tag{2.28b}$$

$$\left. \frac{\partial \mu}{\partial u} \right|_{V} = -(1/\mathscr{L})(\mu - \mu^{ss}) \tag{2.28c}$$

where u is slip displacement. Equation 2.28b defines the direct velocity effect on friction and Equation 2.28c describes the evolution of friction over the critical slip distance $\mathscr{L}$ following a velocity change.

If $\mathbf{a} - \mathbf{b} < 0$, the behavior is velocity weakening, if greater than zero, velocity strengthening. This constitutive law, when coupled to an elastic loading system, results in nonlinear behavior for which the stability conditions are complex. However, the results of both linearized and nonlinear stability analyses show the general result that slip will be unstable for $\mathbf{a} - \mathbf{b} < 0$ and stable for $\mathbf{a} - \mathbf{b} > 0$ (Rice and Ruina, 1983; Gu et al., 1984).

Following Dieterich (1979a), we may interpret these several effects by an interaction between time dependencies of the contact and shear strengths discussed earlier. Suppose first that the contact area decreases with sliding rate, as may be inferred from Figure 2.19, which is equivalent to p increasing with rate. Suppose also that the shear strength s also has a similar rate dependence, but with different governing coefficients. Steady-state sliding will then be determined by a friction coefficient μ^{ss} which depends on the equilibrium values of p, s, and A_r, all of which are functions of the sliding velocity. If a sudden increase in sliding velocity is imposed, there will be an immediate increase in friction caused by the consequent increase in s, because A_r requires some finite slip $\mathscr{L}$ to reestablish a new equilibrium contact at the higher velocity. Because p also increases with V, A_r and hence μ, will decrease with slip until new steady-state values are established. The direct effect, Equation 2.28b, can be identified with the rate effect on s; the steady-state velocity effect, Equation 2.28a, is determined by the difference between the p and s rate effects; and the evolution between these states, Equation 2.28c, is determined by the characteristic sliding distance $\mathscr{L}$, which reflects a

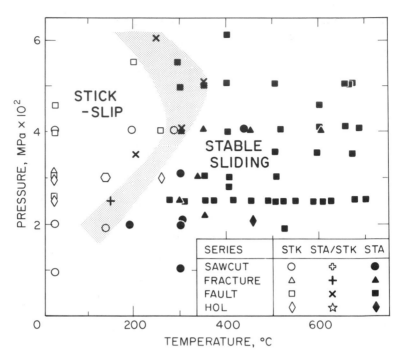

Fig. 2.21 Fields of stability and instability for a granite. (After Stesky et al., 1974.)

fading memory of the old contact state. In this interpretation then, the state variable ψ defines the contact state between the two surfaces, which we have described simplistically as A_r.

2.3.3 *Frictional stability transitions*

The stability of frictional sliding of rock therefore depends on the relative size of two competing effects that are two aspects of the time dependence of rock strength, both of which are environmentally dependent. The variety of parameters that effect stability, as summarized in Section 2.3.1, should not be surprizing. However, the nature of the constitutive law that underlies this behavior, given in Equation 2.27, has been worked out for only a few materials, most notably granite. At present it is only possible to generalize by inference to the other cases.

Because **a** and **b** are material properties that must depend on such parameters as pressure and temperature, one might also expect frictional sliding of a given material to be characterized by fields of stability or instability, depending on the sign of **a** − **b**. The earliest attempt to map out the stability field was done for granite by Brace and Byerlee (1970),

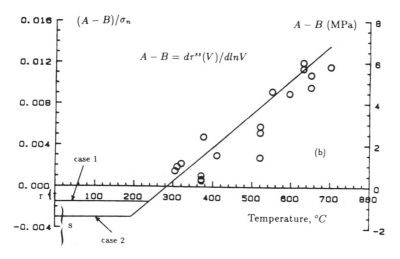

Fig. 2.22 The steady-state rate parameter $\mathbf{a} - \mathbf{b} = (A - B)/\sigma_n$, shown as a function of temperature for granite. (From Tse and Rice, 1986; data from Stesky, 1975.)

Byerlee and Brace (1968), and Stesky et al. (1974). Their results are shown in Figure 2.21, where it can be seen that stability is favored by high temperature and low normal stress and that there is a fairly sharp transition between the fields of stick slip and stable sliding.

Stesky (1975, 1978) measured the velocity dependence of friction in his high-temperature experiments. Tse and Rice (1986) were able to recast these results in terms of the constitutive law parameters $\mathbf{a}$ and $\mathbf{b}$, as shown in Figure 2.22. The parameter $\mathbf{a}$ increases sharply above $400°C$ and $\mathbf{a} - \mathbf{b}$ changes sign from negative to positive at about $300°C$. A transition from unstable to stable slip is therefore expected at $300°C$. Brace and Byerlee (1970) concluded that this transition marked the depth limit of tectonic earthquakes, a conclusion supported more quantitatively by Tse and Rice (1986). In terms of crustal faulting, we refer to this as the *lower stability transition*.

Although it has been traditionally thought that the base of the seismogenic zone is controlled by the brittle–plastic transition for crustal rocks (Macelwane, 1936), a view that has been revived recently (e.g., Sibson, 1982), it was remarked in Section 1.4.4 that because earthquakes are a frictional instability, their depth limitation must be controlled by a stability transition, rather than by a transition in bulk rock rheology. Not unexpectedly, however, there is a connection between the brittle–plastic transition and the lower stability transition.

In Section 2.2.2 it was pointed out that temperature and ductility have little effect on the base frictional strength (μ_0 in Eq. 2.27) until the

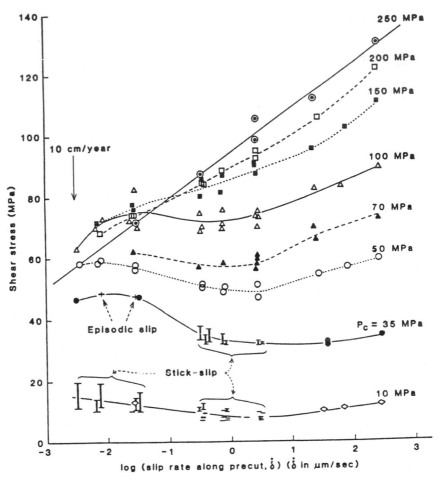

Fig. 2.23 Velocity dependence for frictional sliding of halite gouge at a variety of confining pressures. (From Shimamoto, 1986.)

brittle–plastic transition is complete, the surfaces become welded, and there is a transition from friction to bulk flow. Ductility does have a profound influence on the rate parameters, however, and hence on stability. The velocity dependence of friction for the halite experiments of Shimamoto (1986) is shown in Figure 2.23. At the lowest normal stresses there is a region in which velocity weakening, and hence stick slip, is observed. At higher normal stress, where the behavior of the halite is in the semibrittle regime (see Fig. 2.11), velocity strengthening is observed at all rates and the sliding is stable. At the highest pressure, where the brittle–plastic transition is complete, the rate dependence is uniformly positive and is governed by the halite flow law.

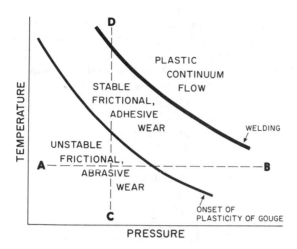

Fig. 2.24 Schematic diagram showing the stages in the brittle–plastic transition for frictional sliding.

By analogy with Figure 1.21, we may schematically represent the brittle–plastic transition for frictional sliding as shown in Figure 2.24. The experiments of Shimamoto follow the path A–B in the figure. The onset of ductility in the halite corresponds to a stability boundary: The transition from friction to bulk flow occurs at higher pressure where the brittle–plastic transition for the halite is complete. The experiments on granite of Stesky (1975, 1978), on the other hand, followed path C–D. He observed the stability transition at 300°C, which, from the results discussed in Section 1.4, corresponds to the temperature of the onset of quartz plasticity. We also expect (Sec. 2.2.3) a change from abrasive to adhesive wear to occur at the lower transition boundary, and hence the type of rocks that might occur in faults. The lower stability transition is therefore of profound significance in the mechanics of faulting, and its role, together with that of the brittle–plastic transition, will be discussed in that context in Section 3.4.

The transition that occurs at low normal stress, also seen in Figure 2.21, may be called the *upper stability transition*, because it would be expected to occur at a shallower depth in the earth. This transition arises for entirely different reasons than the lower transition (Dieterich, 1978). If the change in friction from a static to a dynamic value $\Delta\mu$ occurs over a characteristic slip distance $\mathscr{L}$, application of Equation 2.24 shows that slip will change from stable to unstable at a normal stress given by

$$\sigma_n = \frac{K\mathscr{L}}{\Delta\mu} \qquad (2.29)$$

where K is the stiffness of the system. Therefore the normal stress at which this transition occurs is not an intrinsic property of the constitutive law but depends also on the stiffness of the loading system, as was demonstrated experimentally by Dieterich (1978). If the slip zone is treated as an elliptical crack in an infinite medium of Young's modulus E and Poisson's ratio ν, the shear stiffness is given by

$$K = \frac{E}{2(1 - \nu^2)\ell} \qquad (2.30)$$

where ℓ is the length of the slipping zone. Combining Equations 2.29 and 2.30 we find that instability occurs when the slip zone reaches a critical length,

$$\ell_c = \frac{E\mathscr{L}}{2(1 - \nu^2)\sigma_n \Delta\mu} \qquad (2.31)$$

This critical length has been referred to as the breakdown length, or *nucleation length*. In laboratory experiments a stability transition will occur at a normal stress when ℓ_c becomes larger than the test surface. Because ℓ_c varies inversely with normal stress it will become large at shallow depths, which will tend to inhibit earthquake nucleation there (Sec. 3.4.1). It is also of central importance to one class of theories relevant to earthquake prediction (Sec. 7.4.1).

A one-dimensional block-slider analysis shows that three stability states are possible (Rice and Ruina, 1983; Gu et al., 1984). The critical stiffness may be reexpressed in terms of the parameters of Equation 2.27 as

$$K_c = -(\mathbf{a} - \mathbf{b})\sigma_n/\mathscr{L} \qquad (2.32)$$

Notice, by comparison with Equation 2.29, that $(\mathbf{b} - \mathbf{a}) = \Delta\mu$. If $(\mathbf{a} - \mathbf{b}) > 0$, the system is stable under any velocity perturbation. If $(\mathbf{a} - \mathbf{b}) < 0$, then two types of behavior are possible, as shown in Figure 2.25. If $K < K_c$ the system is intrinsically unstable, but if $K > K_c$ it will be conditionally stable, that is, stable under quasistatic loading but unstable if subjected to a sufficiently large velocity jump, as may occur under dynamic loading during an earthquake. Recognizing these three stability states is central to understanding seismic coupling of faults (Sec. 6.4.3).

With constitutive laws that employ a single state variable, there is a sharp boundary between the unstable and conditionally stable states. However, with a two-state-variable law, Gu et al. (1984) found a transitional region characterized by self-driven oscillatory or episodic slip and, close to the stability boundary, by chaotic behavior (Fig. 2.25; see also Hobbs, 1988). This episodic behavior near the stability transition has been observed in the laboratory; an example is shown in Figure 2.26. A two-dimensional numerical analysis shows that this behavior may be

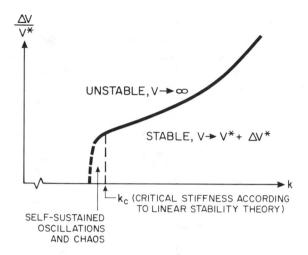

Fig. 2.25 The stability fields for a spring-slider model with rate–state-variable friction and velocity weakening. When $K < K_c$ the behavior is always unstable. When $K > K_c$ it becomes unstable only when subjected to a velocity jump ΔV greater than a critical value. With a single-state-variable law there is a sharp transition. With a two-state-variable law, there is a region of oscillatory and chaotic behavior at the transition. (After Gu et al., 1984.)

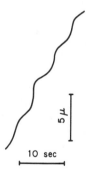

Fig. 2.26 Episodic stable sliding near the stability transition for granite. (From Scholz et al., 1972.)

very complex, due to the scale dependence of the critical state (Horowitz and Ruina, 1985). Their analysis brought out a rich variety of slip behavior, including "creep waves" that may become critical or damp out. Similar behavior is seen in the aseismic slip of faults, discussed in Sections 5.3.1 and 6.4.1, and the possible significance of the chaotic behavior will be discussed in Section 5.4.

This discussion has considered only transitions within a single lithology. On a real fault, stability transitions may occur because of lithological changes. One such change that may be common is from consolidated to unconsolidated fault gouge at shallow depth, which may produce a change in the sign of (a − b) from negative to positive and thus a stability transition at shallow depth by a mechanism different than that discussed above (Marone and Scholz, 1988; see also Sec. 3.4.1). This change in sign of a − b for granular materials occurs because they exhibit dilatancy during shear that has a positive rate dependence that overwhelms the negative rate dependence intrinsic to the grain-to-grain contact friction (Marone et al., 1990).

2.3.4   *Dynamics of stick slip*

An evaluation of the motion of the block-slider model of Figure 2.16a provides both a good description of stick-slip motion observed in the laboratory and a one-dimensional analog of earthquake motion. Consider a slider of mass m loaded through a spring of constant K that is extended at a load point velocity v. The governing differential equation will be

$$m\ddot{u} + a\dot{u} + F(u, \dot{u}, t, t_0) + K(u - vt) = 0 \qquad (2.33)$$

where the first term gives the inertial force; the second, damping, including seismic radiation; the third is the frictional force during sliding; and the fourth is the force exerted by the spring.

The simplest solution of Equation 2.33 (e.g., Jaeger and Cook, 1976; Nur, 1978), is for the case when there is no damping and friction is assumed to drop from an initial static value μ_s to a lower dynamic value μ_d upon sliding. At the onset of sliding ($t = 0$), the spring has been extended an amount ξ_0 just sufficient to overcome the static friction, that is, $K\xi_0 = \mu_s N$. The driving force is therefore the difference between the static and dynamic frictions, $F = \Delta\mu N$. If we further assume that the load point velocity is negligible compared the average velocity of the slider, Equation 2.33 simplifies to

$$m\ddot{u} + Ku = \Delta\mu N \qquad (2.34)$$

with initial conditions $u(0) = \dot{u} = 0$. The solution is

$$u(t) = \Delta\mu \frac{N}{K}(1 - \cos nt)$$

$$\dot{u}(t) = v = \Delta\mu \frac{N}{\sqrt{Km}} \sin nt$$

$$\ddot{u}(t) = a = \Delta\mu \frac{N}{Km} \cos nt \qquad (2.35)$$

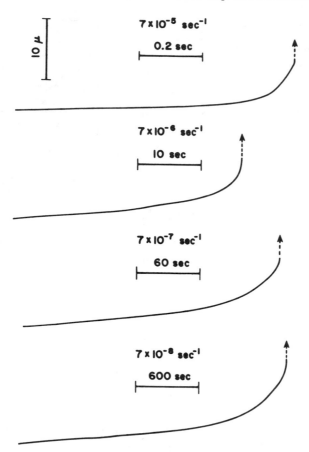

Fig. 2.27 Precursory stable sliding for a number of stick-slip loading cycles at different rates. The initial slope is due to elastic strain. (From Scholz et al., 1972.)

where $n = \sqrt{K/m}$. The slip duration is given by

$$t_{\mathrm{r}} = \pi \sqrt{\frac{m}{K}} \qquad (2.36)$$

after which static friction is reestablished and the loading cycle begins anew. The slip duration, or *rise time*, therefore depends only on the stiffness and mass and is independent of $\Delta\mu$ and N. In contrast, the total slip $\Delta u = 2\,\Delta\mu(N/K)$ and the particle velocities and acceleration are directly proportional to the friction drop. The corresponding force drop is $\Delta F = 2\,\Delta\mu\,N$ and the stress drop is $\Delta\sigma = 2(\mu_{\mathrm{s}} - \mu_{\mathrm{d}})\sigma_{\mathrm{n}}$.

Equation 2.35 provides a good first-order description of stick-slip motion observed in the laboratory (e.g., Johnson, Wu, and Scholz, 1973; Johnson and Scholz, 1976). The velocity v in Equation 2.35 is the

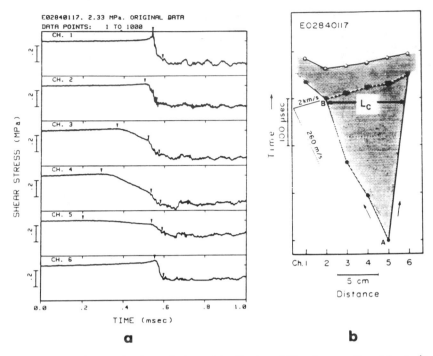

Fig. 2.28 The nucleation of the stick-slip instability in granite: (a) strain records at various points along the length of the sliding surface; (b) slip chronology at the same points. Stable slip begins first on channel 5, indicated by the stress reduction at the first arrow [point A in (b)]. Stable slip propagates at about 260 m/s to channel 2 (point B), when ℓ_c is reached. Then dynamic slip propagates over the surface at about 2 km/s. At the end of the stick slip, healing propagates back across the surface at a near sonic velocity. Normal stress is 2.33 MPa. (From Ohnaka et al., 1986.)

particle velocity, and is distinguished from the *rupture velocity* v_r, which is the propagation velocity of the boundary between unslipped and slipping regions. During dynamic slip v_r usually takes on a value close to an elastic wave velocity of the medium, typically the shear velocity (Johnson and Scholz, 1976).

In this simple model, friction was assumed to drop instantly from a static to dynamic value. However, as stated earlier, the friction drop evolves over a critical slip distance, and so stable slip must occur during nucleation prior to the instability (Dieterich, 1979b). Experimental observations of this precursory stage of stable slip are shown in Figure 2.27. The amount of precursory slip that occurs prior to the instability is independent of normal stress and loading rate, so it probably is also controlled by a critical slip distance. At the end of this stage nucleation and propagation of the instability occurs, shown in detail in Figure 2.28.

Records of shear stress measured at different points along the sliding surfaces are shown in Figure 2.28a, and a chronicle of slipping events in Figure 2.28b. Nucleation begins when the quasistatic slip becomes fast enough to reduce the stress at point A, which propagates along the fault at a velocity much smaller than sonic. Dynamic slip begins at point B and propagates back down the fault with a rupture velocity slightly lower than the shear velocity. Cessation of slip (healing) similarly propagates at near the shear velocity. The nucleation length ℓ_c in this experiment was in good agreement with Equation 2.31. Ohnaka et al. also showed the results for an experiment at about three times higher normal stress in which nucleation was not observed. This is expected from Equation 2.31, which predicts that for this case ℓ_c would be comparable to the spacing between their measurement points.

The magnitude of stress drop typically observed in friction experiments is on the order of 10% of the total shear stress (except for triaxial experiments, which produce unreliable results in this respect). These rather small values of stress drop are consistent with solutions of Equation 2.33 that employ the more complete constitutive law of Equation 2.27 and measured values of **a** and **b**, although the results depend in detail on assumptions about friction at high slip velocities. For example, Okubo and Dieterich (1986) compared the dynamic slip predictions of Equation 2.27 and Dieterich's (1981) version of the constitutive law, in which the velocity effect is assumed to become neutral at high velocities. They found, in comparison with experiment, that the former overpredicted the stress drop whereas the latter was in much better agreement. The actual behavior, however, is probably even more complex. Shimamoto's results (Fig. 2.23) show a high-speed frictional regime characterized by pronounced velocity strengthening. This will have an even stronger effect in reducing the stress drop than that described by Okubo and Dieterich.

Even in their simplest form the friction laws predict a rich variety of behavior. In detail they appear to be much more complex, requiring several state variables and different formulations over different velocity ranges. It remains an open question, however, to what extent these details, worked out in idealized laboratory conditions, apply to natural faults, or conversely, to what extent observations of natural phenomena will allow their confirmation.

2.4   FRICTION UNDER GEOLOGICAL CONDITIONS

The discussion in Section 2.1.2 makes clear why a single friction criterion, such as Byerlee's law (Eq. 2.14) should be essentially independent of lithology and temperature. To further extrapolate these results to geological conditions appears to pose no significant problem. The appli-

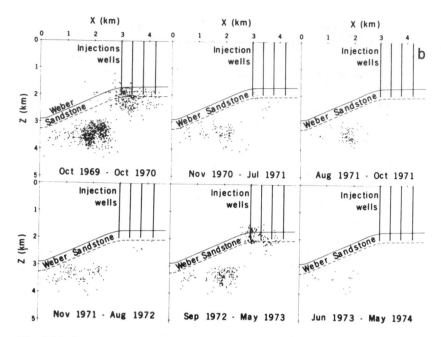

Fig. 2.29 Cross sections through the Rangely oilfield at various times during the earthquake control experiment, showing seismicity and the location of the experimental wells. Fluid was injected during the period of the first frame, withdrawn in the second and third, injected in the fourth and fifth, and withdrawn again in the sixth. The time record is shown in Figure 2.30. (From Raleigh et al., 1976.)

cation of this friction law to geological scales involves nothing more than a reaffirmation of Amontons's first law. For the important case of faults that slip seismically we know that the rate effect on friction must be negative, which says that friction under geologic loading rates must be somewhat higher than that measured in the laboratory. The coefficients of the rate effect measured in the laboratory are small, however, amounting to only a few percent change in friction per order of magnitude change in rate (Fig. 2.18), so that the difference in friction between geological and laboratory rates is not large.

This comment regarding scaling of friction was confirmed in a large-scale field experiment on induced seismicity conducted at Rangely, Colorado (Raleigh, Healy, and Bredehoeft, 1972, 1976). Intensive microearthquake activity had been observed on a fault within an oil field in the vicinity of Rangely. It had been hypothesized that this activity was being induced by overpressurization of the field during secondary recovery operations. To test this, an experiment was devised in which the fluid pressure at the hypocentral depth was controlled with a series of wells

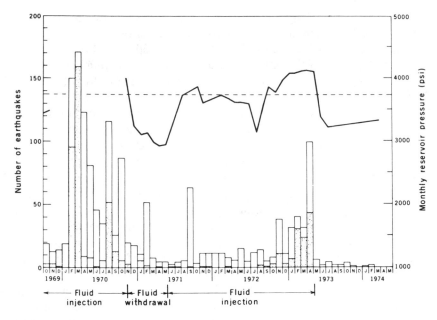

Fig. 2.30 The pressure and seismicity record during the experiment at Rangely. The dashed line indicates the critical pressure for triggering estimated from the measured state of stress and frictional strength of the Weber sandstone. (From Raleigh et al., 1976.)

from which fluid could be injected or withdrawn. The state of stress was measured in situ at the same depth, and from laboratory friction measurements on the reservoir rock, knowledge of the fault orientation from earthquake focal mechanisms, and the application of the effective stress law, the critical fluid pressure for frictional sliding was calculated. Cross sections of the experiment for different time windows are shown in Figure 2.29 and the seismicity and pressure record in Figure 2.30. It was found that reducing the fluid pressure below the critical value brought about an abrupt cessation of seismic activity, and the seismicity was observed to resume as soon as the pressure once again exceeded the critical level. This experiment confirmed, in a natural environment, not only the effective stress law for frictional sliding and its role in induced seismicity (see also Sec. 6.5), but also the applicability of the laboratory-determined value of friction.

These considerations lead to confidence in the use of Byerlee's law as a lower bound for the strength of the schizosphere. One cannot be as sanguine regarding the rate and state variables that control stability, however. The constitutive law (Eq. 2.27) contains critical sliding distances $\mathscr{L}_i$ that presumably scale with some characteristic distance of the mechanical contact. Rabinowicz (1951, 1956, 1958) found that $\mathscr{L}$ for metallic friction corresponds to an average contact junction diameter, which could be determined from direct observation or from the autocorrelation function of the friction force. In rock friction, laboratory values of $\mathscr{L}$ are typically of the order of 10 μm, similar to the roughness of the surface topography, and Okubo and Dieterich (1984) found that $\mathscr{L}$ increases with surface roughness, which is consistent with the model of Rabinowicz. In constructing models that both use this constitutive law and simulate natural faulting phenomena, Tse and Rice (1986) and Cao and Aki (1985) found that they had to assume that $\mathscr{L} \approx 1$ cm, or about three orders of magnitude greater than the laboratory values.

Ground surfaces used in laboratory measurements have corner frequencies in their topographic spectra, such as in Figure 2.3b, which define a characteristic length scale for these surfaces that may give rise to a critical slip distance $\mathscr{L}$. Natural faults, however, have fractal topography, more like that shown in Figure 2.3a, which does not possess a characteristic length scale. This poses the problem of how the parameter $\mathscr{L}$ scales from laboratory measurements to geological scales.

If we consider fractal surfaces in contact under a normal load, however, a characteristic length does appear (Scholz, 1988). Fractal surfaces have asperity amplitudes that increase with wavelength. A fault consists of two such surfaces in contact, which must be geometrically mismatched at all distances less than the net fault slip; yet no large-scale aperture can be observed between fault surfaces in the field. The fault must have closed, but little evidence may be found of inelastic deformation that would allow this closure, and the physical conditions are not always present that would allow such inelastic deformation. This closure must largely have occurred elastically.

Consider the contact between two surfaces to be composed of a network of bridges, of varying length λ, between asperities of height h. Approximate the behavior of the bridges by treating them as cracks, of aspect ratio h/λ. An order-of-magnitude approximation for the pressure required to completely close a crack elastically is (Walsh, 1965)

$$p = E(h/\lambda) \qquad (2.37)$$

For fractal surfaces, h and λ are related. For the special case of Brownian surfaces,

$$h^2 = \kappa_0 \lambda \qquad (2.38)$$

where κ_0 is a reference roughness. Combining Equations 2.37 and 2.38, we obtain an expression for the maximum length of bridge that may remain open under a pressure p,

$$\lambda_c = (E/p)^2 \kappa_0 \qquad (2.39)$$

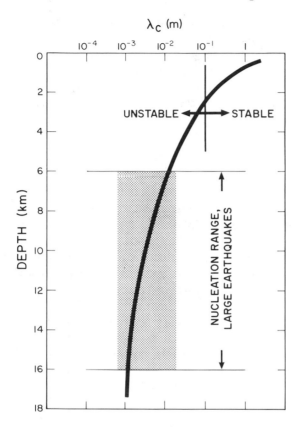

Fig. 2.31 Calculated values of the maximum contact spacing, λ_c, as a function of depth for fractal fault surfaces in contact based on fault topography data from Power, Tullis, and Weeks (1988). Effective normal stress is assumed to increase 15 MPa/km. Maximum $\mathscr{L}$ for fault instability is about 10 cm, so that if $\mathscr{L}$ is equated with λ_c, an upper stability transition is predicted, as shown. (From Scholz, 1988.)

This shows that Brownian surfaces in contact under a pressure p will be mated at all wavelengths longer than λ_c. The mechanical behavior of contacting surfaces depends only on the unmated part of the contact spectrum, so that, in terms of contact properties, λ_c appears as a characteristic length.

Equation 2.39 was evaluated using some fault topography data, with the result shown in Figure 2.31. If we tentatively identify $\mathscr{L}$ with the maximum contact spacing λ_c, we see from Figure 2.31 that this decreases with depth and is in the range 10^{-2}–10^{-3} m at seismogenic depths. This rough estimate agrees with the value of $\mathscr{L}$ that Tse and Rice (1986) had to assume in order for their model to produce geophysically realistic results. Their work also shows that the fault will become stable if $\mathscr{L} > 10$ cm, so, as indicated in Figure 2.31, this predicts a stability transition at about 2.5 km.

The depth to this stability transition can be estimated in a different way. Equation 2.31 gives the maximum size of a patch size that will remain stable. Assuming $\mathscr{L}$ is given by Equation 2.39, and substituting that result into Equation 2.31, with the recognition that $(\mathbf{b} - \mathbf{a}) = \Delta\mu$ (Eq. 2.32), we obtain

$$\ell_c = \frac{4E^3\kappa_0}{7p^3(\mathbf{b} - \mathbf{a})} \tag{2.40}$$

The transition will occur at a depth equal to this maximum stable dimension. Assuming a representative value of $(\mathbf{b} - \mathbf{a}) = 0.005$, Equation 2.40 indicates this depth is about 3.5 km, in reasonable agreement with our first estimate.

An additional scaling problem is that the parameters $\mathbf{a}$ and $\mathbf{b}$ depend on the micromechanisms of friction, and it is not clear if these are the same in nature as in the laboratory. There is evidence, discussed in Section 4.3.2, that the strength of natural faults is much more rate dependent than observed in laboratory friction experiments, suggesting that other healing mechanisms, such as chemically enhanced healing, may operate in nature. Our only guide, in pursuing this question, is to see how well experience gained in the laboratory explains natural faulting phenomena, and to pay close attention to discrepancies that arise.

3 Mechanics of Faulting

Having completed the preliminaries, we now begin to address the central topic. For organizational reasons, we have divided the topic of faulting into two parts: that concerned with statics is presented here; that concerned exclusively with dynamics, and hence seismic faulting, is reserved for the next chapter. We begin with a discussion of the elementary theory of faulting, followed by a more modern treatment of the formation and growth of faults and a description of the rocks and structures formed by faulting. Here we rely more heavily on geological observations than elsewhere. We summarize with a discussion of the strength and rheology of faults, finishing with the topic of heterogeneity and its role in faulting, which continues a subtheme to be found throughout this book.

3.1 MECHANICAL FRAMEWORK

3.1.1 *Anderson's theory of faulting*

In his seminal paper of 1905 and in his memoir of 1942, E. M. Anderson developed the modern mechanical concepts of the origin of faults and emphasized their important role in tectonics. His key contribution was to recognize that faults result from brittle fracture and to apply the Coulomb criterion to this problem. This led him to expect that faults should sometimes form conjugate sets, with planes inclined at acute angles on either side of the maximum principal stress and which include the intermediate principal stress direction (Sec. 1.1.4, Eq. 1.32). By applying the condition that near the free surface one of the principal stresses is vertical, Anderson showed that the three major classes of faults – reverse, normal, and strike-slip – result from the three principal classes of inequality that may exist between the principal stresses.

Anderson's concept can be grasped readily by examining a section through a Coulomb fracture surface, as shown in Figure 3.1a. For convenience, we assume that the three principal stresses are oriented north–south, east–west, and vertically, denoted σ_{NS}, σ_{EW}, and σ_V. The

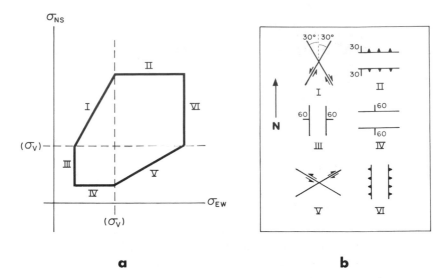

Fig. 3.1 (a) Section through a Coulomb fracture surface at σ_v = constant. Each side of the resulting figure, labeled in roman numerals, defines the relations between the horizontal stresses for a different type of faulting. (b) Shows these different faulting types in map view (ϕ = 30°).

section shown is at some arbitrary value of σ_V so as to illustrate, say, faulting at some fixed depth. A six-sided figure, symmetrical about the line $\sigma_{EW} = \sigma_{NS}$, is obtained that completely describes the fracture criterion at this value of σ_V. Side I of the figure is the locus of failure for the conditions $\sigma_{NS} > \sigma_V > \sigma_{EW}$, described by the line of Equation 1.33. This side therefore defines the conditions for strike-slip faulting on conjugate planes with orientations labeled I in Figure 3.1b. (We assume that the angle of internal friction is ϕ = 30°.) At the corner between sides I and II, $\sigma_{EW} = \sigma_V$. Beyond this, σ_{EW} becomes the intermediate principal stress, so that side II defines the conditions for thrust faulting, on faults dipping 30° with an EW strike, labeled II in Figure 3.1b. Side III defines normal faulting under the conditions $\sigma_V > \sigma_{NS} > \sigma_{EW}$, producing north–south-striking faults dipping 60°. Similarly, the faults produced under the conditions defined by the other three sides of the fracture surface section also are shown in Figure 3.1b.

In support of his hypothesis, Anderson (1942) described many conjugate fault systems, with examples from Britain in which the angular relationships conform to the expectations of the Coulomb criterion. The theory, though, only relates the orientations of faults to the stress field that existed at the time of their formation and it is usually not possible to prove the simultaneity of origin of the old faults he discussed. A more illuminating example is given in Figure 3.2. There we see the Izu

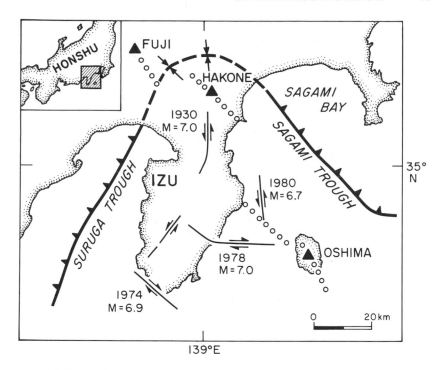

Fig. 3.2 Map of the Izu Peninsula, Japan, showing the major active tectonic elements. Curves with dates are rupture zones of large historic strike-slip earthquakes. Rows of circles indicate the alignment of parasitic eruptions emanating from active volcanoes. The Suruga and Sagami troughs are subduction boundaries between the Philippine Sea plate, to the south, and the Eurasian plate, to the north. (After Somerville, 1978.)

Peninsula, Japan, in which the northernmost part of the Izu–Bonin arc is presently colliding, in a NNW direction, with the Japan arc (Somerville, 1978; Nakamura, Shimazaki, and Yonekura, 1984). Folding occurs in the northernmost part of the peninsula, but the main tectonic features are conjugate systems of strike-slip faults. These Holocene faults are presently active: Major earthquakes, which have ruptured these faults during the twentieth century, are indicated by dates and sense of slip. In this case there is little question as to the simultaneity of the origin and period of activity of these faults.

It is also of interest to note that the alignments of volcanic cones and flank eruption vents locally bisect the angle between the conjugate faults and lie parallel to the maximum compressive stress (σ_1) direction (Nakamura, 1969). This observation conforms with Anderson's other major contribution to structural geology, which has to do with the formation of dikes by the mechanism of hydraulic fracturing (Anderson,

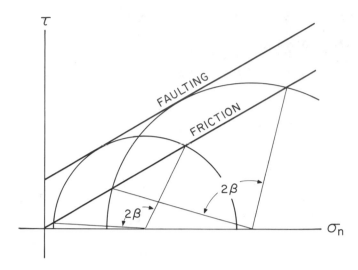

Fig. 3.3 Mohr diagram with two failure envelopes, one for fault formation (Coulomb failure) and one for slip on a preexisting fault (friction
criterion). The Mohr circles are drawn giving the stresses for fault
formation. Preexisting faults with a range of orientation given by β will
be activated first.

1936, 1942). In this and similar tectonic environments both of these
features, the orientation of strike-slip faults and the orientation of flank
eruptions, have been used to infer the direction of σ_1 (Lensen, 1981;
Nakamura, 1969; Nakamura, Jacob, and Davies, 1977; see also
Sec. 6.2.1).

This example also provides several warnings about the limitations of
this analysis. One may notice in Figure 3.2 spatial variations in both the
strikes of the faults and of the flank eruptions, so that even in this rather
small region the orientation of the stress field is not constant; extrapolation of stress orientations into regions in which data are lacking is
consequently not always justified. A second point is that the inferred σ_1
direction in the northern part of the peninsula is not normal to the strike
of the adjacent subduction zones in the Suruga and Sagami bays. The
oblique slip of these zones is confirmed by other data, so that one cannot
use unambiguously the orientation of such structures, which are not
fractures (see Sec. 6.3.2) and needn't conform to a criterion of fracture in
a homogeneous stress field, to infer the orientation of stresses (or even
plate motions).

This approach has other limitations, because the frictional strength of
faults is less than the stress necessary to form them and once formed
they constitute planes of weakness that may be reactivated in stress fields
that are not optimally oriented. This leads to ambiguity in inferring the

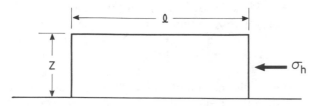

Fig. 3.4 Diagram illustrating the force balance in overthrust faulting.

stress-field orientation from the orientation of active faults or earthquake focal mechanisms. This problem is illustrated in Figure 3.3 with a Mohr diagram with two strength envelopes: a frictional one that represents the strength of a fault; and a Coulomb envelope that represents the strength of the surrounding rock. Mohr circles are drawn for the condition necessary for the formation of a new fault. It is clear that old faults of a wide range of orientations will be reactivated before this condition is met. The range β in possible fault orientations translates directly into the same angular uncertainty in the inferred direction of σ_1. Notice that this uncertainty increases as the stress magnitude decreases. This problem, can be generalized easily to three dimensions (McKenzie, 1969; Gephardt and Forsyth, 1984), and is a significant problem in seismotectonic interpretations (Sec. 6.2.1). Similar problems are produced by the presence of preexisting weakness planes, since Anderson's theory also assumes that strength is homogeneous and isotropic. A review of this problem and a discussion of ways to overcome it are given by Angellier (1984).

3.1.2 *Hubbert–Rubey theory of overthrust faulting*

A major class of faults not readily explained by Anderson's scheme is low-angle overthrusts. These are very shallow-dipping faults upon which thin sheets of rock, commonly of broad lateral extent, have been transported horizontally, often over considerable distances. They do not have the orientation predicted by Anderson's theory and they also present a fundamental mechanical problem. The difficulty is illustrated in Figure 3.4, where we consider the forces involved in pushing a block along a plane, resisted by a basal friction traction. Summing forces in the horizontal direction for a block of unit width, length l, and thickness z, we obtain, assuming the simple effective stress law,

$$F = \sigma_h z = (\sigma_v - p)l\mu \tag{3.1}$$

where σ_h and σ_v are the horizontal and vertical stresses and μ is the

friction coefficient. Let $\sigma_v = \rho g z$, the lithostatic load, and let the pore pressure $p = \lambda \rho g z$, where λ is some constant. Then

$$\sigma_h = (1 - \lambda)\rho g l/\mu \qquad (3.2)$$

and we can estimate the maximum length of block that can be pushed without σ_h exceeding the strength of the rock, C. If we suppose that p is hydrostatic and $\rho = 2.5$, then $\lambda = 0.4$; $C = 200$ MPa, and $\mu = 0.85$, which we might consider to be normal conditions, we find from equation 3.2 that $l = 16$ km. More realistic situations produce essentially the same answer (Jaeger and Cook, 1969; Suppe, 1985).

The problem is that the length of many overthrust sheets can be observed to be in the range 50–100 km. In the case of *decollements*, where the thrust moves on a layer of a very weak and ductile material, like gypsum or salt, the right-hand side of Equation 3.1 may be replaced by $\tau_{max}l$, where τ_{max} is the shear strength of the ductile material. For these materials, $\tau_{max} \ll C$, and the observation that $l \gg z$ therefore does not present a mechanical problem. Decollements are special cases, however, so that this does not provide a general solution. An alternative loading possibility is that the fault is driven by gravity. If the basal tractions are the same as given in Equation 3.2, this would require that the fault be inclined at an angle $\theta = \tan^{-1}[(1 - \lambda)\mu]$ (Jaeger and Cook, 1969, page 417). Under hydrostatic pore pressure conditions, $\lambda = 0.4$ and this requires a slope of 30°!

Hubbert and Rubey (1959) hypothesized that under some conditions the pore pressure may greatly exceed the hydrostatic head, and even approach a value close to the lithostatic load. This offers a solution because, as λ approaches 1, Equation 3.2 shows that the frictional resistance to sliding (and thus the constraint on the length of the overthrust sheet) vanishes. This proposition created a major controversy. Reprints of the key papers and discussions of the relevant issues have been collected in Voight (1976).

Although Hubbert and Rubey's application of the effective stress law to this problem was novel at the time, this was not the central issue of the debate, which revolved around the mechanism of the generation and maintenance of the "overpressure." Two mechanisms have emerged as likely candidates. The first, proposed by Hubbert and Rubey and supported by abundant oilfield data, results from the compaction of saturated sediments capped by low-permeability rock such as shale. This possibility is therefore limited to sedimentary basins, but as these are typical geological sites for overthrusts, this mechanism is a likely explanation for many cases. The major seismotectonic environments that may be affected by this process are foreland fold and thrust belts and shallow

parts of subduction zones. In the latter case, discussed further in Section 6.3.2, this overpressurization mechanism is augmented in the accretionary wedge by tectonic transport and compaction of sediments by the subduction process.

A second mechanism, less frequently discussed, is the generation of high pore pressure through dehydration reactions in metamorphism. This has been suggested as a way of generating high pore pressure locally on the San Andreas fault (Irwin and Barnes, 1975), but has been more commonly discussed in terms of regional metamorphism at midcrustal depths (e.g., Fyfe, Price, and Thompson, 1978; Etheridge et al., 1984). In a study of fluid–rock interactions in detachment zones, for example, Reynolds and Lister (1987) concluded that high fluid pressures existed within the ductile zone but were vented above the brittle–plastic transition. The depth at which prograde metamorphism is likely to occur overlaps the lower part of the schizosphere, however, so this mechanism may play an important role in the strength of faults (taken up in Section 3.4.3).

The term *detachment* is also applied to overthrusts, but it is more commonly used to describe near-horizontal ductile faults in the crystalline basement. In the latter case, Anderson's theory does not apply, because ductile shearing will occur on planes of maximum shear stress rather than at the Coulomb orientation and because at those depths the near-surface boundary condition need not apply. If the flow is fully plastic, the Hubbert–Rubey concept does not apply either, because the strength is then independent of pressure.

The departure from Anderson's idealized fault orientations with depth from the free surface is illustrated in the loading models of Hafner (1951) and Sanford (1959) which produce stress trajectories predicting listric faults for Coulomb failure. This is because at the base of their models, which represents some depth horizon in the earth, they assume shear stress or displacement boundary conditions, which require that the maximum principal stress dips 45° or shallower. In tectonic terranes dominated by horizontal movements, it is likely that at sufficient depth beneath the schizosphere the horizontal plane is one of maximum shear stress. This favors the formation of detachments. If a detachment occurs within the schizosphere, however, the Hubbert–Rubey constraint applies. This presents a particularly severe mechanical problem for detachments in extensional terranes, such as proposed by Wernicke and Burchfiel (1982), because then $\sigma_h < \sigma_v$. It is difficult to imagine how pore pressure could become high enough under these conditions to reduce friction sufficiently on the detachment without ordinary normal faulting or hydraulic fracturing interceding. The problem is akin to trying to pull the block in Figure 3.4 (but see Sec. 6.3.4 for a possible solution).

Fig. 3.5 Photograph of a horsetail fan terminating a Mode II fracture in granite. Sense of shear on the principal crack is right lateral. Lateral field of view is about 2 m. (Photograph by Therese Granier.)

3.2   THE FORMATION AND GROWTH OF FAULTS

3.2.1   *The problem of fault formation*

Whereas Anderson's application of the Coulomb criterion to faulting provides an adequate and useful phenomenological understanding, we are still left with the difficulty discussed in Section 1.2.3 concerning the inability of shear cracks to propagate within their own planes. How, then, do faults form and grow to their often great lengths?

Several field studies have documented cases in which faults formed by the linking together of joints (Segall and Pollard, 1983; Granier, 1985; Segall and Simpson, 1986). The interpretation is that σ_1 rotated from a direction parallel to the joints (which formed as Mode I fractures) to a new orientation favoring shear. Linkage occurs by formation of arrays of Mode I cracks in "horsetail" structures (Fig 3.5). Notice the similarity of these cracks to those shown schematically, propagating from the Mode II end of the shear crack in Figure 1.16.

These cases, however, do not provide a satisfactory general explanation for fault formation. If this were always the mechanism of fault initiation, every case of faulting would have to be preceded by an early stress field that has an appropriate orientation to form the tensile fractures later reactivated in shear (and in the case of conjugate sets, by two). Furthermore, in these examples, the length of the reactivated joints is limited; for general applicability the initial tensile fractures would have to be persistent enough to form the long fault systems observed.

From the discussion of Section 1.2.3 we expect that fault initiation must consist, on some scale, as the coalescence of crack arrays. Pollard, Segall, and Delaney (1982) and Etchecopar, Granier, and Larroque (1986) describe observations in which en echelon arrays of tensile cracks were found to emanate from the Mode III edges of faults, similar to that shown in Figure 1.16. Knipe and White (1979) found that the formation of such an array was the earliest stage in the formation of a shear zone in a low-grade metamorphic terrane (see Fig. 3.15 for an example).

In an experimental study, Cox and Scholz (1988a, b) also observed the generation of an array of tensile cracks from the tip of a Mode III shear crack. Upon further shearing, however, they found that the cracks in the array became linked by a series of shear-parallel cracks, which formed a shear process zone, and resulted in propagation of the shear crack as a narrow rubble zone. They envisioned the growth of the shear zone as a progressive process. Only the first stage, the formation of the crack array, is amenable to stress analysis, because it results in a very heterogeneous stress field (Pollard et al., 1982; Cox and Scholz, 1988b).

Several examples of fault zones at different scales are shown in Figure 3.6. These faults are not single shear fractures, but complex zones of brittle deformation. In all the cases, arrays of fractures can be seen. Many of these are correctly oriented for Mode I cracks. In some cases they may be relics of the formation of the fault zone, although many may continue to form as deformation progresses. Even well-developed fault zones are surrounded by zones of pervasively fractured rock (Chester and Logan, 1986); often such secondary faulting may be used to identify the fault (Gamond and Giraud, 1982). *Shear joints*, faults with little accrued slip, may often be seen with a pinnate set of cracks emanating from them, which may be left over from echelon crack arrays developed during formation.

The shear zone shown in Figure 3.6e was produced by a single rockburst in a deep mine in South Africa. It is thought to be due to fracture of previously unbroken rock. McGarr et al. (1979) interpreted the tensile crack array as resulting from the shear fracture, although it could just as well have preceded it. En echelon crack arrays are found as

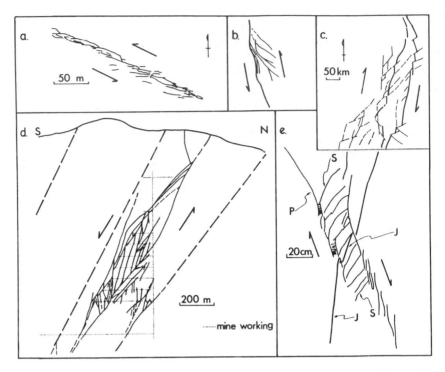

Fig. 3.6 A variety of fault zones mapped at different scales and viewed approximately normal to slip. (a) Dasht-e-Bayaz, Iran; (b) Ales, France, and (c) Taranaki graben, New Zealand – (a)–(c) are map views from Naylor et al. (1986). (d) Cross section through a fault array in the Coeur d'Alene mine, Idaho (thin dotted lines are mine workings, from Wallace and Morris, 1986). (e) Fresh fracture in a deep gold mine, South Africa (from McGarr et al., 1979). J is a preexisting joint that has been offset, and P and S are primary and secondary fractures, according to the interpretation of McGarr et al. (1979) (but see text for a differing interpretation). (Compilation from Cox and Scholz, 1988b.)

a common form of ground breakage produced by strike-slip earthquakes. They are sometimes thought to be near-surface or soil features, because they are usually observed in soil, forming characteristic "moletracks" with en echelon tension gashes separated by compressional push-ups (Sharp, 1979). Sometimes double en echelon sets can be seen, in which the larger cracks in an en echelon set are themselves comprized of an en echelon set on a smaller scale (Tsuneishi, Ito, and Kano, 1978).

   These features are not always restricted to the soil horizons, however. In the South Iceland seismic zone there are several cases of fractures in recent lava beds that seem to have resulted from single strike-slip

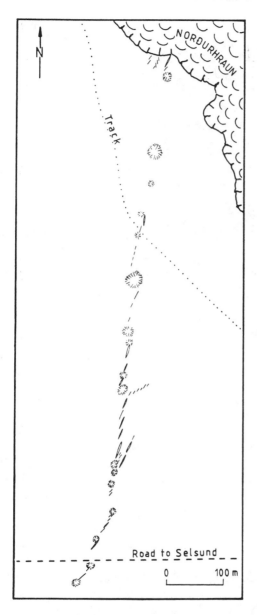

Fig. 3.7 Map of part of the rupture zone of an earthquake of 1912, South Iceland seismic zone. Movement was right-lateral strike slip on a N–S plane, and cuts a recent basalt flow in which there was no preexisting fault. (From Einarsson and Eiriksson, 1982.)

Fig. 3.8 Photograph of an array of push-ups and tension fissures cutting a recent pahoehoe lava flow and forming a strike-slip rupture produced in a single earthquake, South Iceland seismic zone. The push-up in the foreground is about 1.5 m high. Sense of shear is given by the arrows, and tension fissures highlighted by dashed lines and push-up in distance by dotted line. These features can be followed for about 10 km. (Photograph by author.)

earthquakes (Einarsson and Eiriksson, 1982). A sketch and a photograph are shown in Figures 3.7 and 3.8, respectively (for location, see map, Fig. 4.28). There we see the same en echelon structures fracturing solid rock. In these examples, then, we also see the pattern of nascent faulting beginning with the development of an en echelon array of Mode I cracks that later coalesce to form a shear zone. If these strike-slip earthquakes nucleated at depth, they would have propagated up through the near-surface lava beds as Mode III edges. The crack array observed can be compared with that shown adjacent to the Mode III edge in Figure 1.16. In the South Iceland cases, the en echelon crack segments are primarily extensional in their movement. Deng et al. (1986) describe a number of cases of strike-slip faulting in China in which the en echelon segments range from extensional to transtensional, depending on their orientation with respect to the shear zone. The scale of the en echelon segments they described, several kilometers, do not suggest that they are limited to the unconsolidated sediment layer, as may be the case so often in other examples (e.g., Tsuneishi et al., 1978). In the latter case, sandbox models indicate that such en echelon arrays originate as "Reidel shears" resulting from strike-slip faulting in the basement (Naylor, Mandl, and Sijpesteijn, 1986; Tchalenko, 1970). In this case, however, the basement

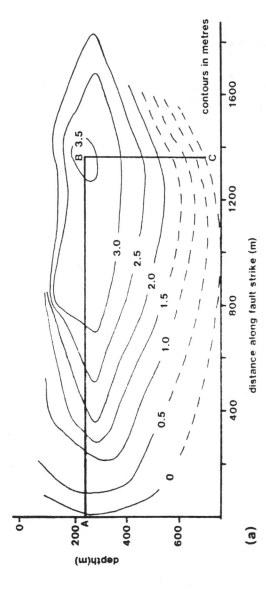

Fig. 3.9 (a) Cumulative slip (throw) distribution on the plane of a normal fault, Coal Measures, Derbyshire, UK (from Rippon, 1985). (b) Profile of slip distribution, from the center to the end of the fault for a similar case. (From Barnett et al., 1987.)

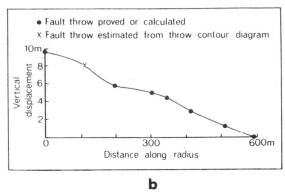

b

Fig. 3.9 *(cont.)*

faults are already well developed, so it is not truly a case of fault nucleation.

3.2.2   *Growth and development of faults*

The cumulative slip distribution on a fault is shown in Figure 3.9a. The pattern is similar to that produced in a single earthquake, with a maximum slip near the center of the fault and slip tapering off to zero at the edges. Slip must taper off in such a way that the stresses at the crack tip remain finite. To maintain this condition, as slip accumulates a fault will be expected to grow in its lateral dimensions. The fault therefore may be envisioned as having originated at a point, growing with progressive slip and gradually developing its characteristic features.

To consider this growth process, we examine the relation between maximum net fault slip, D, and fault length L, as shown in Figure 3.10. Lines of constant strain, $\varepsilon = D/L$, are shown for reference. Elliott (1976) concluded that his data for the Moine and other thrusts in crystalline rock indicated that D was linearly proportional to L, that is, that the growth occurred at a constant critical strain. Watterson (1986) and Walsh and Watterson (1988) combined their data for normal faults in the Coal Measures with Elliott's data and found that the two data sets together defined an $L \propto \sqrt{D}$ relation.

Fracture mechanics would lead us to expect yet a different relationship. From Equation 1.25 the stress intensity factor $\mathbf{K} \propto \Delta\sigma\sqrt{L}$, where $\Delta\sigma \propto D/L$, so if $\mathbf{K}$ is a constant material property and slip builds up by a succession of constant–stress-drop earthquakes, we would expect that $L \propto D^2$. This possibility does not seem admissible by the data. The application of linear elastic fracture mechanics to this problem seems doubtful for other reasons. Recall that this implies a stress singularity at

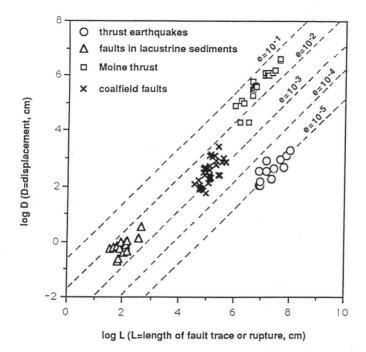

Fig. 3.10 Maximum net slip D plotted versus fault length L for several data sets from different tectonic settings. Dashed lines are lines of constant strain. Data sources are: Moine thrust, Scotland, Elliott (1976); coalfield faults, UK, Walsh and Watterson (1988); faults in lacustrine sediments, Japan, Muraoka and Kamata (1983). Also shown are data from individual thrust earthquakes, from Scholz (1982).

the crack tip, which in the real case is likely to be relaxed by inelastic deformation. In the example shown in Figure 3.9b, the slip tapers off towards the edge much more gradually than given by the static solution for a crack in an elastic medium (Eq. 4.24), which suggests that such relaxation has occurred.

More data such as that shown in Figure 3.10 need to be collected before a firm conclusion can be reached. The data in each individual data set do seem consistent with growth at a critical strain, but if this is the case, the difference between the data sets indicates that the critical strain must be a function of rock type, and perhaps of faulting type as well.

As slip on a fault progressively accumulates and the fault grows in lateral extent, the thickness of gouge that separates the fault walls also expands. Figure 3.11 exhibits data that demonstrate the relationship between thickness of the gouge zone, T, and total slip D for brittle faults in crystalline silicate rock. In this case, $T \propto D$, with a proportionality

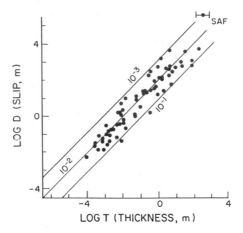

Fig. 3.11 Thickness of the gouge zone T plotted versus total slip D for a collection of faults, mainly in crystalline rock. Point SAF is for the San Andreas fault. (After Scholz, 1987; data mainly from Robertson, 1982.)

constant of approximately 10^{-2}. This relationship can be interpreted to indicate that the gouge was developed by wear (Robertson, 1982; Scholz, 1987). This idea is consistent with the prediction of the steady-state wear law given in Equation 2.22. The data shown in Figure 3.11 imply much higher wear rates than observed in laboratory friction experiments on smooth surfaces (e.g., Fig. 2.13). The explanation of these high wear rates favored by the author is that they result from the much greater roughness of faults. Smooth faults would soon be separated completely by gouge. Then the wear rate would decrease greatly, because more of the slip occurs by rolling and other deformation within the gouge than by abrasion of the surfaces (so-called three-body abrasion, Rabinowicz, 1965). Because faults are fractal (see Sec. 3.5.1) the amplitude of their asperities increases with their wavelength. No matter how thick the gouge becomes there will always be places where asperities directly abut. In these regions wear can be expected to be high. This suggestion has been taken up by Power, Tullis, and Weeks (1988) who developed a model of wear of fractal surfaces. In their model the surfaces are continually "running in" because a steady-state smoothness is never achieved. Sibson (1986a) emphasizes that most brecciation occurs at such fault "jogs." He distinguishes between wear that is produced on slip-vector–parallel walls of faults from the mechanisms of brecciation that occurs at fault jogs or asperities. He describes mechanisms and examples of "distributed crush brecciation" at compressional jogs and "implosion brecciation" at extensional jogs (see Sec. 3.5.2). These are all considered

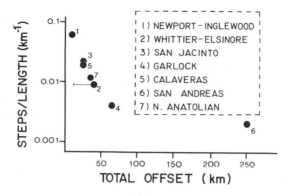

Fig. 3.12 Plot of fault steps per unit length versus net slip for active strike-slip faults. (From Wesnousky, 1988.)

to be modes of abrasive wear and will contribute their products to the thickness of the fault zone.

If the fault originates as an en echelon crack array such as described in the previous section, then as slip proceeds wear will be expected to erode the steps and produce a progressively more continuous and smoother fault. This process is illustrated in Figure 3.12, which shows fault steps per unit length as a function of net slip for a number of active strike-slip faults. This figure is very similar to a wear curve (Fig. 2.14), showing the most rapid changes during running in at small displacements. Note, however, that wear is intrinsically an anisotropic process (Sec. 3.5.2) so that such smoothing will be much less severe in the direction normal to slip. As a result, the surface traces of pure dip-slip faults need not exhibit such smoothing.

The wear law given in Equation 2.22 also predicts that gouge zones should increase in thickness with depth, if other things, such as rock type, remain constant. Some confirmation of the thickening of a fault with depth in the schizosphere was obtained in a seismic refraction study across the San Andreas fault by Feng and McEvilly (1983). They found that the fault was marked by a near-vertical low seismic velocity zone thickening with depth to about 10 km. From an interpretation of gravity data from the same locality, Wang et al. (1986) found agreement with this result and also found a secondary zone of low-density rock that increased in thickness towards the surface. They interpreted this using a two zone model of Chester and Logan (1986), in which a central gouge zone thickens with depth and is surrounded by a zone of shattered rock that thins with depth. The thickening of the gouge zone with depth may be interpreted as due to an increase in wear with increasing normal stress, and the thinning of the shattered zone by the increase in strength of rock outside the fault zone. In another locality, Anderson, Osborne, and Palmer (1983) found that the gouge zone thinned with depth. This may be a near-surface result in which weathering increases the friablity of the rock, leading to greater wear rates. These properties, however, are variable along the fault. Mooney and Ginsburg (1986) noted that a pronounced low-velocity zone was associated with the section of the San Andreas fault in central California that is slipping aseismically, but within the resolution of the seismic method (2 km) they could not detect one for sections of the fault that are frictionally locked, suggesting a connection between the gouge-zone structure and the fault's frictional characteristics (as will be discussed in Sec. 6.4.1).

The mechanical work of faulting is expended in three main ways: frictional heating; the surface energy of gouge formation; and, if seismic slip occurs, by elastic radiation. It is possible to estimate the fraction of energy that goes into gouge formation from the wear model discussed in Section 2.2.3. According to the model, each fragmentation produces new surface area $\pi d^2/2$, so proceeding from Equation 2.22, we find that the amount of surface area created per unit sliding distance is

$$A = \frac{2\kappa ND}{hd} \tag{3.3}$$

The inverse dependence on d in Equation 3.3 reflects an increase in area produced by grain size reduction. The frictional work is $W_{fr} = \mu ND$, and the surface energy is $U_s = A\gamma$, where μ is the friction coefficient and γ the specific surface energy. If we define $\vartheta = U_s/W_{fr}$, then

$$\vartheta = \frac{2\kappa\gamma}{hd\mu} \tag{3.4}$$

Using $d = 5$ μm, the dominant gouge grain diameter measured in the experiments of Yoshioka (1986), $\gamma = 10^3$ erg/cm^2 (Brace and Walsh, 1962), $\mu = 0.5$, and κ and h as in Section 2.2.3, we find, for Yoshioka's experiments, that $\vartheta \approx 10^{-4}$. This is in reasonable agreement with the more thorough estimate he made. If the larger wear rate observed for the natural faults is explained by a higher κ (or lower h), and we assume the same d, then we obtain $\vartheta = 10^{-3}$. This is near the upper limits of Yoshioka's results, but still a negligible energy sink.

The thickness of mylonites in a ductile shear zone also increases approximately linearly with slip (Hull, 1988). In this case, the mechanism is more likely to involve strain hardening of the mylonites (White et al., 1980; Means, 1984). The constant of proportionality between T and D is similar to that found for the brittle faults, except for the case of superplastic zones, such as the Glarus in Switzerland (Schmid, Boland, and Paterson, 1977), where it is several orders of magnitude smaller. In lower-grade terranes, thin (relative to slip) faults can also be observed, particularly in phyllitic rocks and carbonates. In these cases, it is likely that the wear mechanism is not abrasion, but adhesion or some other mechanism associated with a much smaller wear coefficient. Equation 2.22 may still apply, but with a much smaller value of κ. As we will see in Section 3.3.2, some mylonites may also result from wear.

Another such scaling relationship is that between D and T for pseudotachylytes (Sibson, 1975), for which it was found that $D \propto T^2$. He made a case that these rocks were melts formed during seismic slip, and so were dimensionally controlled by frictional heating rather than wear.

These macroscopic scaling relationships show that faults progressively grow in lateral extent and thickness as slip accumulates on them. At the same time, deformation increases within the fault zone, leading to intensification of strain and associated structures within the rocks of the fault zone, and to an overall smoothing of the fault zone topography.

3.3   FAULT ROCKS AND STRUCTURES

Geologists have associated crushed rock with faults for centuries, but the first description of a rock intimately related to faulting was by Lapworth (1885), who coined the term *mylonite* to describe certain types of rocks associated with the Moine thrust of northwestern Scotland. Although the term means literally milled rock, the type of rock to which Lapworth applied it is now recognized as a product of plastic deformation processes. From Lapworth's day to the present, a great deal of confusion has arisen over both genetic understanding of fault rocks and the terminology used to describe them, not the least of which is caused by their great variety. It is possible, for example, to view Lapworth's "type" mylonite (flinty and finely laminated, with a single foliation and lineation) at the Stack of Glencoul on the Moine thrust, then walk in an updip direction for a few hours to Knocken Crag, where one finds the fault filled with an unfoliated finely pulverized rock flour without a trace of recrystallization.

The confusion over the genesis of such rocks has a rich history. Christie (1960), also working with the Scottish Highland rocks, was the first to recognize clearly the importance of recrystallization in addition to cataclasis in forming these rocks and summarized the debate up to that time. It wasn't until recently, however, that it was shown that the grain size reduction typical of mylonites could be produced entirely by plastic flow, not just by brittle comminution (Bell and Etheridge, 1973). Although fault rocks must form a continuous series for a given starting material (protolith), arguments over their subdivision and the preferred

Table 3.1. *Textural classification of fault rocks*

				Random fabric	Foliated		Percent of matrix
Incohesive			Fault breccia (visible fragments > 30% of rock mass)		?		
			Fault gouge (visible fragments < 30% of rock mass)		Foliated gouge		
Cohesive	Nature of matrix	Glass–devitrified glass		Pseudotachylyte	?		
		Tectonic reduction in grain size dominates grain growth by recrystallization and neomineralization	Cataclasite series	Crush breccia (fragments > 0.5 cm) / Fine crush breccia (0.1 < fragments < 0.5 cm) / Crush microbreccia (fragments < 0.1 cm)			0–10
				Protocataclasite	Mylonite series	Protomylonite	10–50
				Cataclasite		Mylonite (Phyllonite varieties)	50–90
				Ultracataclasite		Ultramylonite (Phyllonite varieties)	90–100
		Grain growth pronounced		?		Blastomylonite	

nomenclature continues to the present. (See, for example, Wise et al., 1984.)

3.3.1   *Fault rocks and deformation mechanisms*

Rocks that occur within fault zones provide primary evidence for the processes that occur there. Because faults pass entirely through the schizosphere and, as ductile shear zones, much of the plastosphere as well, there is a broad suite of deformation mechanisms involved in their deformation and a rich variety of rocks result. Furthermore, many such rocks may have been subject to multiple periods of deformation, often under different conditions. In studying fault rocks from deeper levels one must rely on the examination of ancient faults that have been uplifted and exposed, and if fault movement continued throughout that process there will be exposed in the fault zone a variety of fault rocks formed under different conditions. Later structures and rock types may obliterate earlier ones, and rocks formed during rapid seismic shear may be geologically ephemeral. The task of reconstructing the deformation mechanisms and structural history of fault zones at different structural levels is difficult and only a simplified account is possible at present.

Sibson (1977) discusses the various types of fault rocks and deformation mechanisms associated with them. His textural classification, with amendments, is given in Table 3.1, and photomicrographs of some representative types are shown in Figure 3.13. The main textural divisions are between random fabric and foliated types, and between cohesive and incohesive rocks. Subdivisions within the incohesive types are based on grain size, from breccia to gouge. The cohesive types are divided based on the tectonic reduction of grain size from that of the host rock and the fraction of fine-grained matrix relative to lithic fragments and residual, coarse crystals. The latter, in the mylonite series, are referred to as *porphyroclasts*. They are distinct from *porphyroblasts*, which occur in blastomylonites and have experienced growth during or after plastic deformation.

In Table 3.1, foliated gouge has been added to Sibson's original list, because it occurs in nature and can be produced in the laboratory (Chester, Friedman, and Logan, 1985). One excellent exposure of this kind of rock known to the author is in the Linfield Falls (North Carolina) window of the Blue Ridge thrust. There the fault is marked by 1 m of fine quartzo-feldspathic sand, which though very friable, shows tabular splitting and a mylonitic foliation and lineation (a photomicrograph of a similar, nearby rock is shown in Fig. 3.13a). In the Linfield Falls locality also one can see a weak mylonitic foliation in the wall rock (a granite gneiss) within a meter of the gouge zone.

Sibson classifies the mechanisms that produce fault rocks into two main types, "elastico-frictional," in which the micromechanism is primarily brittle fracture, and "quasi-plastic," in which the mechanism involves some degree of crystalline plasticity. In the latter category he includes the various solution- and diffusion-aided processes.

The elastico-frictional processes include brittle fracture of the host rock, frictional (abrasive) wear, and cataclastic deformation of the gouge or breccia. In this context, the term cataclastic flow refers to the deformation of a granular aggregate involving grain rolling, frictional sliding at grain boundaries, and comminution by brittle fracture, rather

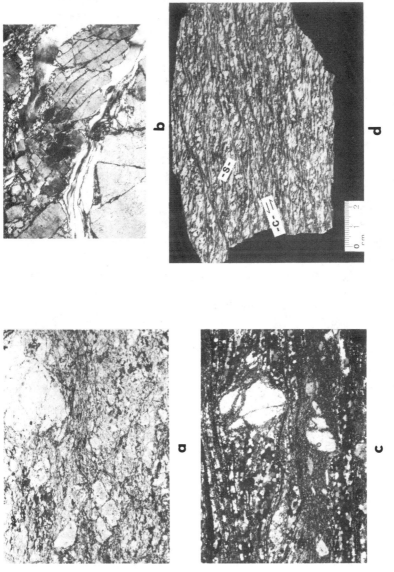

Fig. 3.13 Photomicrographs of fault rocks, all formed from a granitic protolith. (a) Foliated cataclasite in Striped Rock granite, in the Blue Ridge thrust below the Fries thrust zone, Independence, Va. Field of view (FOV) 3.3 mm, plane light. Clasts of quartz, feldspar and lithic fragments in a fine-grained foliated matrix of phyllosilicates. (b) Granitic mylonite from the semibrittle field (upper greenschist grade). Fractured microcline surrounded by plastically deforming mica and quartz ribbons. FOV 3.3 mm, crossed nicols. Borrego Springs shear zone section of the Peninsular Ranges Mylonite Zone, southern California. (c) Granitic mylonite from the fully plastic field (amphibolite grade). Dynamic recrystallization of Kspar with well-developed quartz ribbons. FOV 3.3 mm, crossed nicols. Same locality as (b). (d) Hand specimen of an S-C mylonite, lower amphibolite facies. From the Santa Rosa mylonite zone, Eastern Peninsular Ranges

than its more general usage for semibrittle behavior noted in Section 1.4. The quasiplastic processes include the various types of dislocation creep with associated syntectonic recrystallization, pressure solution processes, and, in certain cases, superplasticity (see Sibson, 1986b; see also Sec. 1.4).

Sibson attributes the formation of the incohesive rocks and the cataclasite series to the elastico-frictional processes and the mylonite series to quasiplastic ones. In general it might be expected that, with increasing depth (hence increasing pressure and temperature), the rocks encountered in a fault zone will follow some path from upper left to lower right in Table 3.1. However, transitions within a given series do not necessarily reflect different ambient temperature and pressure conditions. Sibson (1977) notes that as strain increases towards the center of a shear zone one may find a transition from protomylonite to mylonite to ultramylonite. Even for near-surface cases, where the strain profile is highly discontinuous, one may see a transition from brecciated rock to gouge to foliated gouge (Chester and Logan, 1986). Increasing strain may effect a transition from the unfoliated types to the foliated types for both the cohesive and incohesive rocks. A progressive reduction of grain size by cataclastic or plastic flow may also enhance the solution- and diffusion-aided processes, because they are grain-size dependent (Mitra, 1984).

Beyond the general association of the cataclasites with the schizo-sphere and mylonites with the plastosphere, one cannot estimate un-equivocally depth of formation of these rocks based solely on textural evidence. Nonetheless, it is clear that such a progression exists. Simpson (1985), for example, studied mylonites formed from a granitic protolith over a range of conditions from lower greenschist to amphibolite meta-morphic grade (approximately 300–450°C, see Sec. 1.4). The sequence of rocks is shown in Figure 3.13b–d. Lower greenschist mylonites were typified by plastic deformation of quartz, brittle fracture of feldspar, and kinking of biotite. By midupper greenschist grade, recrystallization of quartz, biotite, and orthoclase was evident, but there was only minor low-temperature plasticity in the plagioclase. In the amphibolite grade and above, recovery and recrystallization could be seen in all minerals. This sequence thus spans the brittle–plastic transition, with the green-schist grade boundaries defining the limits of the semibrittle field, much in the way described in Section 1.4. Using Sibson's textural classification it is only possible to say that the mylonitic series lie within the semibrittle field and the blastomylonites within the plastic field. The transition between them is at approximately the greenschist–amphibolite boundary (~ 450°C).

Sibson (1977, 1986b) has also tried to differentiate those types of fault rocks generated by rapid seismic slip from those produced during slow deformation during other parts of the seismic cycle. Prominent among

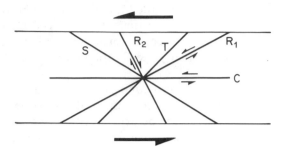

Fig. 3.14 Schematic diagram of a shear zone, showing the geometrical relationships of the main structural elements.

the former are pseudotachylytes. These are very fine grained, usually dark rocks, often containing various amounts of lithic fragments, and usually found in fault zones. Based primarily on veining relationships, Sibson (1975) argues that they are the result of frictional melting during seismic slip. Although they can sometimes contain glass, they are usually devitrified so that their origin must be inferred from veining relationships, texture, and flow structures (Maddock, 1983). These serve to distinguish these rocks from similarly appearing rocks that were formed by very fine pulverization, rather than by melting (Wenk, 1978). Formerly thought to be uncommon, the more frequent recognition of pseudotachylytes now suggests that their supposed rarity may have been from a lack of recognition. Their occurrence, even when recognized and mapped, may underestimate the rate at which they are produced, because they can be obliterated easily by any further deformation. Grocott (1981) suggested that they form preferentially in special structures within fault zones.

Other structures result from rapid changes in pore pressure induced by seismic slip, such as gouge-filled clastic dikes (Gretener, 1977). Sibson (1986a, 1987) has identified "implosion breccias" which he infers to have resulted from sudden coseismic reductions of pore pressure.

3.3.2   *Fabrics and surfaces*

Mylonitic rocks are characterized by an anastomosing shape fabric with a single lineation and foliation. This forms the "fluxion" structure first described by Lapworth (1885). In addition, there are a number of typical nonpenetrative structural elements of shear zones, all of which have a distinctive geometrical relationship with the shear zone itself (Fig. 3.14). Two nomenclatures are used for these fabric elements. One was originated by Berthe, Choukroune, and Jegouzo (1979), and the other, based on similar structures produced in the laboratory, by Logan et al. (1979), following the terminology of Riedel (1929). Where they overlap, we use the former, indicating the latter in brackets.

The primary foliation, referred to as the S [p] foliation ("schistosite") by Berthe et al. (1979) is found at an angle of 135–180° from the shear zone. It is identified with the plane of maximum flattening in the finite strain ellipsoid. In the simple shear case it forms initially at 135° and with progressive shear may be rotated to near parallelism with the shear zone. The lineation lies in the S plane in the direction of shear. Secondary Reidel shears are often seen in the R_1 orientation, with the same sense of shear as the zone itself, and antithetic shears, usually less well developed, can sometimes be found in the R_2 orientation. The amount of slip on such planes is necessarily quite limited by their geometry, but discrete slip surfaces, with the same sense of shear as that of the overall shear zone, may also occur parallel to the shear zone. In this way, they are not limited in the amount of slip that can take place on them. These are referred to as C [y] ("cisaillement") surfaces, and mylonites in which this fabric element is well developed are called S-C mylonites (Fig. 3.13d). Many of these fabric elements can be seen in the rocks shown in Figure 3.13, and in the simulated gouges deformed in the laboratory (Fig. 2.15).

These fabrics reflect the geometry of the strain field in the shear zone, commonly one of simple shear parallel to the zone, and are not diagnostic of a particular deformation mechanism. The S foliation, for example, may result from the preferred orientation of platy minerals due to mechanical rotation, particularly in clay-rich gouges (Rutter et al., 1986). In high-grade mylonites it results from preferred crystallographic orientation due to plastic deformation and recrystallization (Simpson and Schmid, 1983; Lister and Williams, 1979). Similarly, the C surfaces may be zones of concentrated slip in unconsolidated gouge (Chester and Logan, 1986), or thin zones marked by recrystallized, finer grain size material in high-grade mylonites (Simpson and Schmid, 1983; Lister and Snoke, 1984). Reidel shears were originally defined for brittle deformation with which they are usually associated, but in ductile deformation a foliation or cleavage, called C′, sometimes is developed in the R_1 orientation (see Fig. 3.17).

Tensile fracture (gash) (T) arrays are also common. Initially, they form normal to the least principal strain direction (inferred from a simple shear configuration). Subsequently, they are rotated by finite strain into sigmoidal shapes. These are noticed most often in low- and medium-grade metamorphic terranes, where vein-filling material makes them prominent. They often occur early in the history of the shear zone, and arrays of tension gashes can be seen to mark incipient shear zone formation (Knipe and White, 1979). An example is shown in Figure 3.15.

Fig. 3.15 Photograph of a tension gash array in an incipient shear zone, Hartland Quay, Cornwall. Sense of slip: left lateral. (Photograph by Simon Cox.)

The presence of such extension fractures in shear zones is sometimes taken to mean that anomalously high fluid pressures were present, and that such fractures formed by hydraulic fracturing (e.g., Reynolds and Lister, 1987). However, this need not be the case for arrays of tension gashes formed within the zone of strain localization associated with a shear zone, because the local stress field there may deviate considerably from the ambient stress and the tension crack arrays may form in a zone of stress concentration similar to their formation along the edges of shear cracks (Secs. 1.2.3 and 3.2.1). Thus, in cases such as shown in Figure 3.15, where veining is localized to tension gash arrays in incipient shear zones, an argument that pore pressure exceeds one of the principal stresses in the surrounding rock is not justified.

The strain in shear zones often increases towards the center of the zone. In this case, the strain profile may be either continuous or discontinuous. Schematic examples are shown in Figure 3.16. If the strain is continuous (Fig. 3.16a), the S foliation is deformed into a sigmoidal form, from which the strain profile may be measured (Ramsay and Graham, 1970). The limits of strain measurement are met in cases of extreme strain, where the S foliation approaches parallelism with the shear zone. When C surfaces are present (Fig. 3.16b), which may sometimes develop progressively with shear (Fig. 3.17; see also

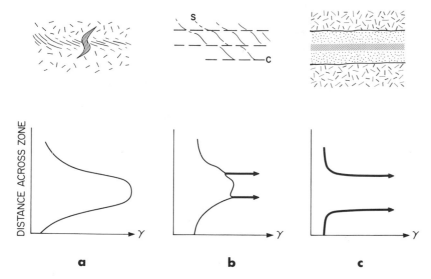

Fig. 3.16 Schematic sketches and strain profiles for three types of shear zones: (a) continuous; (b) discontinuous (S-C mylonite); (c) highly discontinuous (brittle).

Simpson, 1984), strain is discontinuous. The total strain across the shear zone can no longer be estimated without the use of offset markers. The discontinuity of shear is most pronounced in the case of purely brittle faulting (Fig. 3.16c). The presence of discrete frictional offsets renders the usual concept of strain meaningless, but one can still often discern an increase in the intensity of deformation towards the center of the fault zone (Chester and Logan, 1986).

Another principal characteristic of shear zones, both brittle and ductile, are lenticular-shaped masses of almost undeformed rock, which may occur on scales ranging from kilometers to millimeters. The long axis of the lenticels lies roughly parallel to the slip direction. The actively deforming finer mass, whether it be gouge zones on the map scale or the mylonitic matrix on the microscale, appears to flow around them in an anastomosing pattern. On the grain scale, these are called porphyroclasts, and they are usually composed of the materials most resistive to deformation, typically feldspars in quartzo-feldspathic rocks. In mylonites, porphyroclasts form asymmetric *augen* structures, often with trails of finer-grained recrystallized material, that may be used to infer the sense of shear (Simpson and Schmid, 1983).

Slip surfaces in fault zones are typically (but not always) smooth and shiny, and commonly striated in the direction of slip. These are called *slickensides*, although some authors have used this term to describe the

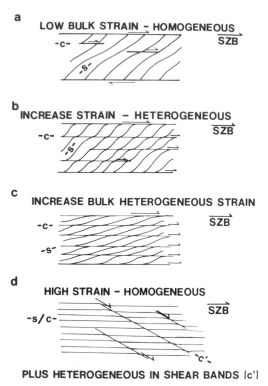

Fig. 3.17 Diagram showing the development of the S and C fabric elements in mylonite with progressive strain. All case show right-lateral shear parallel to the shear zone boundary (SZB). The development of a cleavage (C′) as a shear band in the Reidel direction is shown at high strain. (After Simpson, 1984.)

striations themselves (a somewhat objectionable usage, see Fleuty, 1975). The striations have long been used as an indication of the most recent slip direction and, sometimes, small steps perpendicular to the striae also can be used reliably to determine the sense of slip (downstepped in the direction of slip).

Slickensides may form by a variety of mechanisms. Classical slickensides are polished surfaces covered with grooves and debris tracks and are most certainly the result of abrasive wear. Surfaces of laminated and lineated gouge may also be found (Engelder, 1974). Other fault surfaces called slickensides are covered with mats of slip-parallel fibrous growths of crystals or lineated vein material (Durney and Ramsay, 1973). In this case the slickensides are formed by a growth process rather than a wear process. Solution and dissolution may play a role in the formation of these and other features seen on fault surfaces.

3.4   STRENGTH AND RHEOLOGY OF FAULTS

More than 50 years ago, when developments in seismology allowed the accurate determination of earthquake depths, it was discovered that most earthquakes were of shallow origin. It was recognized then that this reflects the different rheological properties of the schizosphere and plastosphere. Macelwane (1936), in an address on progress on the "geologico-seismological frontier," stated the current understanding and a salient problem succinctly:

Geologists have been accustomed to speak of the zone of fracture near the surface of the earth and of the underlying zone of flow. Now it has been found that the depth of first yielding in most destructive earthquakes is not, as might be expected, near the top of the zone of fracture but is at or somewhat below its base. The *normal* depth of focus of earthquakes or depth of first significant radiation of earthquake waves seems to be between ten and fifteen kilometers. [italics in the original]

These ideas and observations have not much changed to date. Only recently, however, can more definitive statements be made concerning the zones of fracture and of flow, and the question brought out by Macelwane regarding the initiation depth of earthquakes can be understood also. From an understanding of the fracture and friction of rock, geological observations of faults, and relevant seismological observations, it is now possible to assemble a reasonably detailed picture of the rheology and strength of faults in the schizosphere and much of the plastosphere. Yet when it comes to the quantitative aspects of this model, there remains a fundamental difference of opinion.

3.4.1   *A synoptic shear zone model*

The concept, expressed by Macelwane, that there exists a zone of fracture and a zone of flow, was not greatly improved upon until 1980. Brace and Kohlstedt (1980) and Kirby (1980) assembled laboratory data to produce a simple model of the rheology of the continental lithosphere (Fig. 3.18). The upper part of the model uses Byerlee's friction law for the limiting strength and assumes a hydrostatic pore-pressure gradient. An Andersonian stress state and fault orientation is also assumed, so the stress required for thrust faulting is much greater than that for normal faulting. The stress for strike-slip faulting lies somewhere in between. The lower part of the model is based on the extrapolation to these low temperatures of a high-temperature steady-state flow law for an appropriate rock (wet quartzite is shown). This model gives a lower limit of the strength of the lithosphere, because it describes the stress levels required to drive a suitably oriented fault. The strength predicted in the upper,

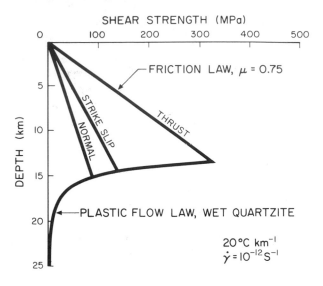

Fig. 3.18 Simple model of the strength of the lithosphere. In the upper part, optimally oriented faults are assumed with Coulomb friction and constant $\mu = 0.75$ and hydrostatic pore pressure. For the lower part, an experimentally determined flow law for wet quartzite (Jaoul, Tullis, and Kronenberg, 1984), has been extrapolated, assuming the strain rate and temperature gradient as indicated.

frictional part depends only on the pore pressure assumed, whereas that of the lower part depends strongly on the assumed rock type, temperature, and strain rate.

The intersection of these two laws was taken to mark the brittle–plastic transition in the earth. Sibson (1982) and Meissner and Strelau (1982) compare this model with observations of the deformation mechanisms of fault zone rocks and earthquake hypocentral depth distributions. They find a rough correspondence between the predicted brittle–plastic transition, the depth at which evidence of ductility can first be found in fault rocks, and the maximum depth of earthquakes. This seemed to provide strong confirmation of this model and so detailed refinements and applications were made, in which often minor variations in the depth distribution of seismicity were, for example, ascribed to variations in heat flow (Chen and Molnar, 1983; Sibson, 1984; Smith and Bruhn, 1984; Doser and Kanamori, 1986).

However, as discussed in Section 1.4, the brittle–plastic transition cannot occur at a point, as depicted in this simple model. It must occur over a range of pressure and temperature in which semibrittle behavior gradually changes from the fully brittle to fully plastic modes of deformation. The steady-state flow law used in the lower part of the model cannot be extrapolated into the semibrittle field and therefore cannot be extrapolated to intersect the friction law. The strength of the rock in the semibrittle field will always be higher than predicted by extrapolation of such a flow law because, as the semibrittle field is entered, plastic flow mechanisms will gradually be eliminated, first in one mineral component and then in another, which will increase the flow strength. Microscopic brittle processes will become increasingly necessary to accommodate the strain, and which carry with them a pressure sensitivity of strength that was not included in this extrapolation.

This is what we observed in the structure of mylonites (Sec. 3.3). In quartzo-feldspathic mylonites, quartz flows readily throughout the semibrittle field, but the feldspar remains resistant to deformation other than brittle fracture and forms porphyroclasts. The rock deforms as a composite material and cannot be characterized with a quartzite flow law. Because a typical granite contains only about 10% quartz, it resembles, in that respect, concrete at moderate confining pressure, which is bound together by a relatively ductile lime-rich matrix but which owes its strength to the hard silicate aggregate it contains. This problem has been analyzed by Jordan (1988), who showed that for small fractions of the weaker material, the strength is determined by that of the more resistant component. It is only in a case when throughgoing compositional banding has been produced at high strain that the strength approaches that of the weaker component.

As the rock progresses from protomylonite to ultramylonite, and through the semibrittle field in the way described by Simpson (1985), the nondeforming fraction is reduced and the compositional foliation increased. The rock then may approach steady-state plastic flow. This is not to say that the mylonites do not obey some flow law before this state is reached, but that the law cannot be estimated from extrapolation of a steady-state plastic flow law derived from high-temperature experiments.

Another assumption implicit in the interpretation of the model shown in Figure 3.18 is that there is a sudden transition, at the brittle–semibrittle transition, between discrete frictional slip on a fault and bulk, continuum flow in a ductile shear zone. It is assumed here that friction is synonymous with brittle processes and excluded by plastic deformation. [This is implied, for example, by the names of the two classes of deformation used by Sibson (1977), "elastico-frictional" and "quasi-plastic."] In Chapter 2 we noted that friction is independent of the micromechanism involved in deformation of the asperities, which may be brittle fracture or plastic flow. Plastic flow is the mechanism of friction of metals and of rocks at high temperatures. The first occurrence of mylonites in a fault zones cannot be used to mark the base of the frictional layer. Indeed, the abundance of slip discontinuities in mylonites (C surfaces) suggests otherwise. Frictional slip, with its attendant linear dependence of strength with normal stress, will only give way to bulk deformation when the frictional surfaces become welded, that is, $A_r \approx A$, which is likely to occur at a depth greater than that of the first onset of ductility.

In Section 2.3.3 we saw that although the onset of ductility does not change the frictional character of the fault nor the base value of friction, which determines the strength, it does result in a sign change in the velocity dependence of slip, from velocity weakening to velocity strengthening. The lower friction stability transition occurs at the onset of ductility and defines the greatest depth at which earthquakes can nucleate, that is, the base of the seismogenic zone. Also, there will be a change in the wear mechanism at that depth, from abrasive wear to adhesive wear. The latter wear mechanism is typical of metallic friction and results from plastic shearing out of asperities, often with material transferred from one surface to the other. Mylonites formed at this structural level may be partially the products of this type of wear, rather than the products of bulk ductile deformation of rock in place (Scholz, 1988).

A synoptic model of a shear zone that embodies these characteristics is shown in Figure 3.19. The particular example uses a temperature gradient based on a heat flow model for the region of the San Andreas fault (Lachenbruch and Sass, 1980, model B), so the depths indicated on the left of the figure are specific to that model. This model is for a quartzo-feldspathic crust, so the important fiducial points are 300°C, marking the onset of quartz plasticity and hence the brittle–semibrittle transition, and 450°C, marking the onset of feldspar plasticity and the semibrittle–plastic transition. The first of these transitions is denoted T_1 and the second T_2.

At T_1 there is a change in fault rocks from cataclasites to mylonites and in wear mechanisms from abrasion to adhesion. If there is no change in the protolith, Equation 2.22 indicates that the fault thickness should increase with depth, as shown. At T_1 it is shown as narrowing, because the wear coefficient for adhesive wear is usually much smaller than that for abrasive wear (Rabinowicz, 1965). The asymmetric hourglass shape of the fault zone shown is highly idealized, and only applies, at best, to a pure strike-slip fault in rock of uniform properties.

The seismic properties of the fault are described with the rate variable $A - B = (a - b)\sigma_n$ from Equation 2.27. The fault is velocity weakening and hence unstable where this variable is negative and velocity strengthening and stable where it is positive. The model used is the same as that shown in Figure 2.21, except near the surface, where $a - b$ is assumed to become positive, resulting in an upper stability transition T_4. The presence of this upper transition was first proposed by Scholz, Wyss, and Smith (1969) and is based on the observation that there are both upper

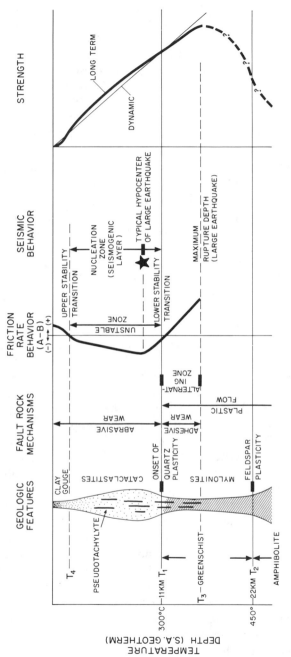

Fig. 3.19 Synoptic model of a shear zone. See the text for explanation. (From Scholz, 1988).

and lower seismicity cutoffs on well-developed faults (Fig. 3.20). There are two possible reasons for this transition. The first (see Secs. 2.3.3, and 2.4) is due to the normal stress dependence of the critical stiffness, which could possibly result in very large nucleation lengths at shallow depths. The second explanation is based on the presence of unconsolidated gouge in the fault zone at shallow depth, because that material is generally found to be velocity strengthening (Sec. 2.2.3). The second explanation was argued by Marone and Scholz (1988) because the observation of an upper cutoff in seismicity at about 2–4 km seems to be limited to well-developed fault zones where a thick gouge layer is likely to be present (Fig. 3.20). In regions where there are no well-developed faults, such as the central Adirondacks of New York, earthquakes can be observed to occur at much shallower depths (e.g., Blue Mountain Lake, Fig. 3.20). An example will be given later (Sec. 4.4.1) of a case in the Imperial Valley, California, in which the upper transition clearly seems controlled by the presence of sediments.

The lower stability transition occurs at T_1, so that the seismogenic layer, which is the zone in which earthquakes can nucleate, is bounded by T_1 and T_4. In this model, then, the base of the seismogenic layer is marked by the 300°C isotherm and the first appearance of mylonites, in agreement with the interpretations of the simple model of Figure 3.18, but for different reasons. Though T_1 and T_4 delimit the region in which earthquakes can *nucleate*, they do not bound the region in which they can *propagate*. Clearly a large earthquake that ruptures the entire seismogenic zone can propagate dynamically through T_4 and breach the surface. Similarly, the depth to which that earthquake may propagate is not limited by T_1, but the earthquake will propagate to some greater depth T_3 ($< T_2$) as shown in the models of Das (1982) and Tse and Rice (1986). But the hypocenters of the aftershocks of that earthquake will be limited by T_1, so they will not define the bottom of the rupture zone of the mainshock.

Although it is usually assumed that aftershocks delineate the mainshock rupture zone, including its depth extent, Strelau (1986) points out that the evidence for this is either ambiguous or circular. Standard seismological methods of the past, and particularly geodetic methods, do not have sufficient resolution to define independently the maximum depth of faulting. For example, Thatcher (1975) shows that the geodetic data for the 1906 San Francisco earthquake cannot resolve slip at the 1 m level below 10 km (see also Fig. 5.2). More modern seismological techniques, however, can, in suitable circumstances, resolve this difference. Nabelek (1988), for example, found significant moment release for the 1983 Borah Peak earthquake down to a depth of 16 km, even though most aftershocks cut off at 12 km (Fig. 3.20). However, both the mainshock hypocenter and that of one of the largest aftershocks oc-

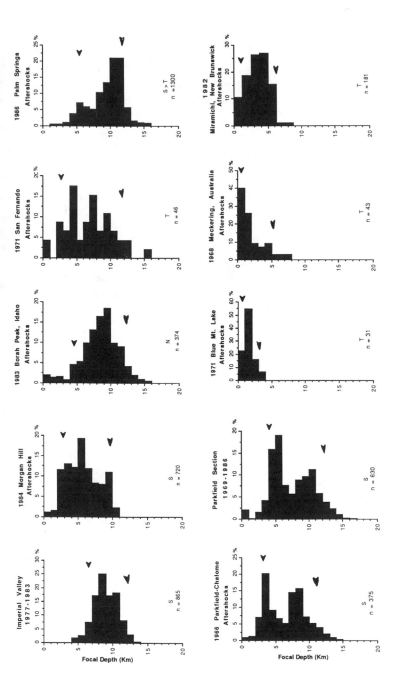

Fig. 3.20 (a) Earthquake depth distributions along well-developed and poorly-developed rocks. The dominant faulting mode (strike-slip, S; thrust, T; normal, N) is shown for each. Arrows indicate depths below which and above which 90% of the earthquakes occur. For well-developed faults on which considerable met slip has occurred, upper and lower cutoffs are apparent, but for midplate faults of very little slip (Blue Mtn. Lake, Meckering, Miramichi) an upper cutoff is not apparent. (After Marone and Scholz, 1988.)

curred at 16 km, so there must at least be patches in the unstable field at that depth (Boatwright, 1985).

We can now distinguish the seismogenic layer, bounded by T_1 and T_4, from the schizosphere, which extends from the surface to T_3. The region between T_1 and T_3 is one of alternating behavior, with coseismic dynamic slip and interseismic semibrittle flow. There is a growing body of geological evidence for such an alternating zone, in observations of shear zones in which mylonites are found interlaced with pseudotachylytes and breccias (Sibson, 1980a; Stel, 1981, 1986; Wenk and Weiss, 1982; Passchier, 1984; Hobbs, Ord, and Teyssier, 1986). Hobbs et al. (1986) try to explain the presence of these pseudotachylytes as a result of a ductile instability. In their model they found that an instability cannot arise in a material with a wet quartz rheology at a temperature greater than about 200°C. However, if the earthquake is assumed to impinge dynamically on T_1 from above, it will impose a coseismic strain rate some 10^8 times greater than the interseismic rate (or about 10^{-3} s^{-1}), which from their model will allow propagation to a much greater depth.

We may now address the other question posed by Macelwane: Why do large earthquakes typically nucleate near the base of the seismogenic zone, and not from near the top? Sibson (1982) pointed out that the base of the schizosphere is the region of highest shear strain energy and so is the most likely origin of large earthquakes. This qualitative idea has been filled out in two ways. In their quasistatic frictional model of a fault zone, Tse and Rice (1986) (see also Sec. 5.2.2) found that nucleation is most likely where $A - B$ is minimum. In their model this occurs in the lower part of the seismogenic zone. A different point was made by Das and Scholz (1983) who, using a dynamic model in which strength and stress drop increase with depth, found that ruptures that initiate in the shallow, low-stress environment are prevented from propagating into the deeper, high-strength regions. Only ruptures that initiate in the deep high-stress region are capable of propagating through the entire schizosphere.

The strength profile that results from this model is shown in the right-hand frame of Figure 3.19. The most significant difference between Figure 3.19 and Figure 3.18 is that in Figure 3.19 the strength continues to follow the friction law to at least T_3 with no strength discontinuity at T_1. The peak in strength has been placed approximately at T_3. Below this depth the strength will drop, joining a high temperature flow law at T_2. Although full plasticity is not reached until T_2, we expect that the fault becomes welded and the strength reduced above that point by flow aided by *pressure solution* mechanisms resulting from the intrusion of fluids into the fault zone (Etheridge et al., 1984; Kerrich, 1986).

This fault zone model is limited to a quartzo-feldspathic crust, and the depth scale is, of course, dependent on the assumed thermal gradient.

The results can be generalized, though with many fewer constraints, to two other principal seismotectonic terranes: subduction zones and faults in oceanic lithosphere. These cases will be taken up in Sections 6.3.3 and 6.4.3.

3.4.2 *Thermomechanical effects of faulting*

If we neglect work against gravity, almost all mechanical work done during deformation must be converted into heat. Because faults are sites of concentrated tectonic work, they are favorable places to look for evidence of mechanically generated heat. A general energy balance for faulting can be written as

$$W_f = Q + E_s + U_s \qquad (3.5)$$

W_f is the mechanical work done in faulting, including both friction and ductile deformation; Q is heat; U_s is surface energy, which from the evaluation of Equation 3.4 was found to be negligible for faulting, and E_s is the energy radiated in earthquakes. The partition of energy into elastic radiation will be taken up again in Section 4.2; for the present discussion this term will be neglected.

Because we usually do not observe heat, but rather temperature or heat flow, it is more convenient to consider the time derivative of Equation 3.5. Neglecting the last two terms, this is

$$\tau v \geq q \qquad (3.6)$$

In this inequality, τ is the mean shear stress acting on a fault sliding at velocity v and q is the heat flow generated by the fault. The equality holds when the last two terms on the right-hand side of Equation 3.5 are truly negligible.

There are two classes of thermomechanical effects in faulting. There will be a steady-state thermal signature resulting from the work of faulting averaged over geologic time. Here the appropriate velocity to use in Equation 3.6 is the geologic slip rate, 1–10 cm yr^{-1}, for major plate-bounding faults. A transient heat pulse may also occur from the coseismic slip during earthquakes. In that case the value of v to be used in Equation 3.6 is a particle velocity appropriate for dynamic slip, 10–100 cm s^{-1}.

Shear heating The first case is known as *shear heating*. Evidence for this phenomenon may be found in the Alpine schists, a suite of rocks that lie just to the east of the Alpine fault of New Zealand (Fig. 3.21). The Alpine fault is a major continental transcurrent fault that has undergone about 450 km of dextral movement in the late Tertiary (see

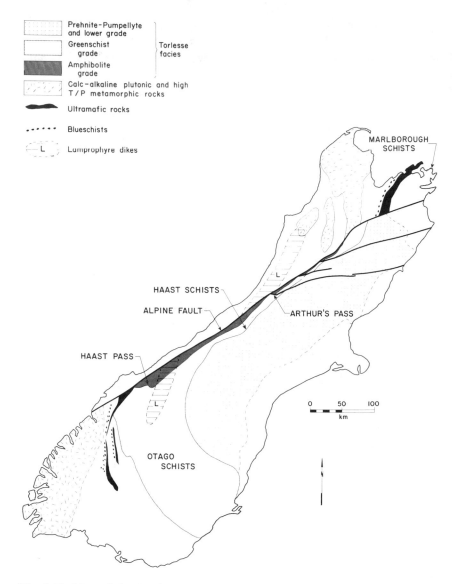

Fig. 3.21 Map of the South Island of New Zealand, showing the Alpine schists and their association with the Alpine fault. (From Scholz et al., 1979.)

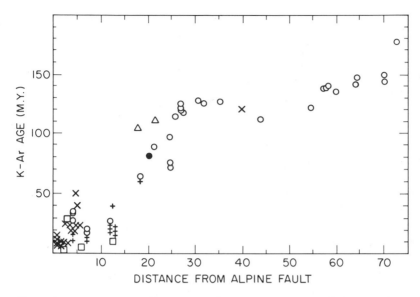

Fig. 3.22 K–Ar age of the Alpine schists as a function of distance from the Alpine fault. (From Scholz et al., 1979.)

Sec. 6.3.1). Its motion has taken on an oblique component since the Pliocene, producing about 10 km of uplift on the southeast side, resulting in the uplift of the Southern Alps. Subsequent erosion allows us a view of a suite of rocks that were adjacent, at midcrustal depths, to an active strike-slip fault in the recent geological past.

The deformation and metamorphism of the schists is of Mesozoic age. They exhibit a distinctive pattern of K–Ar ages, however, that shows that they have undergone a more recent period of heating that is both temporally and geometrically related to the more recent period of fault motion. The K–Ar ages of the schists, shown in Figure 3.22, monotonically decrease as the fault is approached, with a depletion aureole some 20 km wide and approaching zero age within 1 km of the fault. Scholz, Beavan, and Hanks (1979) modeled this by using the known slip and uplift history of the fault to calculate the heat generated by faulting and the resulting diffusion of radiogenic argon from the adjacent rocks. They concluded that if the depth-averaged shear stress on the fault was in excess of 50 MPa, consistent with strength models such as shown in Figure 3.18, then shear heating, together with the uplift of the schists, could adequately explain this aureole, which could not be explained by uplift alone.

There is also an intimate relation between the metamorphism of the schists and their deformation. As is common in metamorphic belts, the maximum in deformation preceded the maximum in heating, so that the most intense structures are overprinted by the metamorphism, and the highest metamorphic grades of the schists are associated with the highest strained rocks. This suggests that an internal source of heat, perhaps deriving from the deformation itself, resulted in the metamorphism. Because the strain pattern of the schists is similar to that of a dextral ductile shear zone, Scholz et al. (1979) suggested that the metamorphism resulted from shear heating during an earlier Mesozoic period of dextral shear. They showed that with a similar choice of parameters used to explain the K–Ar depletion, sufficient heat could be generated by this mechanism. However, because the deformation history of the schists in the Mesozoic is poorly known, this case is much more speculative than the more recent heating episode recorded in the K–Ar ratios.

Scholz (1980) reviews a number of other cases of shear heating revealed by metamorphic aureoles around faults. The most geologically definitive and obvious examples are those in which the hanging walls of thrusts have inverted metamorphic gradients. These cases, of course, all occur in ductile-deforming terrane where the strong temperature sensitivity of the strength of the rocks results in thermal softening that provides a buffer for the shear heating (for models of this see Brun and Cobbold, 1980; Fleitout and Froidevaux, 1980). Shear heating in carbonate sequences is limited to 300–400°C by rapid thermal softening of calcite in that temperature range. In faults associated with ultramafic rocks, shear heating appears to be buffered by serpentinite dehydration at 550°C and in granite by melting at 650°C, so that in the latter cases sillimanite grade metamorphism may be achieved, as in the Main Central thrust of the Himalayas. In the latter case, anatectic granites are found along the fault. This example was analyzed in some detail by Molnar, Chen, and Padovani (1983), who concluded that if the granites were produced near the fault, then shear heating, resulting from shear stresses in excess of 50 MPa, would have been required. However, in this case the constraints on both the thermal properties of the crust and the metamorphic history of the rocks are weak enough that other possibilities may be just as likely.

As in the case of the Main Central thrust, evidence for the source of heat in metamorphism is usually ambiguous enough that only a circumstantial case for shear heating may be made. For this reason, in metamorphic petrology usually only static sources of heat are considered. It is, however, an inescapable conclusion of Equation 3.6 that if the rate of fault motion exceeds 1 cm/y and the shear stress 50 MPa, as would be expected from the strength models explored earlier, then the work of faulting will be an important source of heat to the adjacent terrane. For a thorough treatment of this problem for fold and thrust belts, see Dahlen and Barr (1989) and Barr and Dahlen (1989). As we shall see later, however, the viability of such strength models has been challenged in at least one important case: the San Andreas fault of California.

Coseismic transient heating   Because of the high slip velocities involved, and the low thermal conductivity of rock, transient heating during rapid coseismic slip occurs essentially without conduction, and so if the fault is thin enough, very high temperatures may be reached locally. With the assumption of reasonable values for earthquake parameters, one may expect friction melting (McKenzie and Brune, 1972), or the vaporization of any fluid phase present (Sibson, 1973). These effects depend on the slip being concentrated within a zone no thicker than about 1 cm (Cardwell et al., 1978; Sibson, 1973). The occurrence of pseudotachylyte is testimony that friction melting sometimes occurs, and fault-associated veins and clastic dikes may also indicate transient high pore pressures.

A question then arises as to what effects these phenomena may have on the dynamic frictional resistance of the fault. If the fault becomes covered with melt, or filled with vapor at superlithostatic pressure, its frictional resistance may drop abruptly so that the earthquake stress drop might approach the total stress, rather than being about 10% of it, as would be the case otherwise.

Melt lubrication is sometimes important in sliding of metals at high velocity and in some other cases, such as the friction of skiis on snow or skates on ice, where the contact pressure induces the melting (Bowden and Tabor, 1964). However, silicate melts have high viscocity when dry, and coseismic flash melting probably does not allow enough time for significant water to enter the melt. If we assume the viscosity of a friction melt to be that of dry granite (10^7–10^8 poise) and a coseismic shear strain rate across a 1-cm layer (10^1–10^2 s^{-1}) the melt layer should retain considerable shear resistance (10–100 MPa).

In an experimental study of friction melting of rock, Spray (1987) found only a slight force drop occurred upon melting and, in an analysis of natural pseudotachylytes, Sibson (1975) also concluded that the residual strength of the melt was significant. We must also consider, in natural faulting, how much of the fault plane will have slip confined to a zone narrower than 1 cm, a prerequisite for these effects to occur. In a study of a very large prehistoric landslide in the Austrian Tyrol, Erismann, Heuberger, and Preuss (1977) showed that much less friction melt occurred at its base than would be predicted from the drop in potential energy calculated for the slide. In that case the friction melt was found in thin, discontinuous slivers distributed through a 10–100-cm-thick rubble zone. They concluded that rapid movements of the particles in the rubble zone resulted in a very high thermal conductivity for that zone, so that the efficiency of producing melt was much less than otherwise would have been expected.

The production of high fluid pressures by transient vaporization probably will be limited similarly to localized patches of intimate asperity contact. If that is the case, then its effect on globally reducing the dynamic fault strength will not be significant, since the real area of asperity contact is likely to be much smaller than the total fault area.

3.4.3   *The debate on the strength of crustal fault zones*

There is a long tradition in seismology that the stress drop in a earthquake (~ 1–10 MPa) completely relaxes the ambient shear stress and hence is a measure of the absolute value of stress in the schizosphere. This dates from Tsuboi (1933), who identified the "ultimate strain" of the earth with a geodetically measured coseismic strain drop. A more recent example is the identification of the strength of the earth's crust with earthquake stress drops by Chinnery (1964). With the advent of the stick-slip theory of earthquakes this position became untenable, though vestigial echoes of this long-held belief still may be found in the seismological literature. Stress sometimes is discussed synonymously with stress drop and sections of faults that exhibit high stress drop are referred to, without qualification, as *strong* (as compared to *weak* sections where smaller stress drops occur).

However, an argument is still alive as to whether shear strength of continental faults are on the order of earthquake stress drops (~ 1–10 MPa) or at the level predicted by friction models such as discussed in Section 3.4.1 (~ 100 MPa). This argument centers on the lack of any evidence for a thermomechanical effect associated with the San Andreas fault (Brune, Henyey, and Roy, 1969; Lachenbruch and Sass, 1973, 1980). With an abundance of surface heat flow measurements in California, no local conductive heat flow anomaly can be found at the fault. Brune et al. considered that the maximum heat flow anomaly that can go undetected by their measurements (assuming a two standard deviation uncertainty) is one produced by a mean shear stress of about 20 MPa. Lachenbruch and Sass, considering a much larger data set, also found no measurable anomaly, and quote the same value of shear stress as a maximum value (twice the mean stress). An example of the data is shown in Figure 3.23, where they are compared with the expected steady-state conductive anomaly for a fault moving at 35 mm/y with a depth-averaged shear stress of 50 MPa. With the exception of one point, discussed later, the data show no evidence of any heat flow anomaly.

On the other hand, there is a broad heat flow anomaly associated with the San Andreas, about 100 km wide with a magnitude of about 0.8 HFU (Fig. 3.24). This anomaly dies out as the age and total slip on the fault decrease to zero at the Mendocino triple junction, so it is most likely that this anomaly is mechanically generated by fault motion. The amount of heat in this anomaly is sufficient to account for heat generated by the fault sliding with 100 MPa of ambient shear stress (Hanks, 1977).

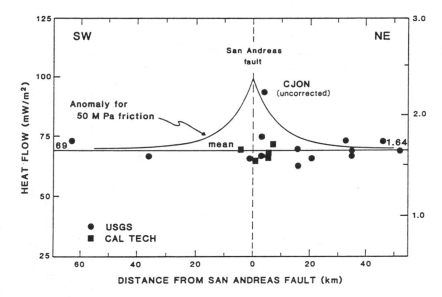

Fig. 3.23 Heat flow in the vicinity of the San Andreas fault, Los Angeles–Mojave region, California, projected on a section perpendicular to the fault. The data point CJON is from a deep drillhole at Cajon Pass. (From Lachenbruch and Sass, 1988.)

However, because the heat flow anomaly is not localized to the fault, Lachenbruch and Sass interpreted this anomaly using a model in which it is generated by basal shear stresses of about 40 MPa on a detachment surface, with negligible stress on the fault.

The heat flow data shown in Figure 3.23 were taken in the region of the bend in the San Andreas fault in southern California where the fault strikes more east–west than elsewhere (see map, Fig. 5.5). In situ stress measurements made by the hydrofracturing technique to a depth of 0.9 km in the same region are shown in Figure 3.25 (Zoback, Tsukahara, and Hickman, 1980; McGarr, Zoback, and Hanks, 1982). These data show that the maximum principal stress is N 20° W, about 45° from the local strike of the fault, and that the shear stress increases with depth at a gradient of 7.9 MPa/km. If extrapolated to depth, these data indicate a depth-averaged shear stress of 56 MPa and a friction coefficient of 0.45. These results, which are similar to stress gradients measured elsewhere in the continental lithosphere to depths of 2.5 km (McGarr and Gay, 1978), contradict the heat flow data of Figure 3.23, but the extrapolation of either data set to seismogenic depths may be questioned.

One may also argue, prompted by the complete lack of any heat flow anomaly, that the heat conduction models upon which the heat flow argument is based are incorrect. This may be because pore fluid convec-

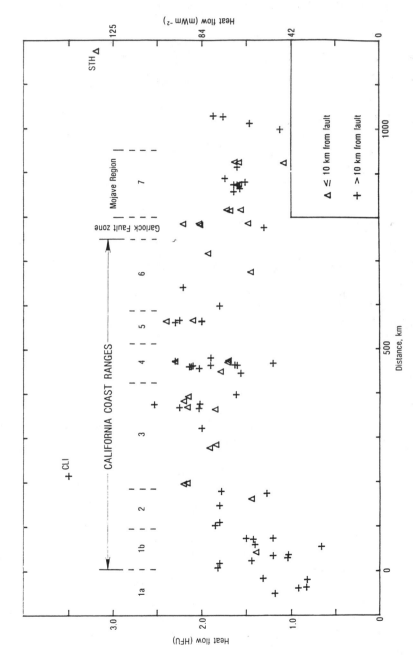

Fig. 3.24 Heat flow data from coastal California, projected parallel to the San Andreas fault. The distance axis is measured from the Mendocino triple junction, where the fault has zero age. (From Lachenbruch and Sass, 1980.)

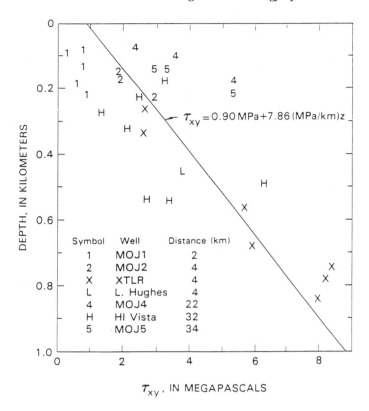

Fig. 3.25 Horizontal shear stress as a function of depth from hydrofracture stress measurements made near the San Andreas fault in the same locality as the heat flow measurements of Figure 3.23. (From McGarr, Zoback and Hanks, 1982.)

tion disperses the heat so that a purely conductive model is not appropriate (O'Neil and Hanks, 1980). Modeling studies have shown, though, that convection driven by a frictionally heated fault is not likely to destroy the near-fault heat flow anomaly expected from purely conductive models (Williams and Narasimhan, 1989). In the same modeling study, however, it is pointed out that much of the San Andreas fault lies in elevated terrain, and in that case gravitationally driven fluid flow could easily remove any heat flow anomaly for measurements made in shallow boreholes. The heat flow measurements shown in Figure 3.23 and elsewhere in California were all made in shallow (~ 100 m) holes. In contrast, the one data point in Figure 3.23 which appears to show the anomaly was made in a 1.8-km borehole near Cajon Pass. This measurement is also equivocal, however; the high heat flow there may be a result of unusually rapid uplift and erosion (Lachenbruch and Sass, 1988).

Mount and Suppe (1987) and Zoback et al. (1987) have assembled evidence, in the form of near-fault fold axes, wellbore breakouts, and off-fault focal mechanisms, that suggest that the maximum principal stress is nearly normal to the San Andreas fault along much of its length. This suggestion that the San Andreas is nearly a principal plane means that its total shear stress could be scarcely more than the stress drops of earthquakes that occur on it. Jones (1988), based on a study of off-fault seismicity in the vicinity of the San Andreas fault in southern California, concluded that the maximum compressive stress is at a constant angle of about 65° from the fault. Oppenheimer, Reasenberg, and Simpson (1988) on the same basis argue that it is in the range of 63–80° from the Calaveras fault. Although these results also suggest that the fault is weak, the difference between them and the Mount–Suppe and Zoback et al. conclusions are significant.

The reasons laboratory-determined values of rock friction should apply on the geological scale have been given in Section 2.4, and the Rangely experiment recounted as an example. If natural faults have significantly lower strengths than given by Byerlee's law, there seem to be only two possible reasons. One is that faults are filled ubiquitously throughout the schizosphere with saturated and unconsolidated clay-rich gouge, which can reduce friction significantly, Wang (1984) (see also Fig. 2.7b). The other reason is that friction is reduced by the presence of elevated pore pressure, which requires that the Hubbert–Rubey condition be much more widespread than originally envisioned. This necessitates a mechanism for the generation and maintenance of such overpressures, only now on a much wider scale than in certain types of overthrust terranes.

Assuming a representative coefficient of friction of $\mu = \tan 30°$, the greatest angle the maximum principal stress can make with the fault is 60°, in the case when pore pressure is equal to the least principal stress. The value reported by Jones and the lower range of values reported by Oppenheimer et al. would thus be consistent with an overpressurization mechanism. The conclusions of Mount–Suppe and Zoback et al. cannot be explained simply by elevated pore pressure, because that would require the pore pressure to nearly equal the maximum principal stress, which could only occur in a hydrostatic stress state. Zoback et al. consequently interpreted it as meaning that the fault is everywhere filled, throughout the seismogenic depth, with clay-rich fault gouge, which has an intrinsically low strength. However, although in undrained laboratory tests (e.g., Wang, Mao, and Wu, 1980) saturated clay has a very low strength, it has a normal frictional strength when dry (Sec. 2.2.4). This suggests that its low saturated strength is not inherent, but that it has the capacity to build up an internal pore pressure nearly equal to the applied load. Throughout the duration of the laboratory tests, the effective stress is very small and deformation may occur with the clay particles sliding past one another with little frictional resistance. For this mechanism to work in the scenario envisioned by Zoback et al., the internal pore pressure must nearly equal the maximum principal stress and be maintained over geologic time. This seems highly unlikely because the internal pore pressure would exceed two of the principal stresses and could only be restrained from escaping by molecular forces. Furthermore, although clay-rich gouge is commonly found in near-surface exposures of faults, it is not generally observed at deeper levels (e.g., Anderson et al., 1983). It is rather to be expected that those parts of the fault that do contain such gouge are aseismic, because frictional sliding in the presence of clay is always stable (see Sec. 6.4.1). Indeed, much of the evidence of Mount–Suppe and Zoback et al. comes from the region of the San Andreas fault where it is stably sliding.

The above discussion provides both the flavor and main arguments in this evolving debate. We need to consider how these results for the San Andreas fault apply to the strength of the lithosphere in general. A special issue of the *Journal of Geophysical Research* (Hanks and Raleigh, 1980) is devoted to this question. If the San Andreas fault is as weak as some of the arguments indicate, it does not seem likely that this result will apply to the lithosphere as a whole. Abundant determinations have been made of the flexural rigidity of both the oceanic and continental lithosphere from their response to loads produced by islands, glaciers, and sediments. These results, which provide an estimate of the long-term elastic thickness of the lithosphere, are consistent with rheological models such as in Figures 3.18 and 3.19 (Kirby, 1983). This shows that regions of the lithosphere that do not contain well-developed faults can support large stresses to depths corresponding to the base of the seismogenic region, which occurs at about $300°C$ for continental and $600–700°C$ for oceanic lithosphere, respectively (see Sec. 6.3.3 for a discussion of the oceanic case).

It does not seem credible that overpressures are developed and maintained universally throughout the schizosphere. If true, the mechanism for doing so, which must be fundamental, has not yet been proposed. Furthermore the flexural studies referred to earlier show that the lithosphere is not universally weak. It is possible, however, that faults themselves can provide a mechanism for internally generating and maintaining overpressures. In the shear heating examples given in the previous section, prograde metamorphic reactions are driven by the heat generated by the shear zone. These reactions typically involve dehydration, so they can provide a source for metamorphic fluids at high fluid pressure. As discussed in Section 3.1.2, rocks deforming in such prograde terranes often show evidence of near lithostatic pore pressure. If these fluids leak into the brittle fault zone above the ductile shear zone and are

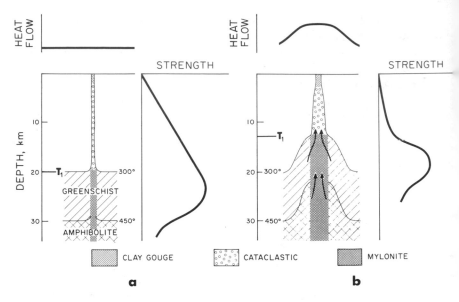

Fig. 3.26 An evolutionary model of a fault zone, in which the fault becomes overpressured and weakened by self-heating.

trapped there by low-permeability near-surface fault zone material, this can provide a source of high fluid pressures localized within the fault zone.

Calculations show that, starting with a strength model of the type described in Section 3.4.1, only faults moving at least 1 cm yr^{-1} or faster will generate sufficient shear heating for this mechanism to operate. They must slip for about 5 My before the thermal system matures (see, e.g., Scholz et al., 1979). With this reasoning we therefore arrive at an evolutionary model of faulting, as shown in Figure 3.26. It applies only to major, rapidly moving faults for which shear heating is significant. The fault starts as a high-strength feature, with a hydrostatic pore-pressure gradient and a frictional strength given by Byerlee's law (Fig. 3.26a). As shear heating progresses, the geotherms are elevated and high pore pressures begin to develop within the frictional regime. This lowers the strength there, but has little or no effect on strength in the ductile regime at depth. After about 5 My of fault movement (at least 50 km slip, from the above figures) the thermal structure becomes mature and the fault is weak within the schizosphere. Within the plastosphere it remains strong and continues to generate heat. [Analyses of microstructures in mylonites from well-developed shear zones indicate that they support steady-state flow stresses in the range 40–200 MPa. This is sufficient to generate considerable shear heating at slip rates of 1 cm yr^{-1} and above (Kohlstedt and Weathers, 1980; Ord and Christie, 1984).] The mature fault zone is shown in Figure 3.26b. The heat flow anomaly over the fault is broad because the heat source is deep, and the heat may also be dispersed by water circulation.

Because the low-stress state of the San Andreas has not yet been established, this model is not only speculative but conditional. It also requires special circumstances for the fault plumbing system. Specifically, high fluid pressures are required to drain from their source in the metamorphic rock by means of high-permeability cataclasites and be capped by low-permeability clay gouge. However, consequences of this model can be explored. It means that some rapidly moving and mature faults may have low strength, but that faults are not universally weak. This agrees with in situ stress measurements made in a variety of active faulting environments that indicate fault friction coefficients in the range 0.6–1.0 (Zoback and Healy, 1984). The model also provides a mechanism for strain softening, in that the fault becomes more of a zone of lithospheric weakness progressively during its deformational history, and thus further localizes strain there. This model may also help to explain why slower moving intraplate faults appear to be stronger than plate boundary faults (Sec. 6.3.5).

These points continue to be debated, but to return to the original point: There is little reason to expect coseismic stress drops to equal the total stress. In the one case where the answer is known, for tremors in deep South African gold mines, the stress drop is a small fraction of the total stress (Sec. 6.5.3). The only mechanisms that could cause stress drop to be total are the coseismic transient heating effects discussed in Section 3.4.2, and, as mentioned, these are only likely to be effective locally. Note that these mechanisms don't predict that faults have low initial strength, which would also have to be true for the equality to hold. On the other hand, from the experimental studies reviewed in Chapter 2 we would expect stress drops to be about 5–10% of total stress. Stress drops for San Andreas earthquakes are typically about 1–3 MPa, which is consistent with a fault strength of 10–60 MPa. On the other hand, stress drops of similar earthquakes on less well-developed faults are about 6–10 times higher, suggesting comparably higher strengths that are consistent with normal values of rock friction and pore pressure (Secs. 4.3.2 and 6.3.5).

3.5   FAULT MORPHOLOGY AND MECHANICAL EFFECTS OF HETEROGENEITY

3.5.1   *Fault topography and morphology*

Although the segmentation of faults into crack arrays has been described in Section 3.2.1, and the morphology of slickensides in Section 3.3.2, we have otherwise treated faults as planar frictional features. Faults are not

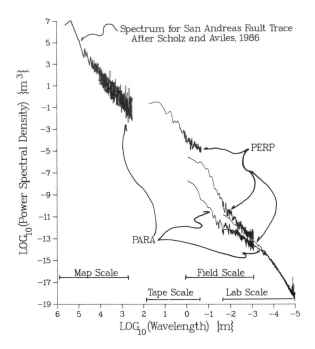

Fig. 3.27 Power spectra for topographic profiles of fault surfaces. Para and perp mean parallel and perpendicular to the slip vector. (From Power et al., 1987.)

perfectly planar on any scale, however, and their deviation from this idealized case provides both useful information on the mechanics of faulting and predictions regarding the heterogeneity of earthquakes (Sec. 4.4). E. M. Anderson appreciated this point, as can be found in the following quote (Anderson, 1951, page 17):

> It is often convenient to use the term fault plane, but on investigation it is seldom that a fault presents an absolutely plane surface. The face is usually fluted on a small scale, giving rise to the appearance known as slickensides. It may be assumed that the fluting shows the direction, at least the latest direction, of movement. A section of such a surface would be a straight line along, but a sinuous line if made across the fluting, and it is probable that what thus happens in miniature is repeated on a much larger scale. A little reflection will show that a fault is likely to take a straight course, if followed along the direction of movement. Otherwise there will be either considerable distortion, or open spaces will develop as faulting proceeds.

This passage shows that Anderson appreciated that the irregular form of faults exists on all scales, and that it is strongly conditioned by its orientation to the fault slip vector. Figure 3.27 shows the power spectra of topographic profiles made along fault surfaces. These data, which extend over the scale range 100 km to 10 μm, show that the faults are irregular surfaces over that entire range, with their surface topography scaling as a generalized fractal. Profiles taken parallel and perpendicular to the slip vector show a strong anisotropy in the topography. Specifically, they are much smoother parallel than perpendicular to slip. This anisotropy extends to map scales: The surface traces of strike-slip faults are much straighter than dip-slip faults. Although several generations of elementary structural geology texts have explained this difference by the interference of a dipping (planar) fault with surface topography, this explanation has missed this essential fact. The anisotropy of fault surfaces reflects the anisotropy of wear, which modifies the surfaces once they are formed, making them smoother in the slip direction.

This description of the fractal nature of fault surface topography does not contain any phase information. It simply tells us that whatever irregularities occur on faults appear at all scales. If the features themselves form a fractal set, they will have a power-law size distribution characteristic of fractals. However, as such structures form by brittle processes, they can be expected to take on distinctive forms as dictated by the mechanics of brittle fracture.

In describing these forms we immediately encounter an almost impenetrable thicket of nomenclature. Geologists have over the years introduced a broad collection of terms to describe such features. Often these terms find only regional usage, and the terminology for strike-slip faults differs considerably from that of thrusts. In recent years a number of workers, recognizing the mechanical role these irregularities take in impeding rupture, have introduced new classification schemes with terminologies descriptive of both the geometry and the mechanical processes they feel are most importantly associated with the feature. Because these systems are usually neither complete nor rigorously defined, none have been embraced universally and so have confused the literature further. A lengthy glossary would be needed to clarify this vocabulary, but such a discussion would not advance our aims.

To avoid this problem we introduce here a descriptive system that has had long use in dislocation mechanics (e.g., Weertman and Weertman, 1964). It has the advantage of being simple, mechanically rational, and of not overly conflicting with existing geological terminology.

Define as the slip plane an idealized plane on which the slip vector lies, and which we take as the local or regional mean of the actual fault plane. Deviations of the fault from the slip plane have two special cases. If the line of deviation of the fault from the slip plane is normal to the slip vector it is called a *jog*, and if parallel, a *step*, as shown in Figure 3.28. The mechanical reason for this classification is obvious: any movement of a jog will require volumetric strain, and that of a step will not. A jog will seriously impede slip whereas the effect of a step will be less.

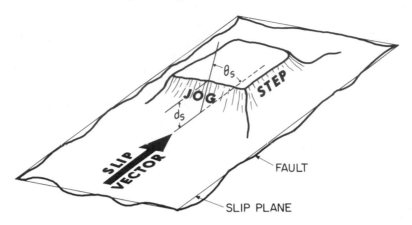

Fig. 3.28 Schematic diagram illustrating the definition of jogs and steps.

Notice also that a rupture will always encounter a jog as a Mode II crack, and a step in Mode III. Just as in the definitions of shear crack modes, the jog and the step are special cases. They will in general be joined by regions that deviate obliquely from the slip vector. To recognize this, and that the slip vector is not always well defined in practice, we retain the geological term *stepover* to refer to any deviation from the slip plane regardless of its orientation with respect to the slip vector.

These definitions are not restricted to any fault type. Deviations from the slip plane of a strike-slip fault will, as seen in map view, be jogs, and will be steps if viewed in cross section. The opposite will be true of dip-slip faults. Jogs have a sign, depending on the sign of the volume strain they induce: compressional ($-$) or extensional ($+$). A step is neutral and has no sign. To find a rule for determining the sign, consider a strike-slip fault, in map view. If the sign of the stepover (right or left) is the same as that of the fault (right lateral or left lateral) then the jog is extensional, if opposite, then it is compressional. The same rule will hold for dip-slip faults except that then one must use a vertical reference frame instead of the customary horizontal one.

Jogs and steps may vary in size, as measured by their deviation from the slip plane (d_s in Fig. 3.28) and in their abruptness, as measured by the angle θ_s they make with the slip plane. Depending on the values of d_s and θ_s and the total slip that has accrued since its formation, jogs may take on different forms, as shown idealized in Figure 3.29. An offset jog is one in which there is no clear connection between the fault strands. They may have overlap or underlap, so θ_s may be larger or smaller than 90°. The other forms, bend jogs and duplex jogs, may correspond to progressive stages of development as slip accumulates. These same forms apply also to steps, except that step duplexes are not known to form.

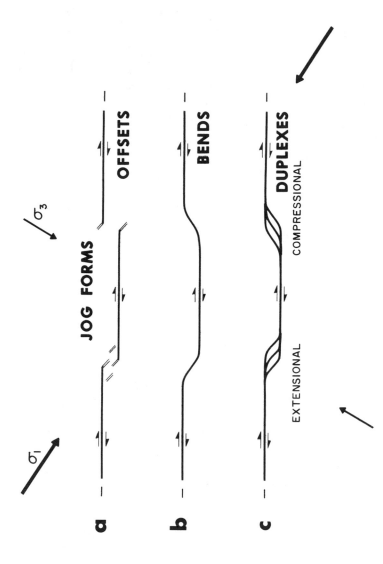

Fig. 3.29 Schematic diagram illustrating different types of stepover structures.

149

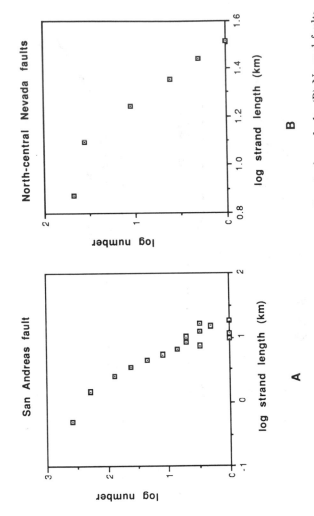

Fig. 3.30 The length distribution of fault strands: (A) San Andreas fault; (B) Normal faults, north-central Nevada. (Data from Wallace, 1989.)

150

We may now note the correspondence of this terminology with that used by others. In the case of strike-slip faults, Sibson (1986c) restricts the term jog to what we call offset jog and uses the terms antidilatational and dilatational for compressional and extensional. Crowell (1984) uses restraining and releasing for the same meanings in the case of fault-bend jogs. King (1986) uses the terms nonconservative and conservative barriers with the same meanings as our jogs and steps. The terminology used here therefore melds these several usages into one systematic treatment, not restricted to a specific fault type.

Although often mapped as continuous features, faults when mapped in detail are found to be composed of discontinuous strands on all scales (e.g., Wallace, 1973). The strand length distribution obeys a power law, like the earthquake size distribution (Sec. 4.2.2), and so probably constitutes a fractal set (Fig. 3.30). This distribution shows an upper limit for strand length that is approximately the same as the thickness of the schizosphere, so this probably corresponds to an upper fractal limit, beyond which fractal scaling does not apply.

As an example, in Figure 3.31 sections of the surface rupture trace of the strike-slip Dasht-e-Bayez (Iran) earthquake are shown at four scales. Similar features may be recognized at the different scales, and may also be recognized in small experimental models constructed of clay and deformed in the laboratory (Tchalenko, 1970). Many jogs may be seen in this figure with forms similar to those idealized in Figure 3.29.

Similar structure may be seen on dip-slip faults if they are viewed in vertical section (Woodcock and Fischer, 1986). The structure of thrust faults often has a very different appearance, however. The slip plane of thrusts is often fairly close to the orientation of bedding, which constitutes a plane of weakness that is frequently taken preferentially as the fault plane. This produces a staircase structure of flats and ramps in which the fault alternately follows bedding and then cuts obliquely across it. The same strength anisotropy favors duplex formation by the propagation of fault-bend folds, so these are much more pronounced on thrusts that cut sedimentary sections than on other faults.

Structures similar to that found in jogs occur at fault terminations, except they are not culminated by an intersection with another strand. Examples are horsetail fans (Fig. 3.5) and their compressional equivalents. Faults in extensional jogs of strike-slip faults have a normal component, and in compressional jogs, a thrust component. As a result, sedimentary basins may be produced in extensional jogs if they are sufficiently developed, the most notable being *pull-apart* basins (Christie-Blick and Biddle, 1985). *Push-ups*, blocks elevated by crustal shortening, are the equivalent features of compressional jogs. Isolated blocks caught within the fault zone are sometimes called horses.

Characteristic splays also occur near the intersection of faults with the free surface, owing to the stress-free boundary condition there. Because

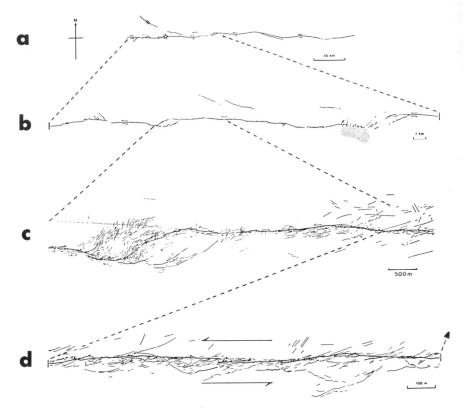

Fig. 3.31 Map of the surface trace of the Dasht-e-Bayez earthquake, shown at four scales. Scale bars are (a) 10 km, (b) 1 km, (c) 500 m, (d) 100 m. (From Tchalenko and Berberian, 1975.)

of the asymmetry of thrusts at that intersection, these are typically restricted to the hanging wall, forming *imbricate fans*. Similar structures in strike-slip faults are more symmetrical and are called *flower* structures.

3.5.2   *Mechanical effects of fault irregularities*

Jogs and steps both present impediments to rupture, the magnitude of which depends upon their severity, as measured by d_s and θ_s. Since these features exist on all scales, they will result in heterogeneity in dynamic rupture at all scales and in extreme cases, rupture termination.

Jogs, because they require for slip either major distortion of the fault plane or pervasive inelastic deformation in the surrounding volume, are much the more serious impediments to rupture. Examples of the deformation within jogs are shown in Figure 3.32. If slip occurs on both sides of the jog, some idea of the stress field within it can be obtained by an

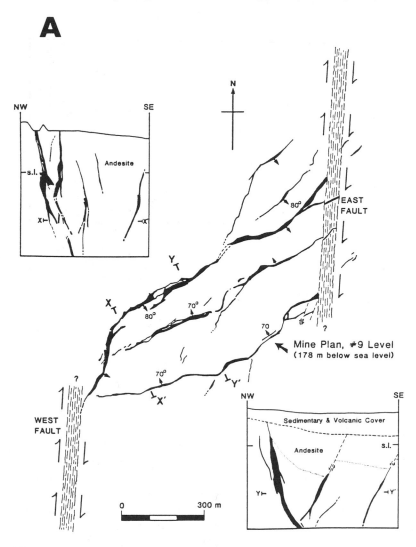

Fig. 3.32 Structures in fault jogs: (a) extensional: a right-stepping jog in a right-lateral strike-slip fault, Martha Mine, New Zealand (from Sibson, 1987); (b) compressional: a left-stepping jog in the right-lateral Coyote Creek fault, Ocotillo Badlands, San Jacinto fault zone, southern California. (After Sharp and Clark, 1972, with additions by N. Brown.)

analysis of the superposition of the stress fields of two cracks (Segall and Pollard, 1980).

The effect of a compressional jog will be greater than that of an extensional one because deformation within the jog will occur at a higher mean normal stress, because the friction on a compressional fault bend jog will be higher than elsewhere, and because gravitational work will be

B

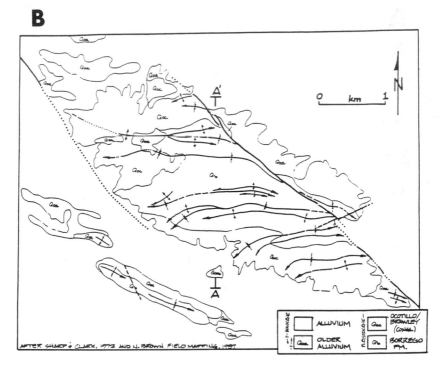

Fig. 3.32 *(cont.)*

done during uplift. However, in the case of dynamic rupture there is an additional effect that increases the efficiency of extensional jogs in impeding rupture (Sibson, 1985). The coseismic reduction of normal stress in such a region will result, due to the coupled poroelastic effect (Sec. 6.5.2) in a sudden reduction of pore pressure that can result in a significant suction force. This transient effect will disappear as pore pressure infiltrates the jog following rupture termination, so postseismic slip, either dynamic or quasistatic, may later ensue. The same effect will increase pore pressure in compressional jogs, resulting in their transient weakening, and possibly may result in coseismic hydraulic fracturing, producing veins and clastic dikes. Surficial phenomena such as increased stream or spring flow may result (Sibson, 1981).

Although steps do not produce kinematic constraints in faulting, both jogs and steps will be impediments to dynamic slip, because for both Modes II and III the shear driving stresses have a maximum in the plane of the crack ahead of the tip. The crack must propagate past a fault bend jog in Mode II and a fault bend step in Mode III, the driving stresses for which are,

Mode II: $\qquad \sigma_{r\theta} = \dfrac{K_{II}}{(2\pi r)^{1/2}}\left\{\cos(\theta/2)\left[1 - 3\sin^2(\theta/2)\right]\right\}$ $\qquad$ (3.7a)

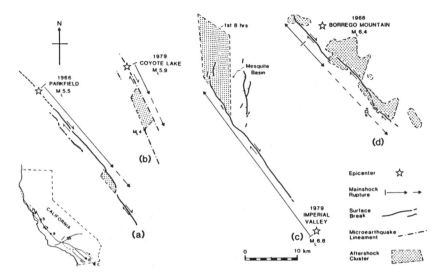

Fig. 3.33 Sketches showing ruptures terminated at jogs in the San Andreas fault zone: (a) Parkfield, 1966; (b) Coyote Lake, 1979; (c) Imperial Valley, 1979; (d) Borrego Mtn., 1968. (From Sibson, 1986c.)

and Mode III:

$$\sigma_{\theta z} = \frac{K_{\mathrm{III}}}{(2\pi r)^{1/2}} \{\cos(\theta/2)\} \tag{3.7b}$$

where r is the distance from the crack tip, z is along its edge, and θ is measured from the plane of the crack (this geometry and these stress functions are illustrated in Fig. 1.6). Since these are both maximum when $\theta = 0$, a bend in either direction and in either mode will reduce the driving stresses.

The presence of a jog on a fault plane may impede or terminate dynamic rupture, and in some cases may also control rupture initiation (King and Nabelek, 1985; Sibson, 1985, 1986c). Several cases of earthquakes stopping at jogs on the San Andreas fault are described by Sibson (1986c) and shown in Figure 3.33. The effect depends on the sign of the jog. In both cases rupture will be impeded because of either the termination of the strand, which represents an increase in G from that necessary for frictional sliding to that necessary to form a new fault, or from an abrupt change in strike, less favorable for slip in the dynamic stress field, as outlined above. In the case of a compressional jog, slip transfer through the jog will be further impeded by the increased compressive stress on either the linking faults or intervening region. In extensional jogs this will not occur, but there may be a suction produced by a dynamic reduction in pore pressure within the jog volume. Slip

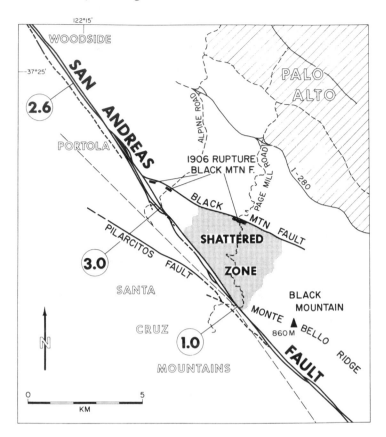

Fig. 3.34 Effects of the compressive fault-bend jog at Black Mountain on the 1906 rupture. Numbers in circles are surface slip, in meters. The rupture propagated from north to south. (From Scholz, 1985.)

transfer may then occur quasistatically after the earthquake when fluid flow equilibrates the pore pressure.

Of the cases shown in Figure 3.33, a, b, and c are extensional jogs. In both the Parkfield and Coyote Lake earthquakes (a and b), rupture propagation on the far side of the jog was delayed and quasistatic, suggestive of the suction mechanism mentioned above. In the Borrego Mountain earthquake (d) the rupture propagated past the compressional jog (marked by a push-up feature, the Ocotillo Badlands, Fig. 3.32b), but slip was much reduced beyond that point. Another similar case, where a fault bend impeded but did not stop an earthquake, is shown in Figure 3.34. There, the 1906 San Francisco earthquake, propagating southwards, encountered an abrupt 9° compressional bend jog at Black Mountain. Coseismic slip was induced on the Black Mountain thrust at the foot of this push-up ridge. The wedge between this thrust and the San Andreas was heavily shattered (Scholz, 1985). The slip as measured at the surface was reduced from 3 m to 1 m at this point, and the depth-averaged slip, deduced from geodetic data, was reduced from 3.5 m to 2 m (Thatcher and Lisowski, 1987). This effect therefore extended through the schizosphere.

Barka and Kadinsky-Cade (1988) made a systematic study of geometric controls on rupture for the North Anatolian and other faults in Turkey. The North Anatolian fault is much more segmented than the San Andreas fault, owing to its smaller net slip (35 km vs. 250 km, see Fig. 3.12). As a result, it tends to rupture in shorter earthquakes. In their study, Barka and Kadinsky-Cade concluded that ruptures always terminated at large jogs, with $d_s > 5$ km (of either sign), but could propagate past smaller features.

The slip distribution produced by an individual earthquake measured at the surface is usually found to be quite irregular (Fig. 3.35). This observation is reenforced by seismic measurement of the moment release distribution in earthquakes (Sec. 4.4). From this we conclude that the previously-mentioned cases of the effects of fault irregularity on rupture are extreme and fairly obvious cases of processes that must occur on many scales, as might be expected from the observation that faults are topographically rough on all scales. Of the two examples shown in Figure 3.35, the surface slip in the dip-slip case (Borah Peak) is much more irregular than in the strike-slip case (Imperial Valley). This observation is fairly typical and may be a result of the much greater irregularity of fault topography in the slip-normal than the slip-parallel direction, as noted earlier. These two earthquakes are described in more detail in Section 4.4.1.

We need to ask how persistent are these features. Because wear is a smoothing process one might expect that these fault irregularities will be gradually removed with progressive slip and that faults will evolve to some smooth, simple geometry (as shown in Fig. 3.12). The fault topography measurements of Figure 3.27, however, indicate that faults remain fractal over all wavelengths, suggesting that there are roughening processes too, that maintain that irregularity. Movement on cross faults, for example, could lead to the growth of jogs.

The San Andreas fault zone near Parkfield (Fig. 3.33a) is a feature some 2 km wide forming the Cholame Valley, which is filled with about 2 km of low-velocity sediments. Yet the active trace of the fault can be defined on the surface to within 1 m (surface ruptures in the earthquakes of 1966 and 1934 were this close together). The jog at the southern end of the 1966 rupture is about 2 km wide and extends to the base of the seismogenic zone (Eaton, O'Neill, and Murdock, 1970; see also Fig. 4.17), and is therefore not a superficial feature. Parallel to the 1966

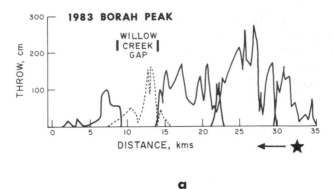

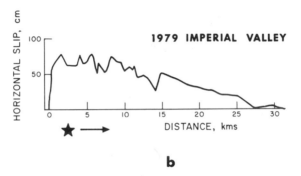

Fig. 3.35 The distribution along strike of surface slip in two earth-quakes of similar fault length. (a) Borah Peak, Idaho, 1983: normal faulting. Data from Crone and Machette (1984), dashed is slip on a secondary fault. (b) Imperial Valley, California, 1979: strike-slip fault-ing. Data from Sharp et al. (1982).

rupture trace and just 2 km southwest of it is another fault trace, also of Holocene age but no longer active (although secondary slip occurred on part of it in 1966, shown in Fig. 3.33a). These two traces extend from either side of a horse (Middle Mountain) in the fault zone at the north end of the valley. If extended to the southeast, this older trace would cut off the jog, so that in the recent geologic past there was a reorganization of the fault zone in which slip switched from the old to the new trace and thus created the jog. This case shows that even when the fault zone is quite wide, the active seismic trace may be much more narrowly defined, but may switch position in time periods of the order 10^4 yr. This conclusion regarding the fineness and durability of the active trace, and by inference the irregularities, was established first by Sharp and Clark (1972) and has been reviewed by Sibson (1986b).

These questions are of considerable importance in understanding earthquake recurrence and in long-range earthquake prediction, to be taken up in Sections 5.4 and 7.4. For example, Schwartz and Coopersmith (1984) have posited that certain faults or fault segments always rupture in "characteristic" earthquakes, perhaps governed by fault irregularities. We need to ask how definitive and durable are such structural controls on rupture, how dependent are they on initial conditions, and how recognizable are they. Many questions need to be considered if these concepts are to find use in hazard assessment and earthquake prediction, and we take this up again specifically in Section 5.3.2.

4   Mechanics of Earthquakes

Friction of faults is often unstable, and slip occurs rapidly as a rupture dynamically propagates over the fault surface. These sudden motions generate seismic waves and this is the mechanism of the most common and important type of earthquake. In this chapter we discuss the dynamics of faulting and review the most important attributes of earthquakes from the point of view of the rupture process.

4.1   HISTORICAL DEVELOPMENT

During most of human history, people's notion of the origin of earthquakes lay within the realm of mythology. Several of the schools of ancient Greek philosophy considered earthquakes to be natural phenomena, although their speculations on the matter relied heavily on imagination and do not bear much relationship with modern theories [see Adams (1938) for an excellent historical account of thinking on this topic from the Greeks up through the Renaissance]. The idea that earthquakes represent an elastic reaction to geological phenomena was promulgated first by Hooke in his *Discourse on Earthquakes*, published in 1668. It was not until the middle and latter part of the nineteenth century that instrumental measurements began to be made and note taken of the geological associations of earthquakes. Lyell (1868) considered earthquakes to be an important agent in earth dynamism, and was aware of both faulting and permanent changes in elevation brought about by them. Although Lyell carefully described the faulting and deformation produced in several earthquakes, like his contemporary Mallet, Lyell believed that the immediate cause of earthquakes was thermal, either a consequence of volcanic activity, or of thermal expansion or contraction.

The first clear connection between earthquakes and dynamic faulting, and its relationship to tectonic processes, was made by a geologist, G. K. Gilbert. He had seen the immediate aftereffects, including the surface faulting, of the 1872 Owens Valley earthquake in California, and in his extensive mapping in the Great Basin had also remarked on the fresh-appearing scarps that so often front the mountain ranges there. He

160

concluded that the elevation of the mountains was produced by repeated sudden ruptures along these faults. Thus, he stated (Gilbert, 1884):

The upthrust produces a local strain in the crust, involving a certain amount of compression and distortion, and this strain increases until it is sufficient to overcome the starting friction along the fractured surface. Suddenly, and almost instantaneously, there is an amount of motion sufficient to relieve the strain, and this is followed by a long period of quiet, during which the strain is gradually reimposed. The motion at the instant of yielding is so swift and so abruptly terminated as to cause a shock, and the shock vibrates through the crust with diminishing force in all directions. ... In this region a majority of the mountain ranges have been upraised by a fracture on one side or the other, and in numerous instances there is evidence that the last increase of height was somewhat recent.

Gilbert claimed only that this theory of earthquakes applied to the Great Basin; but similar connections between earthquakes and faulting soon were made elsewhere. McKay (1890) journeyed to the site of a large earthquake of two years previous in the South Island of New Zealand and discovered there a fresh strike-slip fault scarp on the Hope fault. Soon afterwards, a great oblique-slip scarp was found at the site of the 1891 Nobi earthquake in Japan (Koto, 1893; see also Fig. 4.1). Koto discussed at some length the debate among European geologists as to whether faulting was the cause or effect of earthquakes, quoting extensively from Lyell, and argued cogently for the faulting origin hypothesis. He was evidently unaware of Gilbert's views on the subject. The extensive rupturing of the San Andreas fault during the 1906 San Francisco earthquake, and the geodetic measurements that showed that this ground breakage was not superficial, finally led to the dominance of the faulting theory of earthquakes, as expressed in the analysis of that earthquake by Reid (1910).

That the vast majority of shallow tectonic earthquakes arise from faulting instabilities was proven eventually by seismological observations, but this only occurred after a long delay. Although the theory of radiation from a double-couple source was introduced first by Nakano in 1923, there was slow progress in implementing this development. The determination of an earthquake focal mechanism from its radiated field requires both substantial computation and a widely distributed network of standardized seismometers, so progress awaited developments in instrumentation and computers.

Also, scientific opinion often was divided on the subject. Surface faulting can be observed for only a very small fraction of earthquakes, those large earthquakes that occur on land; even then faulting often may be obscured by heavy vegetation, so it was quite possible to deny this as a general mechanism. There was also a great debate about whether the

Fig. 4.1 The famous photograph by Koto (1893) of the scarp of the 1891 Nobi (Mino–Owari) earthquake, Japan. A recent photograph of the same scene may be found in Bolt (1978, page 41). Faulting in this location was mainly normal, with the north side up.

double-couple or the single-couple is the correct representation of earth-quakes [see Kasahara (1981) for a recounting of this issue]. In retrospect this argument seems futile because the single-couple does not connect two equilibrium states and hence is not physically possible.

The modern era of earthquake source studies began with the installation of the Worldwide Standardized Seismic Network in the early 1960s and with the widespread use of computers. It was only then that dynamic faulting gained widespread acceptance as the origin of the majority of seismic events.

4.2 THEORETICAL BACKGROUND

4.2.1 *The dynamic energy balance*

An earthquake may be considered to be a dynamically running shear crack, so we begin our discussion, as in Chapter 1, with the energy balance for this process. The energy balance, Equation 1.6, may be rewritten by the inclusion of terms for the kinetic energy and the frictional work done on the crack surface behind the tip. Ignoring gravitational energy, we write

$$U = (-W + U_e) + U_s + U_k + U_f \qquad (4.1)$$

As discussed in Section 1.1.2, W is the work done by external forces, U_e the change in internal strain energy, and U_s is the surface energy involved in creation of the crack. U_k is the kinetic energy and U_f the work done against friction.

In the dynamic case we may express the equilibrium condition (Eq. 1.7) as a time derivative and write the equilibrium energy balance as

$$\dot{U} = 0 = (-\dot{W} + \dot{U}_e) + \dot{U}_s + \dot{U}_k + \dot{U}_f \qquad (4.2)$$

where the terms represent the following integrals over the domains shown in Figure 4.2,

$$\dot{U}_k = \frac{\partial}{\partial t} \frac{1}{2} \int_V \rho \dot{u}_i \dot{u}_i \, dV \qquad (4.3\text{a})$$

$$\dot{U}_s = \frac{\partial}{\partial t} \int_{\Sigma_0} 2\gamma \, dS \qquad (4.3\text{b})$$

$$\dot{W} = \int_{S_0} \sigma_{ij} \dot{u}_i n_j \, dS \qquad (4.3\text{c})$$

$$\dot{U}_f = \frac{\partial}{\partial t} \int_\Sigma \sigma_{ij} u_i n_j \, dS \qquad (4.3\text{d})$$

$$\dot{U}_e = \frac{\partial}{\partial t} \frac{1}{2} \int_{V - V_0} \sigma_{ij} \varepsilon_{ij} \, dV \qquad (4.3\text{e})$$

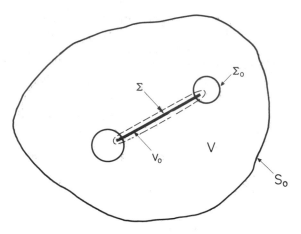

Fig. 4.2 The domains of integration for the dynamic energy balance.

With the assumption of linear elastic fracture mechanics that the crack is cohesionless, the friction term (Eq. 4.3d) vanishes. It is then possible to evaluate the crack tip energy term (Eq. 4.3b) by means of a path-independent integral around the toroid Σ_0 surrounding the crack tip, so in this case one can solve this energy balance (Achenbach, 1973; for examples see Richards, 1976 and Aki and Richards, 1980). However when friction is present, one cannot evaluate the integral over the entire crack surface Σ (Eq. 4.3d) because u_i and σ_{ij} are unknown, so the energy balance becomes intractable. As mentioned in Section 1.1.3, this represents a fundamental problem in fracture mechanics as applied to shear cracks. The result is that dynamic models of shear cracks, such as those discussed in the next section, cannot be solved using Equation 4.2: Instead, we use a stress-fracture criterion that is not tied to the energy balance.

A related problem arises when we try to determine the energy partition for the entire earthquake. Integrating Equation 4.2 over the rupture time, and recognizing that if we chose a large enough volume (say, the whole earth), the term for work at external boundaries (Eq. 4.3c) vanishes, we can write an expression for the radiated seismic energy E_s,

$$E_s \equiv \Delta U_k = -\Delta U_e - \Delta U_f - \Delta U_s \qquad (4.4)$$

If during the rupture process the shear stress drops an amount $\Delta\sigma$ from an initial value σ_1 to a final value σ_2, then the change in internal strain energy will be

$$\Delta U_e = -\tfrac{1}{2}\int_\Sigma (\sigma_1 - \sigma_2)\,\overline{\Delta u}\,dS - \int_\Sigma \sigma_2\,\overline{\Delta u}\,dS$$

$$= -\tfrac{1}{2}(\sigma_1 + \sigma_2)\,\overline{\Delta u}\,A \qquad (4.5)$$

where $\overline{\Delta u}$ is the mean slip and A is the fault area. This expression can be recognized as the work of faulting W_f (Eq. 3.5), with the sign changed. The decrease in strain energy due to the slip on the crack just supplies the work of faulting.

To proceed further we must make some assumption about the behavior of the stress traction acting on the fault surface during slip. If we assume that during sliding the friction stress has some constant value determined by dynamic friction, σ_f, and so define a dynamic stress drop $\Delta\sigma_d = (\sigma_1 - \sigma_f)$, then the seismic energy is

$$E_s = \tfrac{1}{2}(\sigma_1 + \sigma_2)\overline{\Delta u}\,A - \sigma_f\overline{\Delta u}\,A - 2\gamma A \qquad (4.6)$$

which is an explicit form of Equation 3.5. [Note that the integral in Equation 4.3d can be broken into two parts. The second part, $\int \dot{\sigma}_{ij}u_i n_j\,dS$, is assumed to be zero in the present approximation. Kostrov (1974) pointed out that rapid variations in stress must be present during rupture, because that is what gives rise to near-source strong ground motions. In the present approximation we are neglecting the energy in the near-field high-frequency accelerations, which are not measured in E_s.] Making the additional assumption that $\sigma_2 = \sigma_f$ (Kostrov, 1974; Husseini, 1977), and using the result from Section 3.2.2 that surface energy is negligible, we obtain the simple expression

$$E_s \approx \tfrac{1}{2}\Delta\sigma\,\overline{\Delta u}\,A \qquad (4.7)$$

Equation 4.7 expresses the fact that the seismic radiation contains only information concerning the stress change during the earthquake, and none concerning the total stress. This presents a fundamental problem in determining the partitioning of energy because the work in faulting does depend on the total stresses. If we define the seismic efficiency $\eta = E_s/W_f$, we obtain, from Equations 4.5 and 4.7,

$$\eta = \frac{\Delta\sigma}{\sigma_1 + \sigma_2} \qquad (4.8)$$

where we find that the seismic efficiency depends upon the total stresses and cannot be determined from seismological observations alone. If we were to assume, say, the rule of thumb from Chapter 2 that the stress drop is about one-tenth the mean stress, then $\eta \approx 0.05$. On the other hand, if one assumes that stress drop is constant, η must decrease with increasing total stress (Kanamori and Anderson, 1975). In the only case where η has been measured, it is less than 0.01 (Sec. 6.5.3).

The assumption that the final, or "static" stress drop $\Delta\sigma$ is equal to the dynamic stress drop $\Delta\sigma_d$ is also arbitrary. For example, in the case of the spring-slider model given in Section 2.3.4, $\Delta\sigma = 2\,\Delta\sigma_d$ in the absence of any elastic radiation. In the presence of radiation, $\Delta\sigma$ will be less than that, but because of dynamic overshooting (as a result of inertia), we may expect that $\sigma_f > \sigma_2$ (Savage and Wood, 1971; Scholz, Molnar, and Johnson, 1972). We find in the next section that dynamic models usually do predict some overshoot.

4.2.2   *Dynamic shear crack propagation*

Earthquakes are often modeled with kinematic models in which the displacement history of motion is prescribed with some suitably few parameters. Often-used models of this type are the propagating disloca-tion model of Haskell (1964) and the Brune (1970) model, the latter of which, though employing an infinite rupture velocity, has the advantage of being rationalized in terms of the dynamic properties of the source. While such models may provide quite detailed descriptions of earth-quakes they do not yield a physical insight into the rupture process itself, which is of prime concern here. For this purpose, we need to examine some of the results of dynamic models, by which it is meant models that satisfy the dynamic equations of elasticity with the only prescription being that of a fracture criterion. We only discuss some pertinent results of the simplest models of this type. For a more rigorous and detailed mathematical treatment see, for example, Aki and Richards (1980) or Kostrov and Das (1988).

The mechanism of rupture in unstable slip has so far been described in two different ways, as brittle fracture in Chapter 1 and as a stick-slip friction instability in Chapter 2. The two are mathematically equivalent in relating motion in the medium to a drop in shear stress on the fault surface, but have traditionally differed in the way in which the rupture process is considered. In theoretical fracture mechanics it is assumed that a characteristic fracture energy per unit area, a material property, is required for the crack to propagate. In the stick-slip model, on the other hand, rupture is assumed to occur when the stress on the fault reaches the static friction value and the condition for dynamic instability exists. In the fracture model stresses at the crack tip may be arbitrarily high, but in the stick-slip model no energy is dissipated in the crack tip and the stresses there must remain finite. These differences mainly have arisen historically out of the idealizations involved in the two ap-proaches. Later we will discuss a model that merges the two views.

Let us first consider the critical shear crack at the onset of unstable propagation. The simplest case is that of a plane Mode III crack, discussed by Andrews (1976a). Consider such a crack on the $y = 0$ plane extending from $z = -L$ to $z = L$. The shear stress at a large distance from the crack is constant, σ_1, and that on the fault plane is equal to the sliding friction value, σ_f. Displacement on the crack is (Knopoff, 1958)

$$u = \frac{\sigma_1 - \sigma_f}{\mu}(L^2 - z^2)^{1/2} \tag{4.9}$$

and the total offset Δu is twice that value. Incorporating Equation 4.9 in 4.5 and integrating, the change in strain energy is

$$\Delta U_e = -\frac{\pi}{2}\frac{1}{\mu}(\sigma_1 + \sigma_f)(\sigma_1 - \sigma_f)L^2 \tag{4.10}$$

and the net energy available to supply surface energy and be radiated is

$$-\Delta U_e - \Delta U_f = \frac{\pi}{2}\frac{1}{\mu}(\sigma_1 - \sigma_f)^2 L^2 \tag{4.11}$$

As the crack half-length increases an amount dL, the increment of available energy increase is

$$\partial(-\Delta U_e - \Delta U_f) = \frac{\pi}{\mu}(\sigma_1 - \sigma_f)^2 L\, dL \tag{4.12}$$

At the critical half-length L_c this must be just sufficient to supply the fracture energy for both ends of the crack, $2\mathbf{G}_c\, dL$, so

$$L_c = \frac{2}{\pi}\frac{\mu \mathbf{G}_c}{(\sigma_1 - \sigma_f)^2} \tag{4.13}$$

If we use a representative laboratory value for $\mathbf{G}_{\text{IIIc}}$ of 10^2 J m^{-2} (Table 1.1), and assume a stress drop of 1 MPa, $L_c = 1.9$ m. Estimates of $\mathbf{G}_c$ from seismological data yield typical values of 10^6–10^7 J m^{-2}, so values of L_c will be proportionally larger [see Li (1987) for a summary of methods and results of determining $\mathbf{G}_c$ from geophysical data]. Such large values of fracture energy are a consequence of the great crack lengths, and hence stress intensity factors, in large earthquakes. This is usually interpreted as meaning that there is a substantial volume of inelastic deformation around the crack tip.

This model has the nonphysical result discussed in Section 1.1.3 of a stress singularity at the crack tip. One may avoid this by assuming that the crack breakdown occurs over some finite distance, which has the effect of smearing out the stress drop and its associated stress concentration. In brittle fracture this may correspond to the development of the process zone and in friction to the breakdown between the static and dynamic friction values, which, as noted in Section 2.3.2, requires a critical slip distance. Ida (1972, 1973) constructed such a slip-weakening rupture model. One of Ida's models, discussed further by Andrews (1976a), is shown in Figure 4.3. In this model the initial stress σ_1 is less

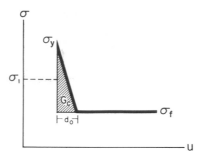

Fig. 4.3 A simple slip-weakening model for stress breakdown at the crack tip.

than a yield stress σ_y, which breaks down to a friction stress σ_f over some critical slip distance d_0.

The fracture energy is the work done in the rupture breakdown in excess of that done against the constant friction σ_f, as shown in the shaded region in the figure, and is

$$G_c = \tfrac{1}{2}(\sigma_y - \sigma_f)d_0 \qquad (4.14)$$

This may be substituted into Equation 4.13 to obtain

$$L_c = \frac{\mu}{\pi} \frac{(\sigma_y - \sigma_f)}{(\sigma_1 - \sigma_f)^2} d_0 \qquad (4.15)$$

A very similar result can be obtained for the Mode II case (Andrews, 1976b). These results may be compared with the friction models discussed in Section 2.3 if σ_y is identified with static friction and σ_f with dynamic friction. In the breakdown model of Section 2.3.3, σ_1 was assumed to be equal to σ_y, and under those conditions, the resulting expression for the critical length, Equation 2.31, differs from Equation 4.15 by only a factor of about 2, even though the bases of the derivations are quite different.

When the crack half-length exceeds L_c the rupture velocity will accelerate up to some limiting value, which is the shear velocity of the medium for the Mode III case (Kostrov, 1966). The stress field of the propagating crack also differs from the static one. To illustrate these points we use the mathematically simpler case of a moving screw dislocation (Cottrell, 1953). The geometry of the field of a screw dislocation is similar to a Mode III crack and the general conclusions will be the same in both cases. Like the Mode III crack, a screw dislocation propagating on the y plane in the z direction has only the u component of displacement, and the equations of motion reduce to

$$\frac{\partial^2 u}{\partial z^2} + \frac{\partial^2 u}{\partial y^2} = \frac{1}{\beta^2} \frac{\partial^2 u}{\partial t^2} \qquad (4.16)$$

where β is the shear velocity. We require a solution in which the dislocation moves with a steady velocity v_r along the z axis and that reduces to the static one, $u = \mathbf{b}(\theta/2\pi)$, when $v_r = 0$ ($\mathbf{b}$ is the Burgers vector, the unit of slip). Equation 4.16 is the wave equation and it is known that steady propagational solutions can be obtained by making a Lorentz transformation. In this case the transformation is made by introducing the variable $z_1 = (z - v_r t)/(1 - v_r^2/\beta^2)^{1/2}$ into the equation. Noting that for steady propagation $\partial^2/\partial t^2 = v_r^2(\partial^2/\partial z^2)$, the equation becomes

$$\frac{\partial^2 u}{\partial z_1^2} + \frac{\partial^2 u}{\partial y^2} = 0 \qquad (4.17)$$

The solution is

$$u = \frac{b}{2\pi} \tan^{-1}(y/z_1) \qquad (4.18)$$

Because z_1 reduces to z as v_r is reduced to zero, the limiting static solution is satisfied. The field of the propagating dislocation is the same as the static one except that z is replaced by $(z - v_r t)/(1 - v_r^2/\beta^2)^{1/2}$. The term $(z - v_r t)$ describes the motion of the dislocation, whereas the factor $(1 - v_r^2/\beta^2)^{1/2}$ indicates that the stress field is foreshortened in the direction of propagation. This "relativistic contraction," a familiar result in electrodynamics, is a consequence of the fact that the motion is governed by the wave equation.

The similarity to electrodynamics can be taken further. If we evaluate the energy density of a dislocation by adding the strain energy density $\frac{1}{2}\mu[(\partial u/\partial z)^2 + (\partial u/\partial y)^2]$ and the kinetic energy density $\frac{1}{2}\rho(\partial u/\partial t)^2$, we can obtain the Einstein equation,

$$U = \frac{U_0}{\left(1 - v_r^2/\beta^2\right)^{1/2}} \qquad (4.19)$$

where U is the energy density of the moving dislocation and U_0 is its rest energy density. As v_r approaches β, U becomes infinite, so the dislocation cannot travel faster than the speed of sound.

The limiting velocity for a Mode III crack can be grasped intuitively by realizing that only shear stresses exist in the crack-tip field in that case, and as they are limited to travel at the shear velocity, so also must be the rupture. Mode II rupture propagation is more complex because both shear and normal stresses are in the crack-tip stress field. The shear components in that case travel at the Rayleigh wave velocity and the normal components at the compressional wave velocity. An example of this complexity is given by a model of Andrews (1976b, 1985) which uses the same slip-weakening model as shown in Figure 4.3. The rupture propagation characteristics are dependent on a dimensionless strength parameter S, defined as the ratio of the stress increase required to initiate slip to the dynamic stress drop that results,

$$S = \frac{(\sigma_y - \sigma_1)}{(\sigma_1 - \sigma_f)} \qquad (4.20)$$

The case for $S = 0.8$ will be illustrated here. A space–time plot of the rupture propagation is shown in Figure 4.4. With models of this type the rupture front is spread over a width, shown as the shaded region in the Figure 4.4a, bounded on one side by the onset of slip and on the other by the completion of the stress drop. Lines with slope equal to the P, S, and Rayleigh wave velocities are shown for reference. The slip behavior is shown in Figure 4.4b. Rupture starts slowly at first, but at a crack length about twice the critical length the Rayleigh speed is approached. As the crack accelerates, the rupture front narrows. The rupture velocity is limited by the Rayleigh velocity for some distance, but then is observed to bifurcate, with some slip initiating supersonically, and eventually the entire rupture front is observed to propagate at a velocity greater than β and approaching the P wave velocity. The temporal development of the shear stress traction on the plane of the crack is shown in Figure 4.4c. When the rupture is propagating just slower than the Rayleigh speed there is a peak in stress before the rupture that is traveling at the shear speed. Eventually this peak becomes large enough to initiate some slip prior to the Rayleigh wave arrival, and eventually slip prior to the Rayleigh wave arrival has become large enough to attenuate the second stress peak. A cusp in the slip function is still observed at the Rayleigh wave arrival even at late times: Although slip can be initiated by the P waves, the driving load, represented by the shear stress components parallel to the direction of slip, arrives at the Rayleigh speed and so the main slip still begins at this time.

Very similar behavior is observed if a rate–state-variable friction law is employed (Okubo, 1988), but the details depend on the rupture resistance parameter S. For example, if S is chosen to be much smaller than 1 so that the fault has a low rupture resistance, the rupture velocity in the Mode II case may very quickly attain the P wave velocity (Burridge, 1973; Das, 1981).

Let us now consider the slip function for a simple three-dimensional model in which we ignore these details. The simplest dynamic model is that of a self-similar shear crack with uniform stress drop that starts at a point and propagates without limit at a constant rupture velocity (Kostrov, 1964). The model is based on the assumptions of fracture mechanics, and because linearity is assumed, the friction stress σ_f may be subtracted from the total stress so that the solution depends only upon

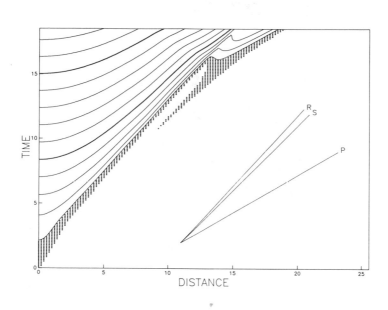

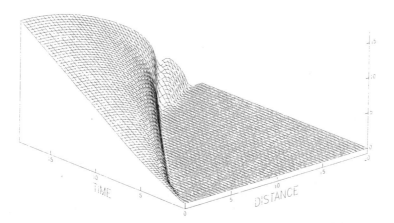

Fig. 4.4 Propagation of a Mode II crack, with slip weakening. (a) Slip contours during propagation of the crack. Dimensionless distance is x/L_c, time is $\beta t/L_c$, and the slip contour interval is $1.3L_c\,\Delta\sigma_d/\mu$. The shaded region indicates the breakdown region where points are slipping with $u < d_0$. Lines denoted P, S, and R indicate the P, S, and Rayleigh speeds for the medium. (b) Mesh perspective of dimensionless slip $(\mu/\Delta\sigma_d)(u/L_c)$ in the same coordinates. Azimuth of view is the Rayleigh wave direction. (c) Mesh perspective, similar to (b), of dimensionless tractions in the crack plane. (From Andrews, 1985.)

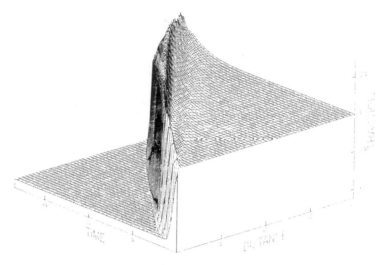

Fig. 4.4 *(cont.)*

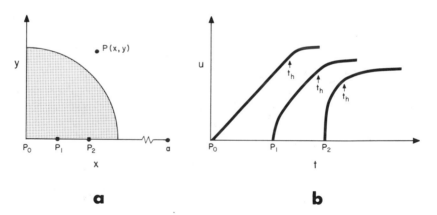

Fig. 4.5 Slip history at point P: (a) The rupture (stippled) approaches *P*. (b) Slip history at three points P_0, P_1, and P_2 at successively greater distance from the point of rupture initiation. t_h denotes the arrival of a healing wave propagating back from the final rupture perimeter at *a*.

the dynamic stress drop $\Delta\sigma_d$. The slip history at a point $P(x, y)$ (Fig. 4.5a) for a rupture propagating at a constant rupture velocity v_r is,

$$u(x, y, t) = v_0\left(t^2 - \frac{(x^2 + y^2)}{v_r^2}\right)^{1/2} \quad \text{for } t_h \geq t \geq \frac{(x^2 + y^2)^{1/2}}{v_r}$$

$$(4.21)$$

where x and y are measured from the point of rupture initiation, v_0 is

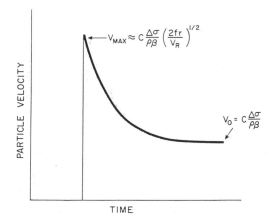

Fig. 4.6 Particle velocity at point P.

an asymptotic particle velocity, and t_h is the time of the onset of healing in the case of a finite crack. The slip history at several distances from the point of initiation is shown in Figure 4.5b. Except for the initiation point, there is a square-root singularity in particle velocity just behind the crack tip, arising because of the singularity in stress in front of the crack tip (Eq. 1.19), and which is, for the same reasons as discussed before, a nonphysical result. The slip velocity history of a point at a distance r from the point of rupture initiation is shown in Fig. 4.6. The singularity has been averaged over a cutoff period $1/f$ to obtain an expression for the maximum slip velocity

$$v_{max} \approx C \frac{\Delta\sigma_d}{\rho\beta} \left(\frac{2 f r}{v_r} \right)^{1/2} \qquad (4.22)$$

in which C is a parameter that depends on rupture velocity v_r (ranging from 1 to $2/\pi$ as v_r increases from 0 to β). From Equation 4.22 we see that v_{max} is proportional to the stress intensity factor $K = \Delta\sigma_d(\pi r)^{1/2}$ (Eq. 1.25; actually, the dynamic stress intensity factor is somewhat less than this, depending on the ratio of v_r to β [Achenbach, 1973]). When the crack tip has propagated a sufficient distance beyond the point so that the influence of the crack-tip stress concentration is absent, the slip velocity drops to the asymptotic level

$$v_0 = C \frac{\Delta\sigma_d}{\rho\beta} \qquad (4.23)$$

The dynamics of motion therefore scale linearly with $\Delta\sigma_d$.

Madariaga (1976) numerically analyzed the same model for the case in which rupture is prescribed to stop at the radius $r = a$. He found that the slip velocity of points within the rupture abruptly decelerate and shortly thereafter come to rest at a time t_h upon the arrival of a Rayleigh wave generated from the termination of rupture at the final perimeter. He called this process *healing* (Fig. 4.5b). The final slip was found to be of the same elliptical form as given by the static solution for a circular crack of radius a (Eshelby, 1957),

$$u(x, y) = \frac{7\pi}{12} \frac{\Delta\sigma_d}{\mu} a \left(1 - \frac{x^2 + y^2}{a^2} \right)^{1/2} \tag{4.24}$$

except that slip was found everywhere to be about 30% greater, due to dynamic overshoot. Thus during healing, as the crack tends from dynamic to static equilibrium as the stress is relieved in the surrounding volume, stress drop increases by 30%, so that the static stress drop is greater than the dynamic.

If we return to the concepts of fault contact discussed in Section 2.1, rupture may be considered to consist of the breaking of a great many asperity contacts, that may themselves exist on all scales. The fracture criteria used in the crack problems described here may be considered spatial averages of the strength of many asperities. On the other hand, for some applications one may wish to consider the mechanics of rupture of a single asperity. This is an "external" crack problem, in which an internal unbroken region is surrounded by a previously broken area. This problem has been idealized by Das and Kostrov (1983) who treated rupture of the unbroken part in the usual way, but required stress on the outer broken part to remain at a constant friction value. Rupture through the asperity occurs in a similar way as crack propagation, the major difference being that when a crack ruptures it causes an increase in stress on the surrounding parts of the fault, whereas when an asperity ruptures it causes an increase in slip in the surrounding area.

4.2.3   *Simple applications to earthquake rupture*

The study of these idealized models of dynamic rupture provides an understanding of the physics underlying the mechanism of earthquakes, but the results usually cannot be applied directly to real cases because of their complexity. Not only are real ruptures likely to obey more complicated rupture criteria, but they are not planar and all the relevant parameters that govern dynamic rupture propagation are likely to be heterogeneous at all scales. We should therefore only expect to observe the general characteristics predicted, and in any case real data are not presently of a quality to allow very detailed investigation of the rupture process.

We may wish to know, for example, what is the rupture velocity for earthquakes. We have found that the Mode III edge cannot propagate

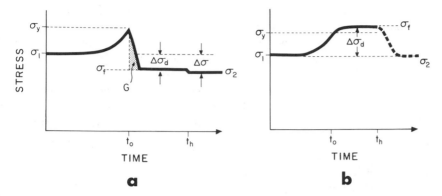

Fig. 4.7 Stress history during rupture at points that exhibit (a) velocity weakening, (b) velocity strengthening, dashed curve is postseismic relaxation.

faster than the shear velocity, but that the Mode II edge can propagate faster than that, although propagation of the main part of the slip is still likely to be limited to the Rayleigh velocity. The quality of real data is such that one is able to identify only propagation of the main slip. So the whole rupture will be seen as propagation at the shear or Rayleigh velocity. This generalization is in accord with reported observations. In exceptional cases, though, one of which is described in Section 4.4.1, variations in rupture velocity can be observed, and one needs to consider the results of this section in interpreting this behavior in terms of fault properties.

It is useful to construct a very simple model in which the dynamic rupture process can be envisioned and with which we can discuss observational results. Consider the stress history at a point P (Fig. 4.5a) as it is being approached by a rupture on the same plane. We discuss two cases, shown schematically in Figure 4.7. At early times, when the rupture is at some distance from P, the applied stress is at its initial value σ_1 and the static strength of the point is σ_y. As the rupture approaches, the stress at P rises due to the dynamic stress concentration ahead of the crack, and when it reaches σ_y at time t_0 slip is initiated. The stress history during slip depends on the frictional constitutive law at P. In Figure 4.7a we consider the case assumed so far, in which there is a reduction in friction during slip giving rise to a positive $\Delta\sigma_d = \sigma_1 - \sigma_f$. It is assumed there that the dynamic friction stress σ_f remains constant during slip but after healing is initiated at t_h the stress drops further due to overshoot so that the final stress drop $\Delta\sigma$ is greater than $\Delta\sigma_d$.

One can see readily from this example why dynamic rupture is governed by the rupture resistance parameter S (Eq. 4.20). Note that S is not a conventional measure of strength because it does not involve total stresses but only the ratio of two stress differences. Furthermore, the parameters σ_y and σ_1 are defined at P, but the parameters in the denominator are spatial averages over the adjacent rupture plane and are also properties of point P only in the case of a perfectly homogeneous fault, as in the idealized models in the previous section. The numerator, which may be called the strength excess is the difference between the static strength and initial applied stress at P, whereas the denominator is what scales the dynamic stress rise caused by the approaching rupture. The initial stress σ_1 will equal σ_y only at the point of rupture initiation.

Any of these factors influence the rupture resistance S. A fault segment with high S may be called strong and with low S weak but the strength, so defined, may be due either to a high value of σ_y or a low value of σ_1, and either case is only relative to the stress drop of adjacent fault segments, which are driving the rupture. An interpretation of fault strength in terms of fault properties from an observation of its rupture behavior clearly must be argued carefully, otherwise confusion is likely to arise, which has happened often in the recent literature (see Sec. 4.5.1).

High relative values of S on a fault can obviously inhibit or terminate rupture, as is illustrated in the models of Das and Aki (1977). Variations in S also can have a strong effect on rupture velocity, as shown by Day (1982). One can envision the latter process with the aid of Figure 4.7a by considering only variations in σ_y. If σ_y at P is lower, rupture will initiate at an earlier time, and if higher, at a later time. So a rupture propagating from a region of high S to a region of low S will show a higher than normal rupture velocity and vice versa.

The case shown in Figure 4.7a is consistent with the fault that obeys a constitutive law of the type given in Equation 2.27 and that exhibits velocity weakening. If velocity strengthening occurs, on the other hand, the behavior will be as shown in Figure 4.7b in which $\sigma_f > \sigma_y$. In this case $\Delta\sigma_d$ is negative and this part of the fault will be a net sink of energy: Slip there can only be driven by drawing energy from adjacent parts of the fault that have positive stress drop. It is clear that rupture cannot propagate far into such a region without being terminated. This was the effect called upon in Section 3.4.1 to determine the base of the schizo-sphere. It is not the only way that a negative stress drop can occur, however. Even where velocity weakening occurs there may be regions in which $\sigma_1 < \sigma_f$, similarly inhibiting rupture. This is an extreme case of what causes "seismic gaps." This will be discussed further with regard to various applications in Section 4.5.

In the particular case sketched in Figure 4.7b, strong velocity strengthening has driven σ_f well above σ_y so that the final stress is higher than the static strength. In this situation friction will relax gradually following the cessation of slip and slip will continue quasistatically until the stress is equal to the static strength value (dashed curve). This process is known as afterslip and will be discussed in detail later (Sec. 6.4.2).

4.3   EARTHQUAKE PHENOMENOLOGY

4.3.1   *Quantification of earthquakes*

Most of what is known about the dynamics of earthquakes has been determined by study of their elastic radiation. This requires careful attention to both effects of the propagation path between source and receiver and the response of the receiver. Many techniques have been developed for these purposes, the description of which is beyond the scope of this book. The reader is directed to one of the many excellent seismological texts on this subject (e.g., Aki and Richards, 1980; Kasahara, 1981). Here we review only what is necessary to understand the dynamics of the rupture process.

The traditional measurement of earthquake size is *magnitude*, which is a logarithmic scale based on the amplitude of a specified seismic wave measured at a particular frequency, suitably corrected for distance and instrument response. There are consequently many types of magnitude (m_L, m_b, M_s, etc.) that are useful under different conditions. Variation in reported magnitudes of an earthquake between stations is to be expected because of the effect of radiation pattern and differences in propagation paths. The different magnitude types reported for the same earthquake also do not always agree, although a number of schemes have been developed to relate the magnitude scales.

While magnitude is a convenient way of measuring earthquake size from a seismogram, a more physically meaningful measurement of earthquake size is given by the *seismic moment*

$$M_{0ij} = \mu\left(\overline{\Delta u_i}\, n_j + \overline{\Delta u_j}\, n_i\right) A \qquad (4.25)$$

where $\overline{\Delta u_i}$ is the mean slip vector averaged over the fault area A, with unit normal n_j, and μ is the shear modulus. (In general, because Δu_i and n_j are functions of position, Eq. 4.25 may be written in integral form.) M_{0ij} is a second rank tensor with a scalar value $M_0 = \mu\,\overline{\Delta u}\,A$ and the two directions define the slip and fault orientations. This latter geometrical information, taken separately, is called the *focal mechanism* (or fault plane solution). For a double-couple source, only the off-diagonal terms in the seismic moment tensor are nonzero.

The relationship between M_0 and magnitude M may be understood by considering the radiation spectra of earthquakes, as shown in Figure 4.8. The spectra shown there (from Aki, 1967) are based on Haskell's kinematic model of an earthquake as a moving dislocation. The spectra

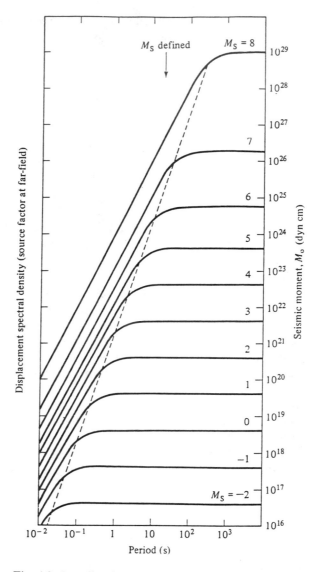

Fig. 4.8 A scaling law for seismic spectra. (From Aki, 1967.)

are flat and proportional to M_0 at long period, and in the case of this particular model fall off as ω^{-3} above a *corner frequency* f_0. Also shown is the period (20 s) at which the surface wave magnitude M_s is traditionally measured. As long as f_0^{-1} is shorter than 20 s, $M_s \propto \log M_0$, so that a magnitude–moment relation may be defined empirically, as for example

$$\log M_0 = 1.5M_s + 16.1 \tag{4.26}$$

(Purcaru and Berckhemer, 1978). For earthquakes large enough that f_0^{-1}

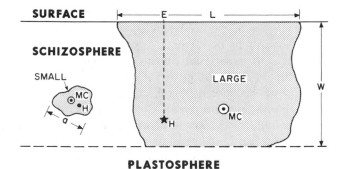

Fig. 4.9 Diagram illustrating the definitions of small and large earthquakes, showing hypocenter (H), epicenter (E), moment centroid (MC), and the dimensions of rupture (a, L, and W).

is longer than 20 s, Equation 4.26 no longer holds and M_s seriously underestimates moment. At this point the M_s scale is said to saturate. This occurs at about $M_s = 7.5$. Because the body wave magnitude m_b is measured at a shorter period ($\sim$ 1 s), it consequently saturates at a lower magnitude. If one wishes to retain the traditional use of magnitude for earthquakes larger than the saturation size, one can calculate a magnitude from the moment with Equation 4.26. This magnitude is denoted M_w (Kanamori, 1977).

Seismic energy can be related to moment by using Equation 4.7 together with the scalar definition of moment to obtain

$$E_s = \frac{\Delta\sigma}{2\mu}M_0 \qquad (4.27)$$

If we combine this with the Gutenberg–Richter empirical relationship

$$\log E_s = 1.5M_s + 11.8 \qquad (4.28)$$

and make the approximation that earthquake stress drops are approximately constant and $\sim$ 3 MPa (Kanamori and Anderson, 1975), we find that Equations 4.28 and 4.26 are consistent (Kanamori, 1977; Hanks and Kanamori, 1979). In view of the approximations and generalizations involved, however, it is always better to use a directly determined moment than a derived magnitude for discussing earthquake rupture processes.

The other principal earthquake parameters that are commonly measured are illustrated in Figure 4.9. The *hypocenter* is the point of initial radiation of seismic waves and hence is the initiation point of dynamic rupture. It may occur anywhere within the rupture: The *epicenter* is simply the projection of the hypocenter on the earth's surface. The *moment centroid*, now often reported, is just that and need bear no relationship to the hypocenter. The earthquake propagates outwards on the fault plane until dynamic rupture ceases at some final perimeter that defines the dimensions of the earthquake.

Qualitative descriptions of earthquake size (moderate, great, etc.) are used often, which, though based roughly on magnitude, are meant to convey the potential destructive power of the earthquake had it occurred in a populated area. Because in this discussion we are considering only the physics of the rupture and not its effects, such terms are not useful. We do, however, find it necessary to divide earthquakes into two classes, called simply, *large* earthquakes and *small* earthquakes. Small earthquakes are all those events whose rupture dimensions are smaller than the width of the schizosphere. They therefore propagate and terminate entirely within the bounds of the schizosphere and their behavior may be described as rupture in an unbounded elastic brittle solid. A large earthquake, in contrast, is one in which the rupture dimensions equal or exceed the width of the schizosphere. Once an earthquake becomes large, it is constrained to propagate only horizontally, with its aspect ratio increasing as it grows and its top edge at the free surface and its bottom at the base of the schizosphere. There are two reasons for making this distinction. The first is that it is found that small and large earthquakes, so defined, obey different scaling relationships and produce radiation with different spectral shapes, which may reflect their different geometries and boundary conditions. The second is that we generally need consider only large earthquakes when quantitatively considering the role of earthquakes in tectonics. Notice that the magnitude level where earthquakes change from small to large depends on the tectonic environment. For the San Andreas fault, say, where the schizospheric width is only about 15 km, this occurs at about $M = 6$–6.5, whereas in a subduction zone, where the downdip width of the schizosphere is much greater, it may be at about $M = 7.5$.

4.3.2   *Earthquake scaling relations*

The analysis of dynamic crack models (Sec. 4.2.2) and of the block-slider model (Sec. 2.3.4) shows that the dynamic characteristics of rupture, slip, velocity, and acceleration, all scale linearly with stress drop, which is therefore the single most fundamental scaling parameter. In principle, measurements of any of these observables can be inverted for stress drop, but this turns out to be difficult in practice.

Geodetic and/or fault slip and length data provide information on static stress drop, and seismic measurements of slip, velocity, and acceleration on dynamic stress drop. Stress drop is defined at a point, but the inversions always yield integral averages over some ill-defined region of the fault. There are several reasons for this. In the general case of a heterogeneous distribution of stress drop on the fault, the slip history of any one point will be a function of the distribution of stress drop over all nearby points, rather than of one point, and both the functional dependence and the definition of "nearby" are not known. Second, the observed seismic waveform is an integral property of the slip distribution, and measurements of the slip parameters are themselves averages; temporal in the case of dynamic and spatial in the case of static measurements. This will result, in the case of dynamic measurements, in the determination being a function of the period and type of seismic waves utilized. Finally, any such inversion is highly model dependent, because a model must be used to relate radiation to source parameters and simplifying assumptions have to be made regarding the geometry of the source.

Consequently, many measures of "stress drop" are reported, but they do not agree with one another, for reasons that are difficult to evaluate. For example, one may measure a value of dynamic stress drop averaged over the rupture duration by an interpretation of body wave spectra (Brune, 1970) or from the rms value of acceleration from strong motion records (Hanks and McGuire, 1981), with results that disagree. Evidently these measurements yield different averages, or are model sensitive in other ways. Some methods determine stress drop over only a part of the rupture event, such as from peak acceleration (Hanks and Johnson, 1976), or slip velocity at the onset of faulting (Boatwright, 1980). As a result, one cannot unequivocally quote a "true" stress drop. Various estimates usually agree within only a factor of four or five; this is not error, but an ambiguity in the underlying physics. Within estimates made by a given method one may sensibly discuss relative difference to a finer degree.

In some respects, the confusion and uncertainty about stress-drop measurement is similar to that which existed with the different magnitude scales some years ago, before they were unified under the concept of seismic moment. In fact, seismic moment may be considered a measure of stress drop, that is, as a stress-drop density integrated over a source volume, but this has never been explicitly related to the notion of a moment tensor as a set of body force equivalents, nor has anyone undertaken to specify what a "source volume" is in a manner that would lead to a proper definition of the parameters of interest. In this respect, the stress-drop problem is more fundamental than the old magnitude problem.

From static crack models we may obtain expressions for the average static stress drop in the general form,

$$\Delta\sigma = C\mu\left(\frac{\overline{\Delta u}}{\Lambda}\right) \tag{4.29}$$

where $\overline{\Delta u}$ is mean slip, Λ is a characteristic rupture dimension and C is a constant that depends on the geometry of the rupture. The term in parentheses is a coseismic strain change averaged over the scale length Λ. Alternatively, we may recognize μ/Λ as a stiffness relating $\Delta\sigma$ and $\overline{\Delta u}$, as in the spring-slider model. Closed-form expressions for the constants in Equation 4.29 are few. For a circular crack, $\Lambda = a$, the crack radius, and $C = 7\pi/16$ (which may be obtained from Equation 4.24, paying attention to the difference between u and $\overline{\Delta u}$). For infinite length strike-slip and dip-slip ruptures of half-width W in a whole space (Knopoff, 1958; Starr, 1928), $\Lambda = W$, and $C = 2/\pi$ and $4(\lambda + \mu)/\pi(\lambda + 2\mu)$, respectively ($\lambda$ is the Lamé constant).

In order to evaluate stress drop from seismic data, a time-history of slip also must be prescribed from a model. In one example (Brune, 1970), the corner frequency in the body wave spectrum, f_0, is interpreted as being proportional to $1/t_R \approx v_r/a$ where t_R is the rupture duration, a the rupture radius, and v_r the rupture velocity. If we take a circular rupture as a model for small earthquakes, we obtain the relation

$$M_0 = \tfrac{16}{7}\Delta\sigma\, a^3 \tag{4.30}$$

where both a and M_0 can then be determined from the seismic spectrum. Data determined for a large number of small earthquakes is shown in Figure 4.10. It is evident that there is a general a^3 dependence on moment over a wide size range, indicating that stress drop is approximately constant and independent of earthquake size. This observation is the most powerful argument for the self-similarity of earthquakes. It is a generality, however, and should not be overemphasized. Although the mean stress drop can be seen from Figure 4.10 to be about 3 MPa, there is clearly a considerable range in the value of $\Delta\sigma$, from about 0.03–30 MPa. Such variations from the mean may have considerable tectonic significance.

When earthquakes become large, they have ruptured the whole of the schizosphere and subsequently propagate along strike to produce events of various length and the same width. The rupture dimensions can be estimated from the length of surface rupture or the areal distribution of aftershocks. If stress drop is constant and the reference length for strain in Equation 4.29 is the fault width, as in the Knopoff or Starr models (W-models), large earthquakes with the same width should have the same mean slip regardless of their length. Instead, it is observed that slip increases nearly linearly with rupture length (Scholz, 1982a). This indicates that $M_0 \propto L^2W$, a relationship shown in Figure 4.11. The empirical proportionality constants between L and $\overline{\Delta u}$ are about 1.25×10^{-5} and 2×10^{-5} for interplate strike-slip and thrust earthquakes, respectively. This observation, according to the previous model would then be interpreted as meaning that stress drop increases with size for large earth-

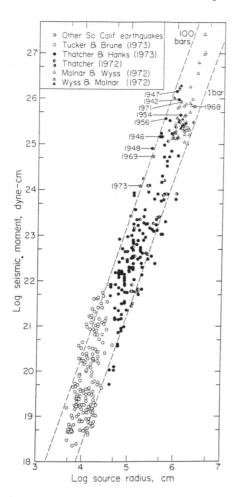

Fig. 4.10 A collection of data for small earthquakes showing the relationship between M_0 and source radius. Dashed lines are of constant stress drop. (From Hanks, 1977.)

quakes, in distinction to the small earthquake scaling just discussed. There are, however, various objections that can be made to this interpretation (Scholz, 1982a). An alternative is that the large earthquake represents the rupturing of the last unbroken patch of lithosphere, the remainder at depth having already failed by aseismic means. It then ruptures like an asperity, and is not constrained on either top or bottom, so the proper scale length in Equation 4.29 is L rather than W (L-model). Then the observations are consistent with stress drop being constant.

This idea may seem confounding at first. If a large earthquake represents a rectangular rupture in an unbroken elastic medium, as in the

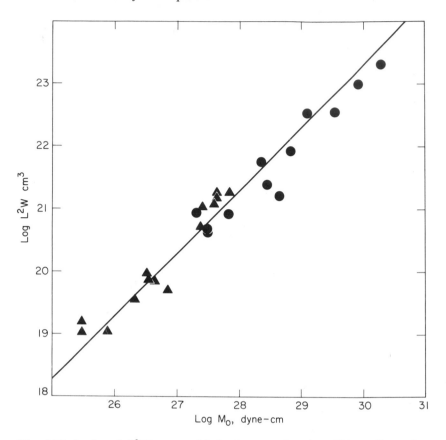

Fig. 4.11 A plot of L^2W versus M_0 for large earthquakes. This scaling reflects the difference between large and small earthquakes, in that for large earthquakes mean slip is proportional to fault length. (From Scholz, 1982a.)

Knopoff and Starr models, then stress drop is clearly controlled by W, the small dimension. But if the deeper part of the lithosphere has already slipped aseismically during the interseismic period, the large earthquake represents the rupture of the last unbroken layer of what is essentially a much deeper crack. Then the controlling stiffness that relates stress drop to slip (and determines the recurrence time) scales with the length of the unbroken segment. A model that works in this way, and which explicitly takes into account slip in the plastosphere, was first suggested by Nur (1981), and further developed by Li and Rice (1983) and Tse and Rice (1986; see also Sec. 5.2.2). The consequences of these static scaling laws on the dynamics of faulting have yet to be firmly established. The end-member cases predict very different behavior, however. The rise time, $t_r \propto W/\beta$ in the W-model and L/β for the L-model (Scholz,

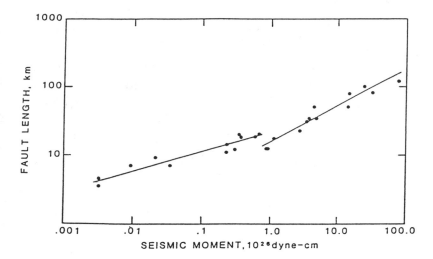

Fig. 4.12 A plot of fault length versus M_0 for intraplate earthquakes in Japan. The break at about 10^{26} dyne-cm is at the change from small to large earthquakes. (From Shimazaki, 1986.)

1982b). These differences should be distinguishable from seismic observations, but neither are likely to be good representations of real earthquakes, because of their heterogeneity.

This scaling break between large and small earthquakes has been demonstrated by Shimazaki (1986). Studying a data set of Japanese intraplate earthquakes, he found that the change from $M_0 \propto L^3$ to $M_0 \propto L^2$ scaling occurred at a moment of about 7.5×10^{25} N-m, where the linear dimension of the rupture is approximately the thickness of the schizosphere (Fig. 4.12). He also observed at this point an offset in the M_0-L relationship, amounting to a doubling of moment. He attributed this to the doubling of slip expected from the breaching of the free surface.

Although there is considerable variation in stress drop among earthquakes, the only systematic effect that has been found is between intraplate and interplate earthquakes (Scholz, Aviles, and Wesnousky, 1986; Kanamori and Allen, 1986). As shown in Figure 4.13, large intraplate earthquakes consistently have greater moment per unit fault length than interplate events, the difference being about a factor of five, so their stress drops are accordingly that much higher. The proportionality constant between mean slip and length for intraplate earthquakes is about 6×10^{-5}. A compilation of parameters of earthquakes in China by Molnar and Deng (1984) reveals the same relationship with a somewhat smaller constant. A discussion of possible reasons for the difference between intraplate and interplate earthquakes is given in Section 6.3.5.

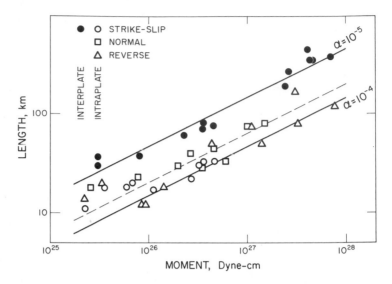

Fig. 4.13 Fault length versus moment for large interplate and intraplate earthquakes. The lines, with slope $\frac{1}{2}$, indicate linear relationships between slip and length. The intraplate events have greater moments per unit fault length, indicating greater stress drops. (From Scholz et al., 1986.)

One might also expect, from a frictional model of faulting, that stress drop would increase with depth for small earthquakes. This has never been found to be a systematically observed effect. However, certain strong ground motion parameters, such as peak acceleration and velocity, which are proportional to dynamic stress drop, do show the expected depth dependence (Figure 4.14; see also McGarr, 1984).

Just as there is a breakdown in self-similarity between small and large earthquakes, there are also problems at the small end of the size spectrum. It may be noticed in some of the data sets in Figure 4.10 that below a certain level M_0 decreases much faster than a^3, suggesting a lower limit to a. There have been two very different explanations of this effect. Hanks (1982) has argued that this is a propagation effect: That. because of very strong attenuation of high frequencies in the immediate vicinity of the earth's surface, there is a maximum frequency f_{max} of waves that may be detected by surface instruments. Aki (1984), on the other hand, has proposed that it represents a true source effect and may represent a minimum source dimension. There is some theoretical justification for this view (Sec. 7.3.1). If this latter interpretation is correct, then because M_0 is not found to have a lower observational limit, $\Delta\sigma$ must decrease without limit for earthquakes at the minimum source

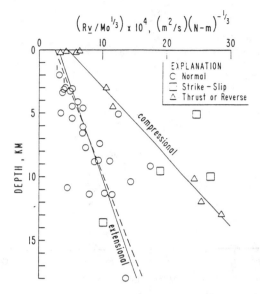

Fig. 4.14 The peak velocity parameter Rv, where R is hypocentral distance and v is peak velocity measured on a seismogram, when normalized by $M_0^{1/3}$ should scale with stress drop. The plot shows this normalized parameter plotted versus depth and sorted into different faulting types. A clear depth dependence is observed, as well as a difference between compressional and extensional events, both of which are expected from a frictional strength law such as shown in Fig. 3.18. (From McGarr, 1984.)

radius. However, the most recent observations favor the attenuation theory of this cutoff (Anderson and Hough, 1984).

The other fundamental earthquake scaling relationship is expressed in their size–frequency distribution. In any region it is found that during a given period the number $N(M_0)$ of earthquakes occurring of moment $\geq M_0$ obeys a relation

$$N(M_0) = aM_0^{-B} \qquad (4.31)$$

where a is a variable in time and space. This is historically known as the Gutenberg–Richter or Ishimoto–Aida relation. This type of power-law size distribution is typical of fractal sets.* It arises simply from the

---

* This type of distribution is observed for widely diverse phenomena and has a rich history. It is often referred to as the Law of Pareto, after V. Pareto (1896) who found that it governed the distributions of firm sizes and personal incomes in a free-market economy. A further tracing of its history may be found in the "Historical sketches" in Mandelbrot (1983).

1964

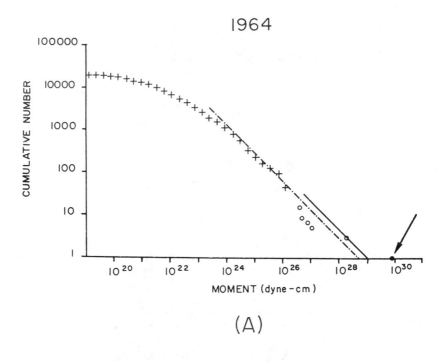

(A)

WORLDWIDE

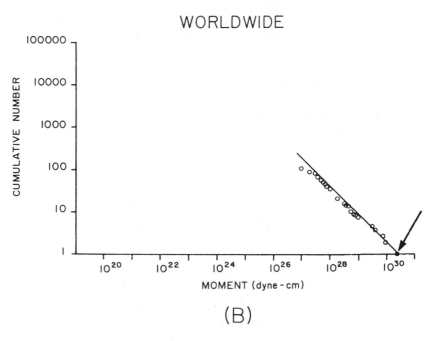

(B)

self-similarity of earthquakes: The exponent $B = 2/3$ for the worldwide distribution and is related to the fractal self-similarity dimension D (Hanks, 1979; Andrews, 1980; Aki, 1981). Because fracture is also a self-similar process, the sizes of faults and joints also obey power law distributions (Wesnousky, Scholz, and Shimazaki, 1983; Segall and Pollard, 1983; Hirata, 1989). These distributions are clearly related.

Mogi (1962) and Scholz (1968a) studied this size distribution as it is observed for the acoustic emissions produced in the deformation of rock samples in the laboratory. Mogi interpreted variations of the B value as being due to variations in the heterogeneity of the rock. In the experiments made by the author it was found that B decreased as the level of applied stress increased. This parameter has been observed to vary from place to place and also temporally. Suyehiro, Asada, and Ohtake (1964), for example, found that the B-value for foreshocks was less than that for aftershocks.

The B-value generally does not depart greatly from the standard value, except for the case of earthquake swarms, when it is sometimes very high. If the size distribution of earthquakes is determined for a single fault or fault segment, however, it is found that the size of the large earthquake that ruptures the entire fault will be greatly underestimated by the extrapolation of the size distribution of small earthquakes for the same fault (Wesnousky et al., 1983; Singh, Rodriguez, and Esteva, 1983; Schwartz and Coppersmith, 1984; Davison and Scholz, 1985). An example of this is shown in Figure 4.15a. This fractal "tear" occurs because of the different scaling relationships for large and small earthquakes, resulting in their belonging to different fractal sets (Scholz and Aviles, 1986). If one considers only a population of large earthquakes, they are self-similar and obey their own size distribution (Fig. 4.15b). This distinction is very important in seismic hazard evaluation (Secs. 5.3 and 7.2). There is also evidence that significant departures from Equation 4.31 occur for earthquakes less than about $M = 3$, suggesting a breakdown of self-similarity at the small end, too (Aki, 1987).

Fig. 4.15 An illustration of the different size distributions for large and small earthquakes. (a) Distribution of small earthquakes within the rupture zone of the 1964 Alaska earthquake, normalized to the recurrence time of that earthquake. The 1964 earthquake is indicated by an arrow. Notice that it is about $1\frac{1}{2}$ orders of magnitude larger than the extrapolation of the small earthquakes would indicate. The rolloff at $M_0 < 3 \times 10^{23}$ dyne-cm is caused by the loss of perceptability of smaller events. (b) Worldwide size distribution of large earthquakes (80 yr). The largest event, that of Chile, 1960 (arrow), is correctly predicted by this relationship. (From Davison and Scholz, 1985.)

4.4   OBSERVATIONS OF EARTHQUAKES

4.4.1   *Case studies*

In this section a number of earthquakes are described in some detail, with particular attention to the rupture process. These cases are not necessarily meant to be taken as typical, since there is considerable variability in detail among earthquakes. The cases chosen represent the three main faulting types, and all occurred in a continental setting (although the crustal structure in the Imperial Valley is distinctly transitional). The three strike-slip earthquakes are interplate events, the others are all intraplate earthquakes. The selections were made primarily of earthquakes that have been unusually well studied.

Three California strike-slip cases: Parkfield, 1966, $M_0 = 1.4 \times 10^{18}$ N-m; Borrego Mountain, 1968, $M_0 = 10 \times 10^{18}$ N-m; and Imperial Valley, 1979, $M_0 = 6 \times 10^{18}$ N-m. The Parkfield earthquake ruptured about 30 km of the San Andreas fault in central California (Fig. 3.33a). This part of the fault is in a stability transition region between the creeping (stable sliding) section of the San Andreas fault to the north and the rupture zone of the great 1857 Ft. Tejon earthquake to the south. The 1966 earthquake propagated unilaterally from north to south and stopped just to the south of the 2-km-wide extensional jog discussed in Section 3.5.2. The mainshock was preceded by an $M_L = 5.0$ foreshock by 17 min, which occurred just to its north. Fresh surface cracks had been noticed on the San Andreas fault in the epicentral area six weeks before the earthquake by K. Kasahara.

A notable feature of this earthquake is that there was evidently little or no coseismic surface breakage, but the earthquake was followed by large amounts of afterslip, at a rate that decayed logarithmically in time (Smith and Wyss, 1968; see also Sec. 6.4.2 for a discussion of afterslip, with an illustration from Parkfield). The net surface slip after two years, as a result of this phenomenon, is shown in Figure 4.16.

The aftershocks of the earthquake are shown in Figure 4.17. They very clearly delineate the fault surface, and even show the jog, where a concentration of aftershock activity occurred. The depth distribution of the aftershocks shows peaks at 4 and 14 km. This corresponds to the top and bottom of the coseismic rupture, as determined from geodetic data (Scholz, Wyss, and Smith, 1969).

The Borrego Mountain earthquake occurred on the Coyote Creek fault of the complex San Jacinto fault system. Surface faulting occurred on 31 km of the fault in three separate breaks (Fig. 4.18). The mainshock epicenter is near the northern break, which also had the greatest coseismic surface slip (38 cm). The Ocotillo Badlands (Fig. 3.32b) occupies a

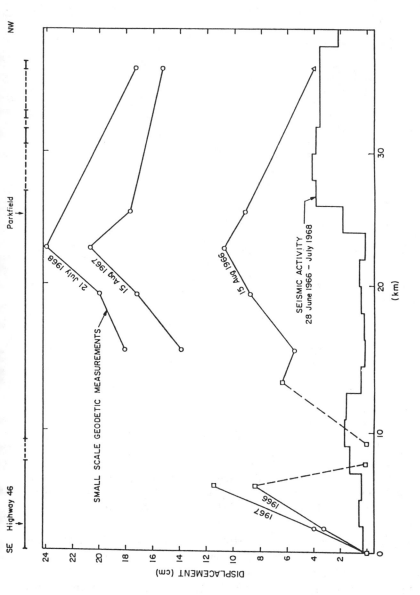

Fig. 4.16 Surface slip for the Parkfield earthquake, measured at three times following the June 27, 1966, earthquake. The minimum in slip 8–10 km from the southern end is at the jog (see map, Fig. 3.33). (From Scholz, Wyss, and Smith, 1969.)

A.

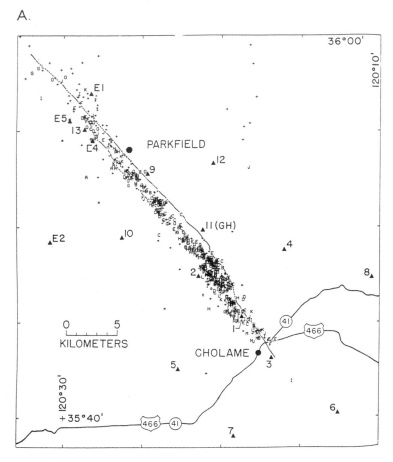

Fig. 4.17 Aftershocks of the 1966 Parkfield earthquake: (a) map view;
(b) a section parallel to the fault plane; (c) histogram showing the depth
distribution. (From Eaton, O'Neill, and Murdock, 1970.)

compressional jog between the northern and central breaks. The maximum slip on the central and southern breaks was 30 and 14 cm,
respectively, with afterslip contributing to these figures (Clark, 1972). No
afterslip occurred on the northern break. A notable feature of this
earthquake was that it triggered slip on the Superstition Hills, Imperial,
and San Andreas faults, located some distance away (Allen et al., 1972).

The mainshock was preceded by 1 min by a single $M_L = 3.7$ foreshock
at the same epicenter. The aftershock pattern was complex. Even ignoring the two groups of aftershocks located at some distance orthogonal to
the rupture (discussed in Sec. 4.4.2), the aftershocks do not define a
simple rupture plane, as they did at Parkfield. They extend over a
distance along the fault of about 55 km, considerably longer than the

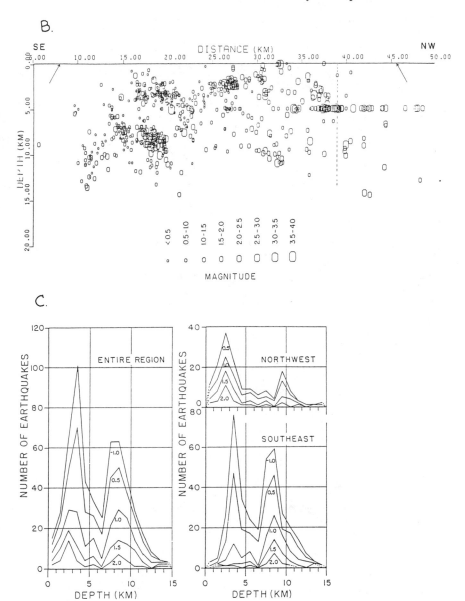

Fig. 4.17 *(cont.)*

region of surface rupture. The main elastic radiation occurred from the region of the northern break. Burdick and Mellman (1976) found that the mainshock was best modeled with a rupture of radius 8 km and stress drop 9.7 MPa centered on the northern break. This contrasts to an average stress drop of only 1.3 MPa obtained for the whole earthquake (Table 6.2). Except for the earliest (and largest) aftershock, most after-

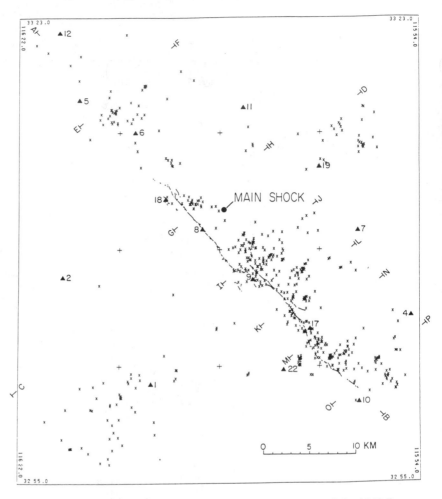

Fig. 4.18 Map of the aftershocks and surface rupture traces of the 1968 Borrego Mountain earthquake. (From Hamilton, 1972.)

shocks occurred on the central and southern breaks. In this earthquake, then, it appears that the main coseismic rupture was confined to a relatively small region, but ruptured past a compressional jog into adjacent parts of the fault, which then experienced considerable after-shock and afterslip activity.

Southeast of the setting of the Borrego Mountain earthquake lies the Imperial fault, which though part of the San Andreas fault system, is quite atypical in its tectonic setting. It is a transform fault right-stepped between the Cerro Prieto fault to the south and the San Andreas fault to the north (Lomnitz et al., 1970). It lies in a transitional region between

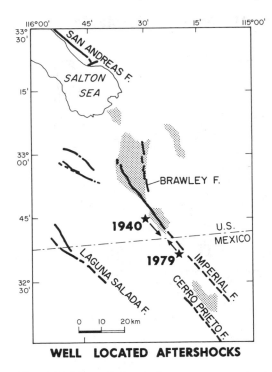

WELL LOCATED AFTERSHOCKS

Fig. 4.19 Map of the surface ruptures and aftershocks of the 1979 Imperial Valley earthquake. Shading indicates zones of concentrated aftershock activity. Stars with arrows indicate epicenters and main propagation directions of the earthquakes of 1940 and 1979. (Data from Johnson and Hutton, 1982.)

the San Andreas fault proper and the sea-floor spreading in the Gulf of California. The extensional jogs at either end of the Imperial fault may be nascent spreading centers. The northern jog is particularly seismic, but the activity does not define an orthogonal offset. Rather it defines a region of high activity, called the Brawley seismic zone, that obliquely connects the Imperial and San Andreas faults. This zone is characterized by swarm activity and is not marked by a throughgoing surface fault, although its eastern edge is paralleled for some distance by the Brawley fault, which splays northwards from the Imperial fault (Fig. 4.19 – note the resemblance of this structure to a horsetail fan, Fig. 3.5). These features lie within the Salton trough, which contains an unusually thick accumulation of sediments. Seismic refraction profiling (Fuis et al., 1982) indicates 4–5 km of sediments, underlain by basement, which they interpret to be metasediments, floored by inferred gabbroic subbasement at a depth of 12–13 km.

KINEMATIC FAULT PARAMETERS

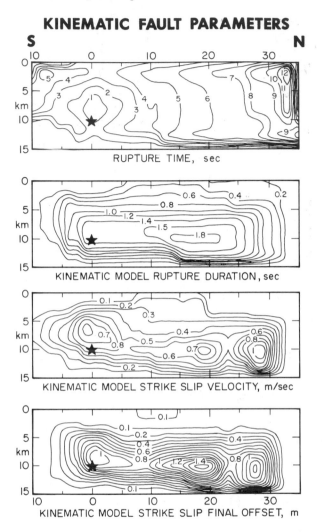

Fig. 4.20 Results of Archuleta's (1984) kinematic model of the Imperial Valley earthquake. The origin of the horizontal scale is at the epicenter, with north to the right. From top to bottom are the positions of the rupture front at different times, the duration of slip, the particle velocity, and the net slip.

The 1979 earthquake initiated just south of the U.S.–Mexico border and propagated almost unilaterally to the north, producing about 30 km of surface rupture on the Imperial and Brawley faults (Fig. 3.31b). It reruptured the northern half of a section of fault that had broken in a larger, $M = 7.1$ earthquake of 1940 (Fig. 4.19). The 1940 earthquake epicenter was north of that of 1979, but it produced its largest surface slip south of the 1979 rupture zone (Sharp et al., 1982). Aftershocks of the 1979 event were mainly in the Brawley seismic zone and in a small zone 12–17 km NW of the mainshock epicenter where one of the largest aftershocks ($M = 5.0$) occurred 3 min after the mainshock. The hypocentral region was notably quiet in aftershock activity (Johnson and Hutton, 1982).

The dynamics of this earthquake can be studied in detail because it occurred within an extensive network of accelerometers that made possible detailed modeling of its rupture characteristics. The results of a kinematic model of Archuleta (1984) are shown in Figure 4.20. The region of high slip velocity and final offset is confined to the basement between the sediments and the mafic subbasement. There are three peaks in slip velocity and offset: in the hypocentral region, at about 12–20 km north, and at 27–31 km north of the epicenter. Rupture velocity was also found to be irregular, averaging 0.94β, but being supersonic at depth between rupture times 6 and 7 s and abruptly decelerating between 7 and 8 s.

These results were simulated with a dynamic model by Quin (1990) which allows us to have some idea of the dynamic parameters of the rupture, as shown in Figure 4.21. It can be seen from the figure that, as might be expected, the regions of high slip velocity correspond roughly to regions of high dynamic stress drop.

There is some correspondence of the dynamic rupture parameters with fault structure. The central patch of high stress drop roughly corresponds with the southern aftershock zone (Fig. 4.19) and the location of the large early aftershock. It may also be notable that the largest offset in the Imperial fault, an 85-m compressional jog, also is in that locality (Sharp et al., 1982). A far more profound effect, however, occurs at the junction of the Brawley fault. Both surface rupture and accelerometer data indicate that the rupture bifurcated at that point, propagating about 10 km up the Brawley fault as well as continuing on the Imperial fault. This bifurcation point also corresponds with the rapid deceleration in slip and rupture velocity (Fig. 4.20), which Quin and Boatwright had to model with a very high value of strength excess ($\sigma_y - \sigma_1$). It also marks the southern edge of the main aftershock zone. The structural complication at the junction of these two faults thus seems to have had a major role in inhibiting this rupture, which finally terminated at the northern end of the Imperial fault where the main fault itself bends to the north. This provides a detailed example of rupture termination at an extensional jog, discussed cursorily in Section 3.5.2.

A pronounced effect of the sediments on this earthquake is also evident. Within the basement, dynamic stress drop was found to increase strongly with depth (Fig. 4.21), as might be expected from a friction model. In order to model the rapid reduction in slip velocities in the

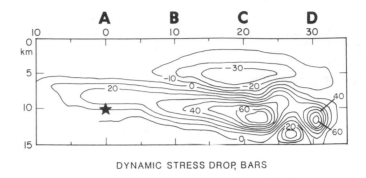

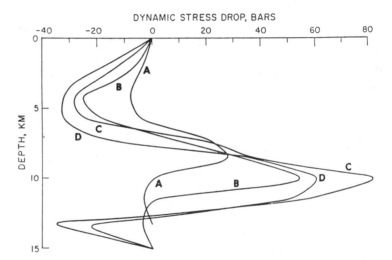

Fig. 4.21 Results of Quin's (1990) dynamic simulation of Archuleta's result. The top figure shows dynamic stress drop contoured on the fault, with the same coordinates as Fig. 4.20. The lower figure shows the depth distribution of stress drop at A–D (0, 10, 20, and 30 km north). Notice that stress drop is large and positive at depth, but is negative at shallow depths.

sediments, however, Quin had to employ a negative dynamic stress drop there (Fig. 4.21). This implies that the sediments are velocity strengthening and that the upper stability transition T_4 occurs at the base of the sediments, in accord with the interpretation of that transition given in Section 3.4.1. This conclusion is supported by the microseismicity: The hypocenters of well-located earthquakes on the Imperial fault, as well as the aftershocks of this earthquake, all lie below the sediments (Doser and Kanamori, 1986). This strong effect of the thick sediments in suppressing dynamic rupture explains why the coseismic surface faulting was so poorly represented (and was generally less than) slip at depth in this earthquake. The coseismic shortfall of slip in the sediments was recovered partially in the pronounced afterslip that followed. Afterslip will be discussed in more detail in Section 6.4.2, but as remarked before, it is likely to occur following rupture of velocity-strengthening regions.

The study of Doser and Kanamori also shows that the lower stability transition occurs at or just below the contact with the mafic subbasement. This may be a case where the lower stability transition is induced by a lithological change rather than by a critical temperature. There is little experimental data on friction of mafic rocks that can be brought to bear on this problem, however.

These three examples illustrate how variable earthquakes are, even among those with the same mechanism on the same fault system. The Parkfield earthquake, except for the behavior at the jog, was a very simple, planar rupture. In contrast, the main energy release in the Borrego Mountain earthquake was from a small region near the epicenter that then produced complex faulting over extensive adjacent regions. The Imperial Valley earthquake, on the other hand, seems to have been relatively simple at the outset, but then became complex at its northern end where an extensional jog was encountered. Three regions of high energy release could be identified in that earthquake, one near the epicenter, one near the jog, and one midway between.

Normal: Borah Peak, Idaho, 1983, $M_s = 7.3$, $M_0 = 32 \times 10^{18}$ N-m. This earthquake ruptured about 36 km of an active normal fault at the base of the Lost River Range, a typical Great Basin structure (Fig. 4.22; see also, Fig. 6.14a), and was of just the type described by G. K. Gilbert. Morphological dating of a preearthquake scarp indicates that the same fault segment had been ruptured last about 6,000–8,000 yr previously (Hanks and Schwartz, 1987).

Rupture initiated at the extreme southwest corner of the rupture zone and propagated unilaterally up and to the north. Aftershocks began immediately after the mainshock, distributed over the entire rupture zone, but there was a progression to the north of the largest aftershocks (Richins et al., 1987). Both the aftershock locations and focal mechanisms indicate that many occurred on subsidiary faults and were not limited to the primary fault plane. The depth distribution of aftershocks of this earthquake is shown in Figure 4.22 and the surface faulting is shown in Figure 3.35a. As was pointed out in Section 3.4.3, moment release for this earthquake occurred to 16-km depth, and although there were a few aftershocks to that depth, they generally showed a sharp cutoff at about 12 km.

The rupture process in this earthquake was affected strongly by structural irregularities. The rupture initiated in a complexly faulted

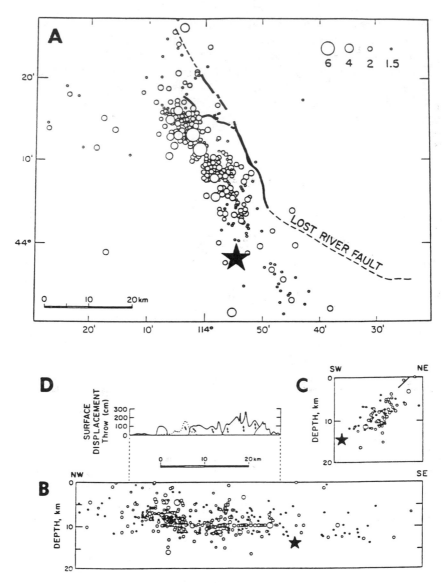

Fig. 4.22 Aftershocks and faulting in the Borah Peak, Idaho, earthquake of 1983: (a) map view; (b) section parallel to the fault; (c) section normal to the fault; (d) surface slip (see Fig. 3.35a for detail). (Data from Richins et al., 1987; Crone and Machette, 1984; figure from Nabelek, in press.)

200

abrupt 55° bend in the Lost River fault, from which it propagated unilaterally to the north (Susong, Bruhn, and Janecke, 1989). Near its northern end, the rupture intersected the Willow Creek Hills, a transverse structural feature that separates the downdropped fault valley into two. At that point, surface faulting diverged from the main fault trace in a secondary splay (Crone et al., 1987). No surface rupturing occurred where the Lost River fault crosses the Willow Creek Hills, but was found to resume to the north after a 4.7-km gap (Fig. 3.35a). Inversion of the geodetic data indicates that slip was sharply reduced at the Willow Creek Hills and that slip north of them was restricted to shallow depths (Ward and Barrientos, 1986). Although the geodetic data are meager, their results also show a fairly irregular slip distribution for this earthquake, with the maximum slip occurring near the hypocentral region.

A comparison of this earthquake with the 1979 Imperial Valley earthquake exemplifies the difference between intraplate and interplate earthquakes shown in Figure 4.13. The two earthquakes had about the same rupture dimensions but the moment of Borah Peak was about five times larger, a difference that could also be seen in the surface faulting for the two cases (Fig. 3.35). We also note again that both the geometry of the main fault trace and the surface slip distribution were much more irregular for Borah Peak than Imperial Valley. The greater fault surface topography normal to the slip vector can occur because it does not produce a kinematic incompatibility, but it does nevertheless produce a dynamic impediment to rupture (Sec. 3.5). However, in the case of the 1983 earthquake the fault trace irregularities did not act as pure steps because the earthquake had a left-lateral component (Nabelek, in press), so that they also may have acted as kinematic barriers.

The most common geological interpretations of the tectonics of extensional regions, of which the Great Basin is most typical, involve listric terminations of normal faults in detachments. The analysis of the seismic and the geodetic data for the Borah Peak earthquake, however, appear to rule out any listric motion in that earthquake (Nabelek, in press; Ward and Barrientos, 1986). This observation of the planar nature of active portions of normal faults within the schizosphere is typical (see Sec. 6.3.4).

Thrust: San Fernando, California, 1971, $M_L = 6.4$; and Tabas, Iran, 1978, $M_s = 7.4$. The San Fernando earthquake occurred on a range-front thrust fault in the Transverse Ranges of southern California. The rupture characteristics of this earthquake, determined primarily from a study of its aftershocks, has been summarized by Whitcomb et al. (1973). Its main features are illustrated in Figure 4.23. The mainshock ($M_0 \approx 10^{18}$ N-m) initiated at 13 km, starting with an initially very high stress drop of 35–140 MPa and propagated updip to the surface, producing about 14 km of surface faulting (Hanks, 1974). The total

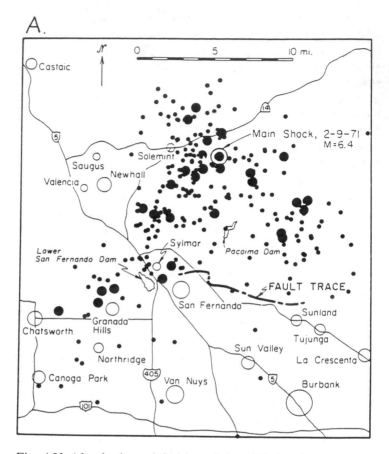

Fig. 4.23 Aftershocks and faulting of the 1971 San Fernando earthquake: (a) map view; (b) simplified schematic diagram of the fault surface. (From Whitcomb et al., 1973.)

rupture duration was about 2.8 s, but 80% of the radiation was produced during the first 0.8 s by the high stress drop region. The average stress drop was at least one order of magnitude smaller than that of the initial phase. The focal mechanism indicated a dip of 52° for the initial motion, but the alignment of aftershocks indicate that above a depth of about 8 km the dip decreased to about 35°. Thus the fault is interpreted as steepening with depth.

An outstanding feature of this earthquake was a downstep of about 3 km of the western portion of the thrust. The northeast strike of this step is essentially parallel to the slip vector for the earthquake so that it is a kinematically compatible feature. Aftershocks on the eastern and western limbs of the thrust primarily had thrust mechanisms similar to that of the mainshock, but the sequence was dominated by aftershocks

B.

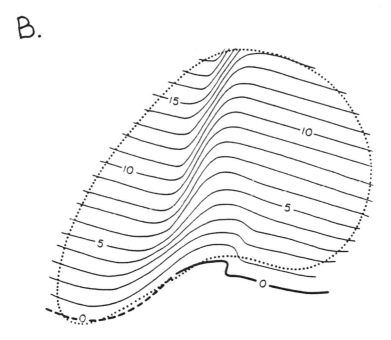

Fig. 4.23 *(cont.)*

with mainly left-lateral strike-slip mechanisms that occurred along the step, indicating slip along the step in a direction nearly parallel with the slip vector for the earthquake as a whole. It is also notable that a number of aftershocks along the step had normal faulting mechanisms, with fault planes oblique to the step. The stress axes for these events have an orientation consistent with the stress field at a Mode III edge, which the step represents, so they may be interpreted as the deep equivalents of the tension gash arrays discussed in Section 3.2.

Aside from the activity associated with the step, the aftershocks for this earthquake were diffusely distributed in space. They do not very well define a principal rupture surface, and they occurred in significant numbers well away from the inferred rupture surface, both on the hanging wall and footwall sides (see also Wesson, Lee, and Gibbs, 1971). The Tabas earthquake (Fig. 4.24), though less well studied, also exhibits this characteristic in its aftershock sequence, as well as some other features of thrust earthquakes. This was also a range-front thrust, and produced about 65 km of surface rupture that broke out in a Neocene monocline along the eastern edge of the Tabas playa (Berberian, 1982). Although from the figure it can be seen that there is a concentration of activity along what may be taken as the principal thrust surface, there is considerable activity on either side, and this is well outside possible

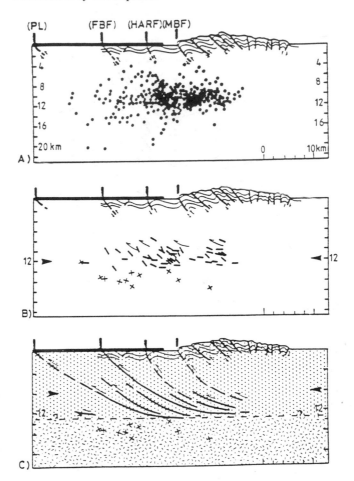

Fig. 4.24 Aftershock and interpretive model of the Tabas-e-Golshan (Iran) earthquake: (a) cross section of well-located aftershocks; (b) inferred fault planes from aftershock focal mechanisms, crosses are events in the basement below the inferred detachment; (c) an interpretation of the relationship between seismicity at depth and surface geology. (From Berberian, 1982.)

relative location errors. Note also the near-horizontal band of activity at depth that extends well back from the mainshock hypocenter on the hanging wall side. This suggests, in contrast to the San Fernando case, that the thrust decreases in dip with depth and may well terminate in a detachment.

In general, aftershocks of large dip-slip earthquakes show greater complexity and diffuseness than for their strike-slip counterparts. Aftershocks for strike-slip earthquakes commonly seem to be restricted to the

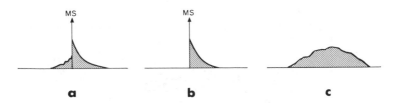

Fig. 4.25 Schematic diagram illustrating the various types of earthquake sequences: (a) Mainshock (MS) with foreshocks and aftershocks;
(b) mainshock–aftershock sequence; (c) swarm.

rupture plane, which they define quite well, and the cases of off-fault
aftershocks that occur can be readily interpreted in terms of slip-induced
stress changes (see next section). Aftershock sequences for thrust earthquakes, in particular, are often much more complex than this – a feature
that may result from the geometric incompatibility of thrust faulting that
requires severe straining in the adjacent volumes of rock.

4.4.2 *Earthquake sequences*

Earthquakes seldom occur as isolated events, but are usually part of a
sequence with variably well-defined characteristics (Fig. 4.25). Foreshock
and aftershock sequences are closely associated with a larger event called
the mainshock, whereas sequences of earthquakes not associated with a
dominant earthquake are called swarms. Occasionally, two or more
mainshocks may be closely associated in time and space. These have
been called doublets and multiplets. We consider these together as
compound earthquakes, discussed in the next section.

Of these sequences, aftershocks are the most ubiquitous, being observed to follow almost all shallow tectonic earthquakes of any significant size. They also have the most well-defined characteristics of any of
the earthquake sequences. In particular, the decay of aftershock sequences follows the Omori law (for his observation of it following the
1891 Nobi earthquake):

$$n = \frac{c}{(1 + t)^p} \tag{4.32}$$

where n is the occurrence frequency of aftershocks at time t after the
mainshock. The exponent p is usually found to be very close to 1, so this
decay law is nearly hyperbolic. The largest aftershock in the sequence is
typically at least one magnitude unit smaller than the mainshock (Utsu,
1971) and the sum of seismic moment for the entire sequence usually
amounts to only about 5% or so of the moment of the mainshock
(Scholz, 1972). Aftershocks are therefore a secondary process.

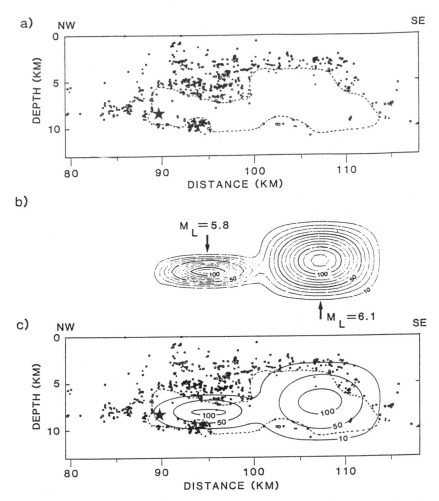

Fig. 4.26 Aftershocks and slip in the 1984 Morgan Hill (California) earthquake, a strike-slip earthquake on the Hayward fault. (a) The distribution of aftershocks on the fault, showing the hypocenter of the mainshock (star) and boundary of the rupture, as inferred by Cockerham and Eaton (1984). (b) Modeled slip distributions in the two principal earthquakes in the sequence. (c) The two superimposed. (From Bakun, King, and Cockerham, 1986.)

Several aftershock sequences have already been described in the previous section. Aftershocks typically begin immediately following the mainshock over the entire rupture area and its surroundings, although they are commonly concentrated in locations where one might expect large stress concentrations to have been produced by the mainshock rupture (cf. Mendoza and Hartzell, 1988). Thus, often they are seen to be concentrated around the rupture perimeter or along structural complexities internal to the rupture, several examples of which have been given earlier. In Figure 4.17, for example, the aftershocks of the Parkfield earthquake delineate the coseismic rupture plane and concentrate along its top and bottom and at the jog that terminated rupture in the south. An even more dramatic case of earthquakes surrounding the rupture zone is shown in Figure 4.26.

The spatial distribution of aftershocks seems almost stationary during the sequence, with only subtle migrations of activity observed, such as was mentioned in the case of the Borah Peak earthquake (see also Whitcomb et al., 1973, for a description of this phenomenon in the San Fernando sequence). However, in the case of subduction zone earthquakes a substantial growth of the size of the aftershock zone with time has been reported. Mogi (1969) noted this phenomenon for large earthquakes offshore of Japan, but found also that it did not occur in the case of earthquakes in the Aleutian arc. Tajima and Kanamori (1985) reported many cases in which the aftershock area increased twofold and more during the months following the earthquake, although they found that the growth was quite variable. These observations have resulted in a dispute about how best to estimate the rupture size of large subduction zone earthquakes from their aftershocks (cf. Kanamori, 1977). This behavior seems to be limited to regions of stability transition, which are common in subduction zones, as discussed in Section 6.4.3.

Aftershocks also sometimes occur in locations distinctly distant from the rupture plane of the mainshock. A number of examples of such off-fault aftershocks in cases of strike-slip earthquakes have been described by Das and Scholz (1982), who pointed out that they occur in specific locations where crack models predict an increase in stress resulting from the mainshock rupture. This is illustrated in Figure 4.27, where the stress changes brought about by the slip on a crack are shown. There are large stress increases near the crack tip, but in addition there are small stress increases on either side of the crack and about one crack length away. These are the regions in which off-fault aftershocks are often seen, as for example, in Figure 4.18. This concept has been developed further by Stein and Lisowski (1983) and Li, Seale, and Cao (1987).

Another interesting example occurred in the case of the 1965 Rat Island earthquake, a large subduction event in the Aleutians. It was followed by an unusual number of large normal faulting earthquakes on the outer wall of the adjacent trench (Stauder, 1968). Presumably this resulted from a decrease in flexure in the outer wall and a decrease in trench-normal stress, both caused by the mainshock.

It seems clear that aftershocks are a process of relaxing stress concentrations produced by the dynamic rupture of the mainshock. However, in order to account for the time delay of aftershocks and the characteristic

Fig. 4.27 Stress changes resulting from slip on a vertical strike-slip fault. Fault-parallel shear stresses are shown; shaded regions have stress increases, elsewhere the stresses are reduced. (From Das and Scholz, 1982.)

decay law, a time-dependent strength must be introduced. Recalling from Section 1.3 that rock strength increases with strain rate, we realize that the dynamic loading during the mainshock may load local regions to stresses much higher than their long-term strength. Such regions will then fail by static fatigue at a later time determined by the imposed stress level. If a static fatigue law of the form of Equation 1.50 is assumed for a large number of such local regions over which the imposed stress levels are assumed to be randomly distributed, it can be shown that the collected rupture behavior of these regions will obey the Omori law (Scholz, 1968b).

An alternative explanation has been given by Nur and Booker (1972). They point out that an earthquake will produce concentrations in volume strain at its ends, which will result in pore-pressure changes there and subsequently induce fluid diffusion. Because of changes in effective stress thereby brought about, this may cause delayed ruptures. They show that under certain conditions, an Omori-law decay may be expected from this process. While earthquake-induced fluid flow undoubtedly occurs (and has been reviewed by Sibson, 1981), the author does not believe this to be the underlying explanation for aftershocks. Aftershocks are observed to "turn on" instantaneously all over the rupture zone after the main shock: They do not migrate in from the ends as would be required if they followed a fluid diffusion front. Secondary migrations of aftershocks, on the other hand, may be associated with fluid flow, such as discussed by Li et al. (1987).

Foreshocks are smaller earthquakes that precede the mainshock. They usually occur in the immediate vicinity of the mainshock hypocenter and are therefore probably a part of the nucleation process. (Earthquakes that precede a mainshock but are not near the hypocenter are probably not causally related to the mainshock and, strictly, should not be considered foreshocks.) Unlike aftershocks, the occurrence of foreshocks is quite variable. Many earthquakes have no foreshocks, and foreshock sequences can range from one or two events to small swarms. Further discussion of foreshocks will be deferred till Section 7.2.3.

Earthquake swarms are sequences of earthquakes that often start and end gradually and in which no single earthquake dominates in size. Sykes (1970) made a global survey of swarm occurrence and found that they commonly are associated with volcanic regions, though this is by no means a universal rule. A major swarm at Matsushiro, Japan, evidently was caused by an upwelling of pore fluid, evidently of plutonic origin (Nur, 1974; Kisslinger, 1975). This seems to have been a natural occurrence of the mechanism responsible for causing earthquake swarms by fluid injection at Rangely and Denver, Colorado (Sec. 2.4). According to this mechanism, the earthquakes are produced by an increase in pore pressure caused by fluid flow. As a result, they occur in a region in which there is an unusually strong strength gradient, so any event in the sequence is prevented from growing very large; strain relief is controlled by the fluid flow, and no dominant large event can occur. For the same reason, the B-value in the earthquake size distribution is often observed to be unusually large in swarms (Scholz, 1968a; Sykes, 1970).

Mogi (1963) has divided these sequences into three types: mainshock–aftershock, foreshock–mainshock–aftershock, and swarm, which he has interpreted as indicating increasing heterogeneity of the source region. He found that these sequence types were variously dominant in different parts of Japan and interpreted this in terms of regional variations of heterogeneity. While this idea has merit, the author feels that it is too generalized. The Adirondacks of New York is a cratonic region in which Precambrian basement outcrops at the surface (Grenville age, 900 My) in a very uniform massif. Several intense earthquake swarms typified by small magnitude events occurred at Blue Mountain Lake in the central Adirondacks in 1972–73, then subsided. In 1975, an $M = 4.0$ earthquake occurred at Raquette Lake, just 10 km away. In spite of the quick installation of portable seismometers, no aftershocks could be found for this event: It was a singleton. In 1985, an $M = 5.1$ earthquake occurred at Goodnow, just 20 km away, and this earthquake had a normal aftershock sequence (Seeber and Armbruster, 1986). Thus, over a very short period of time in a small region of the same structural province, the gamut of earthquake sequences was observed. All the earthquakes had similar thrust mechanisms, the only systematic environmental difference was an increase in depth, from 2–3 km for the swarm, 4 km for the Raquette Lake earthquake, and 6 km for the Goodnow sequence. Whether or not this depth difference is relevant to the difference in sequence type is moot.

4.4.3   *Compound earthquakes*: *Clustering and migration*

Earthquakes are invariably complex in their rupture characteristics because of the geometrical irregularity of faults and heterogeneity in the various parameters in the rupture resistance S (Eq. 4.20). This complexity may be manifested both in irregularity in the propagation of the rupture and in the distribution of moment release within. In many cases this heterogeneity may be pronounced enough to generate distinct seismic waves, in which case various *subevents* can be recognized. High-frequency strong ground motion often appears to be Gaussian noise (Hanks and McGuire, 1981), suggesting random heterogeneity in the faulting process (cf. Andrews, 1980). As long as this complexity occurs within the rupture time, one may consider the earthquake to be a single event. As an example, Wyss and Brune (1967) found that the 1964 Alaska earthquake consisted of a sequence of recognizable subevents that propagated from east to west at a speed near the shear velocity. Sometimes, however, two or more events, often of similar size, occur on nearby but different rupture surfaces close together in time, but with a delay such that their rupture times do not overlap. We call this general class of earthquakes *compound earthquakes*. They are of considerable interest because they imply rupture processes other than those predicted by elastic fracture mechanics.

Perhaps the most common type of compound earthquake is when the rupture surfaces of the two events are contiguous. Well-documented cases of this type have occurred as subduction (underthrusting) earthquakes along the Nankai trough of southwest Japan (see Fig. 5.15). The historical record of earthquakes on this plate boundary dates from 684 a.d. and is particularly well described since 1707 (Ando, 1975). Tsunami and damage records allow the approximate dimension of the earthquakes to be identified, which shows that earthquakes along this plate boundary rupture well-defined segments or groups of segments (A–D, Fig. 5.15). The two most recent earthquake sequences were compound events. In the Ansei earthquakes (1854), segments C and D ruptured first in a single large earthquake, followed 32 hr later by a second earthquake that ruptured segments B and A. This sequence was repeated 90 yr later when the 1944 Tonankai earthquake ruptured segment C, followed by the 1946 Nankaido earthquake that, propagating from the BC boundary, ruptured B and A. This suggests that there is a structural complexity at the BC boundary off the Kii Peninsula that inhibits rupture there. If so, it is not an entirely persistent feature: The Hoei earthquake of 1707 ruptured ABCD in a single, much larger event.

The adjacent rupture planes that fail in this manner need not have the same strike. In the Solomon Islands, where compound earthquakes are common, pairs of such events have been observed to go around the sharp corner of the trench near New Britain (Lay and Kanamori, 1980). The two Gazli, USSR, earthquakes of 1976 occurred 39 days apart on conjugate and contiguous oblique-slip faults (Kristy, Burdick, and Simpson, 1980). The several events in the sequence need not have the same mechanism. The dominant rupture in the 1976 Tangshan earthquake in China, for example, was strike slip, but it was followed closely by a large normal-faulting earthquake that occurred in a right step at one end and a thrust event that occurred in a left step at the other (Nabelek, Chen, and Ye, 1987).

The mechanics of this type of compound earthquake may be understood, like aftershocks, to be a manifestation of the time dependence of strength, discussed later (Sec. 4.5.2). In the case of compound earthquakes that do not occur on contiguous rupture surfaces, however, an explanation is less readily apparent. This type of compound earthquake has occurred repeatedly in the South Iceland seismic zone (Fig. 4.28). In 1896, five strong earthquakes occurred in a three-week period, successively migrating from east to west. A similar sequence occurred in 1784. The first 1896 earthquake produced a north–south-striking right-lateral surface rupture (similar to and just to the west of the one shown in Fig. 3.7), so it is clear that the several mainshocks in each sequence ruptured on different faults, offset in a direction normal to their strikes.

The triggering mechanism for these Icelandic cases may involve the effect shown in Figure 4.27 to explain off-fault aftershocks. Although this explanation is plausible in the Icelandic cases, it is much less likely in the case of a remarkable cluster of earthquakes that occurred in central Nevada in 1954 (Fig. 4.29). There, four large earthquakes occurred over a period of a few months in a region in which the earthquake recurrence time is measured in thousands of years. The last earthquakes in the series, the Dixie Valley–Fairview Peak earthquakes, constitute a typical

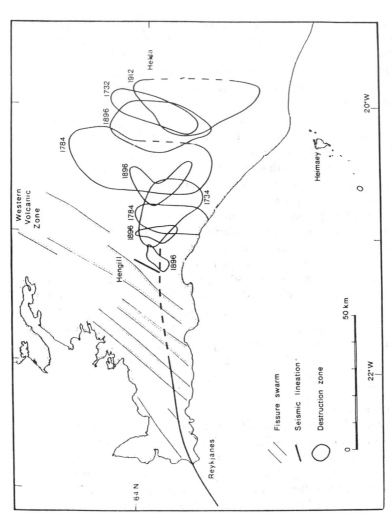

Fig. 4.28 Map of the South Iceland seismic zone, showing the areas of major destruction from large earthquakes since 1700. Notice the migrations of 1784 and 1896. Recall from Fig. 3.7 that earthquakes in this zone are strike-slip, on north–south striking faults. The western rift comes up the Reykjanes peninsula and proceeds to Hengill. The eastern rift begins to die out in the south at about Hekla. (From Einarsson et al., 1981.)

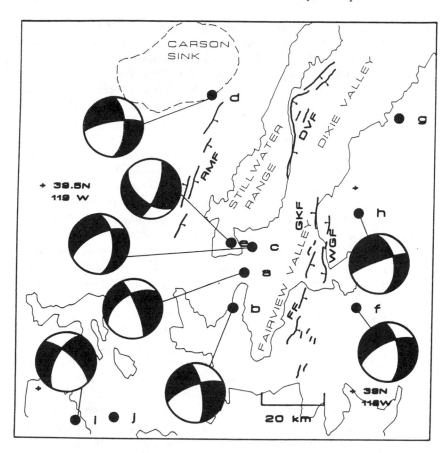

Fig. 4.29 Earthquakes and faulting in central Nevada, 1954. Events a, b, c, d, and e (M = 6.6, 6.4, 6.8, 5.8, 5.5) occurred from July to September and were associated with rupture on the Rainbow Mountain fault (RMF). The Fairview Peak earthquake (f, M = 7.1) occurred at 1107 on December 16 and ruptured bilaterally along the Fairview fault (FF). The Dixie Valley earthquake (g, M = 6.8) occurred at 1111 on December 16 and ruptured the Dixie Valley fault (DVF). Surface rupturing also occurred on the Westgate and Gold King faults (WGF and GKF) [see Slemmons (1957) for details on surface rupturing]. Focal spheres are lower hemisphere, with white areas indicating dilatation. Faults are normal, with barbs on downdropped side. (From Doser, 1986.)

compound event. The Fairview Peak earthquake, propagating bilaterally, ruptured the west-facing normal fault on the east side of the valley. The Dixie Valley earthquake initiated four minutes later on the east-facing fault on the other side of the valley from the northern end of the Fairview Peak earthquake, and ruptured to the north. The earlier Fallon–Stillwater and Rainbow Mountain earthquakes were considerably smaller and occurred on the opposite side of the horst that is bounded on the east by the Dixie Valley fault. Their stress fields, if modeled with a uniform elastic model, could not, however, have contributed a significant triggering stress to the subsequent events.

A recent compound earthquake in the Imperial Valley, California, demonstrates a case of triggering between nonparallel faults in which a rather simple explanation can be conceived. On November 23, 1987, an $M_S = 6.2$ earthquake produced left-lateral motion on a northeast striking cross fault that abuts the right-lateral Superstition Hills fault (Fig. 4.30a). An $M_S = 6.6$ earthquake 12 hr later initiated from the abutment point and propagated to the southeast (Fig. 4.30b). The triggering mechanism can be envisioned easily because the first earthquake would have reduced the normal stress across the Superstition Hills fault, thus bringing it closer to the failure condition. This effect would have been greatest near the abutment point but would not achieve its maximum effect instantly. The movement in the first shock would also reduce the pore pressure there, in proportion to the reduction of normal stress (because of the coupled poroelastic effect, see Sec. 6.5.2). Pore pressure would then commence recovering, at a rate determined by the local hydraulic diffusivity, so that the Superstition Hills fault would be brought progressively closer to failure. This poroelastic effect can explain the time delay between the earthquakes, though other delay mechanisms, such as discussed in the next section, also may have played a role (Hudnut, Seeber, and Pacheco, 1989). The 1927 Tango, Japan, earthquake was very similar to this case. There, two abutting, nearly orthogonal conjugate strike-slip faults ruptured in a single earthquake, but there was evidently no time delay between the events (Kasahara, 1981).

As in the Icelandic cases, compound earthquakes sometimes progress in a single direction, and thereby define an earthquake migration. The most frequently cited case of this was the rupturing on the North Anatolian fault in a sequence of six large earthquakes that progressed from east to west over the period 1939–67 (Fig. 4.31). At first glance, this whole sequence can be considered a single compound earthquake, with each rupture triggering the next, in a delayed domino effect. However, the epicenter of the 1943 earthquake was at the western end of its rupture, nearly 300 km from the end of the previous earthquake (Barka and Kadinsky-Cade, 1988), so the mechanism for the migration of this series of earthquakes is not so straightforward.

Many other examples of earthquake migrations have been cited in the literature. Often, however, these supposed migrations connect earthquakes that are many rupture dimensions distant from one another, and so cannot be directly linked mechanically, as in most of the cases discussed above. It has been suggested, for example, that these migrations are induced by a separate process, such as an episode of astheno-

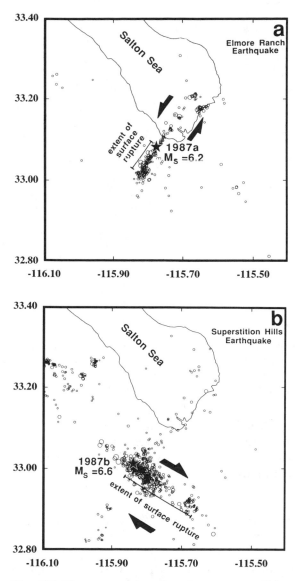

Fig. 4.30 The compound earthquake sequence of November, 1987 near Superstition Hills, California. The left-lateral Elmore Ranch fault ruptured first in an $M_s = 6.2$ event (a), followed 12 hr later by rupture of the right-lateral Superstition Hills fault in an $M_s = 6.6$ event (b). The second initiated from a point near the abutment of the two faults. Stars are the mainshock epicenters. (From Hudnut et al., 1989.)

215

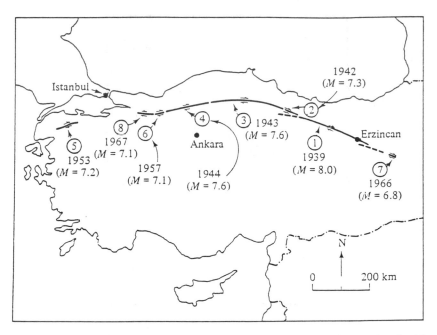

Fig. 4.31 Fault displacements along the North Anatolian fault associated with large earthquakes since 1939. Although the western progression of earthquakes from 1939–67 is generally remarked on, the 1943 earthquake initiated at the western end of its rupture and propagated to the east. (From Allen, 1969.)

spheric flow (e.g., Scholz, 1977). Kasahara (1981) has reviewed these types of migrations and discussed the several mechanisms that have been proposed to explain them, which all involve viscous coupling between the asthenosphere and lithosphere.

4.5   MECHANICS OF COMPLEX AND COMPOUND EARTHQUAKES

The heterogeneity of faults and faulting has been pointed out already in several contexts. This heterogeneity exists on all scales: The particular features discussed are just those evident at the scale of mapping. As would be expected, dynamic faulting is also heterogeneous at all scales, ranging from the apparently random character of high-frequency ground motion (Hanks and McGuire, 1981) to the large-scale irregularities, called asperities, that are recognizable from long-period waves (Lay, Ruff, and Kanamori, 1980).

A considerable amount of attention has been given to the larger scale of these irregularities on faults, because they result in spatial variations in seismic hazard on a scale shorter than the fault itself. These hetero-geneities in dynamic rupture result from variations of strength due to the geometric complexity of faults combined with heterogeneity in the applied stresses, which reflects the most recent rupturing history of the fault.

4.5.1 Heterogeneity in dynamic rupture

The subject of heterogeneity in faulting has been discussed already in several contexts. In seismology the terms *barrier* and *asperity* are often used to describe such heterogeneity, and the usage has spread to other fields as well. The first term was introduced by Das and Aki (1977) who constructed a two-dimensional fault model that contained regions with varying S value. They found that regions of high S, which they called barriers, may arrest rupture and even remain unbroken when the rupture skipped over them. The term asperity, in this context, seems to have appeared first in Lay and Kanamori (1981), who used it to describe regions within earthquakes where relatively high moment release oc-curred. They interpreted this as being due to rupture of a "strong" region, which they called an asperity. They adopted this term by vague allusion to the term used in friction, where it means a contacting point at a protrusion of the surface, between which there is no contact. The reader is cautioned not to confuse these two usages.

Subsequently, there has been great use of the terms barrier and asperity. Various authors seem to have used them differently and a great many diverse phenomena allegedly have been explained with them. The adjective strong is frequently applied to both asperities and barriers, without the required qualification of what strength refers to in this case. Because of both their popularity and vagueness, these concepts need fuller discussion.

A barrier will be defined observationally as a place where rupture during an earthquake is impeded or arrested. Let us take it conceptually to be a region of exceptionally high S relative to the surrounding regions. An asperity will be taken observationally as a region of excep-tionally high moment release and hence a region in which the stress drop is much greater than the surrounding regions on the fault.

Examining the expression for S (Eq. 4.20), we immediately see that there can be more than one explanation for barriers. They may result from a high value of σ_y and be called a strength barrier, or from a low value of σ_1 and be called a relaxation barrier (Aki, 1979). The first case can be a permanent feature of a fault, the second need not be. The relaxation barrier is a major component of what is called seismic gap theory: It explains why there is not significant overlap of the rupture zones of adjacent recent earthquakes. During the seismic cycle, a barrier

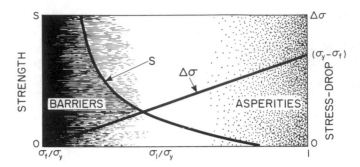

Fig. 4.32 Sketch showing the evolution of a barrier to an asperity during the seismic cycle.

may evolve into an asperity, as shown in Figure 4.32. As σ_1 increases from σ_f to σ_y there is a decrease in S and an increase in $\Delta\sigma$, so this region may act as a barrier or an asperity, depending on which time in its seismic cycle it is impinged upon by rupture of the adjoining region. If it does rupture through, the moment density that it will release, which is proportional to its stress drop (its "strength" as an asperity) will also depend on its position within its seismic cycle.

We may, using these terms, reinterpret the observations and dynamic modeling of the 1979 Imperial Valley earthquake (Sec. 4.4.1). The three peaks in slip and dynamic stress drop along the fault can be called asperities (Figs. 4.20–4.21). There is evidently a trough in σ_y between the second and third asperities because the rupture velocity became supersonic on leaving the second asperity and then abruptly decelerated going into the third, which acted as a barrier with a high strength excess, as the Quin model indicated. There is little basis for determining whether the first asperity corresponded to a strong point (except for the presence of the small jog there) or simply a region that slipped less in the previous earthquake of 1940 and hence had a higher value of σ_1 in 1979. Because the barrier before the third asperity corresponds approximately to the intersection of the Brawley fault, on the other hand, one does have some reason for suggesting that it might indicate a profound strong point.

Structures on faults that may result in strong strength variations have been reviewed in Section 3.5. In many cases earthquakes have been observed to stop or suffer reduced slip at such complications. However, rupture in a given earthquake will also depend on the initial conditions and hence on the most recent rupture history, and so the role of a structural complication like a jog may vary temporally. In the previous section, an example was cited of a structure off the Kii Peninsula in Japan which separated the last two pairs of Nankai trough earthquakes but was propagated through by the earthquake of 1707. Similarly, although the Willow Creek Hills acted as a barrier in the 1983 Borah Peak earthquake, they have not always acted so: the 6,000-yr-old Holocene scarp crosses the 1983 slip gap (Crone et al., 1987). In seismic hazard analysis it is critical to recognize if active faults are *segmented* into strands that rupture in separate earthquakes, which requires that the role of such structures in the rupture process be a permanent feature of them. This subject will be taken up further in Sections 5.3.2, 5.4, and 7.4.3.

So far we have imagined cases in which the rupture behavior of faults is wholly seismic in slip characteristics. However, the way in which Lay et al. (1980) employed their asperity model in describing the rupture characteristics of subduction zones is quite different. There they evidently define asperities as regions that slip seismically with large moment release and assume that the intervening regions normally slip aseismically. These intervening regions are only ruptured seismically when an adjoining asperity ruptures and dynamically propagates into them. This is an entirely different sort of mechanism. In this case an asperity is simply a part of the subduction zone interface that exhibits an unstable constitutive law, such as velocity weakening, whereas the intervening regions have stable, or velocity-strengthening, behavior. In this case no inference regarding the relative strength of the different regions necessarily can be made and the asperity, by virtue of its constitutive law, is a permanent feature. This behavior and definition of an asperity should not be confused with what has been described earlier. In this case what distinguishes the overall behavior is not so much the asperities but the stable regions between. The role of stable slip in faulting is a separate topic and will be taken up in Section 6.4.3.

4.5.2   *Mechanics of compound earthquakes*

The compound earthquakes discussed in Section 4.4.3 cannot be explained with linear elastic fracture mechanics because the time delays between the individual events are too long to result from elastic processes. Because an earthquake dynamically loads the surrounding region, compound earthquakes can result from viscoelastic relaxation in the immediate postseismic period, resulting in a redistribution of loads. There may be many mechanisms for such a viscoelastic response: relaxation of the asthenosphere is one, and pore fluid coupling within the schizosphere another. Asthenospheric relaxation has much too long a time constant to play a role in most compound earthquakes, which often have delay times of minutes, hours, or days. In some cases pore fluid coupling may be important, as in the case of the 1987 Superstition Hills earthquakes, mentioned earlier.

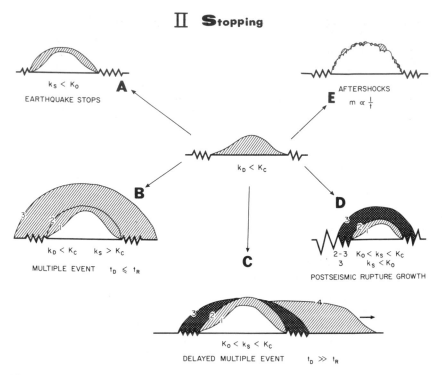

Fig. 4.33 Scenarios illustrating the stopping of a rupture under different conditions of the strength parameters. Light shading indicates dynamic rupture, dark shading indicates subcritical growth. (From Das and Scholz, 1981.)

On the other hand it was noted in Section 1.3 that brittle fracture is inherently time dependent, which results in viscoelasticity in rock due entirely to brittle processes. Das and Scholz (1981) examined the consequences of including subcritical crack growth (Eq. 1.46) in the fracture mechanics formulation of the rupture problem. In this case there are two critical stress intensity factors: $\mathbf{K}_c$, which determines the onset of dynamic rupture; and $\mathbf{K}_0$, which determines the onset of subcritical crack growth (Fig. 1.19). Recall also that the dynamic stress intensity factor K_d is less than the static one K_s.

If one envisions the situation in which an earthquake stops at a barrier because $K_d < \mathbf{K}_c$, there are now a number of possible outcomes, illustrated in Figure 4.33. If, when static equilibrium is reached, $K_s < \mathbf{K}_0$, rupture simply stops (Fig. 4.33a). If $K_s > \mathbf{K}_c$, rupture will be reinitiated before slip has stopped in the interior. This has been called a multiple event (Fig. 4.33b). An example is the 1964 Alaska earthquake, noted earlier.

If now $\mathbf{K}_c > K_s > \mathbf{K}_0$, the crack will grow subcritically into the barrier. If it breaches the barrier and $K_s > \mathbf{K}_c$ on the other side, dynamic rupture will ensue and produce a compound earthquake, with a time delay between the two events being determined by the initial conditions on Equation 1.46 (Fig. 4.33c). On the other hand the rupture may grow subcritically into a region in which $K_s < \mathbf{K}_0$ and stop, so that what is observed is aseismic postearthquake rupture growth (Fig. 4.33d).

This mechanism is the same as that discussed earlier for the generation of aftershocks, except that here we have considered only subcritical growth in the principal rupture tip, rather than in many sites of dynamically enhanced stress (Fig. 4.33e). Similarly, this mechanism need not be limited to contiguous rupture cases. In any nearby area where the stress is increased by an earthquake (e.g., Fig. 4.27) there is the potential that K_s can be raised above $\mathbf{K}_0$ so that the same process can ensue. This limits the distance over which one rupture can trigger another, by this or any other direct mechanism, to about one rupture length.

Environmental effects on strength are important in two ways. They result in the time dependence of friction that causes the earthquake instability. They also allow subcritical crack growth to occur, which can be important in both preearthquake and postearthquake behavior, and can be important in earthquake triggering.

The discussion in this chapter has focused on a description of earthquakes as dynamically propagating shear cracks on a frictional interface. In the next chapter we consider these processes when imbedded in the loading system that drives them.

5 The Seismic Cycle

In the previous two chapters we discussed the statics and dynamics of faulting by treating the fault as an isolated system. We now place the fault into the tectonic engine and consider its behavior when coupled to the loading system. Observational results and models are combined to determine the nature of this loading system, and we explore the question as to whether the seismic cycle is periodic.

5.1 HISTORICAL

G. K. Gilbert understood that slip on faults must accrue through a repetition of earthquakes, and his writings show that he devoted considerable thought to just how such repetition must occur. In Gilbert (1909), he wrote both of rhythm in the recurrence of earthquakes, and of alternation, by which he meant that earthquakes alternate in position along major seismic zones – anticipating the concept of the seismic gap, as it is phrased today.

The credit for a full conceptualization of the loading cycle, however, is usually given to H. F. Reid. In summarizing the mechanism of the California earthquake of 1906, Reid (1910) presented his *elastic rebound theory* in which the earthquake was the result of a sudden relaxation of elastic strains through rupture along the San Andreas fault. The causative strains, according to his theory, were accumulated over a long period of time by the steady motion of the regions on either side of the fault, which under normal conditions remains locked by friction.

Several crucial pieces of information were available to Reid that allowed him to develop his theory. The earthquake ruptured 450 km of the fault with an average right-lateral slip of about 4.5 m. Of this, some 360 km were on land where the ground breakage could be observed readily and the offset measured in many places. At the time, the San Andreas fault was already known as a profound geomorphic feature extending much of the length of California. A triangulation network had been completed some twenty years before the earthquake and was resurveyed just afterwards. Comparison of the two surveys revealed, for the first time, the displacement field produced by an earthquake. It showed that points on the southwest side of the fault had moved to the northwest with respect to those on the other side, and that these displacements, when extrapolated to the fault, were consistent with the offsets measured at the surface break. The displacements fell off rapidly away from the fault, reaching quite small values 10 or 20 km out.

An examination of the differences between two earlier surveys, one done in the 1880s and one in the 1860s, showed that the Farallon lighthouse, a point off the coast well to the southwest of the fault, prior to the earthquake had been moving to the northwest with respect to inland points. This observation formed the basis for Reid's postulation of the loading of the fault by the steady movement of distant points and allowed him to formulate the elastic rebound theory, which was a model for the whole of the loading cycle (Fig. 5.1).

No cause for this steady strain accumulation was then known, although Reid cited in footnote a prescient idea of Bailey Willis' that the floor of the Pacific Ocean must be spreading [sic] and moving to the northwest with respect to its bordering continents. The theory of plate tectonics has validated this idea and the behavior of plate boundaries such as the San Andreas fault is now understood, at least to first order, in this context.

Reid's conception was of a linear elastic system. He consequently proposed that the strain accumulation pattern was just opposite that of the strain release, so that the net result of one earthquake cycle was a block offset of the fault with no net strain (Fig. 5.1). He consequently had difficulty in explaining why the accumulation of strain was so tightly concentrated on the fault. As we shall see, the actual loading behavior is now known to be a great deal more complex and cannot be explained by a totally elastic system with steady strain accumulation.

As a corollary of his theory, Reid proposed a method of earthquake prediction. Immediately following the earthquake, if one were to commence geodetic measurements across the fault, one would know when to expect the next earthquake: It would occur when the strains released in the previous earthquake had reaccumulated. Unfortunately, Reid's recommendation for this systematic measuring program was not taken up in California until 60 yr later, so at present we have only a fragmentary knowledge of the strain buildup in the interim and of the strain accumulation pattern over time. At about the same time as Reid was making his investigation, a geodetic monitoring program was initiated in Japan and continues to the present. This has provided the most complete and reliable data set on this problem, but as it concerns subduction zone loading, the results indicate behavior that is different and more complex than might be expected at a transform boundary, such as the San Andreas fault.

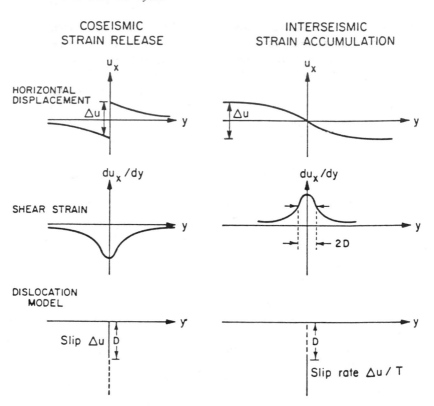

Fig. 5.1 Schematic diagram illustrating a simple elastic rebound model. On the left, the coseismic displacement and strain fields are modeled with a slip on a fault of depth D. On the right, interseismic strain accumulation is modeled with a steady slip rate on the fault below the depth D. (From Thatcher, 1986.)

In Reid's concept the recurrence of large earthquakes is periodic, or at least, to use a modern phrase, time predictable. However, the evidence for this is still inconclusive, if not contradictory, and whether this is true has important consequences both in terms of understanding the mechanics and in earthquake hazard analysis. In this chapter we discuss the present understanding of this loading and relaxation oscillation known as the seismic cycle. It involves study of the entire coupled system: schizosphere, plastosphere, and asthenosphere.

5.2   THE CRUSTAL DEFORMATION CYCLE

In terms of crustal deformation, the loading cycle is often divided into four phases: *preseismic*, *coseismic*, *postseismic*, and *interseismic*. This

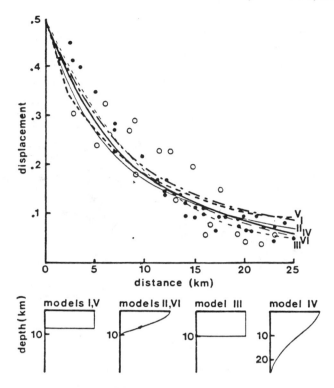

Fig. 5.2 Illustration of the coseismic deformation field of a large strike-slip earthquake. Fault-parallel displacement is plotted as a function of distance from the fault for the case of the Tango, Japan, earthquake of 1927. Curves are from various dislocation models that assume different depth distributions of slip, as shown below. Because the different curves fit the data equally well, the details of the slip distribution at depth cannot be resolved. (From Mavko, 1981.)

fourfold structure has been assembled from geodetic observations in many places: A complete cycle has never been observed at any one location.

Typical coseismic deformation fields are shown in Figures 5.2 and 5.3 for strike-slip and subduction zone thrust earthquakes, respectively. This deformation, as Reid understood, is produced by the strain relief upon dynamic faulting, a conclusion that has been verified with the successful application of elastic dislocation models to explain these fields, beginning with Chinnery (1961) and Savage and Hastie (1966). From modeling these data it is possible to determine the depth, attitude, and extent of faulting, the slip and hence moment, and, if the data are of sufficient quality and quantity, the variation of slip on the fault plane. Examples are shown in Figures 5.2 and 5.3 of the displacement fields predicted by

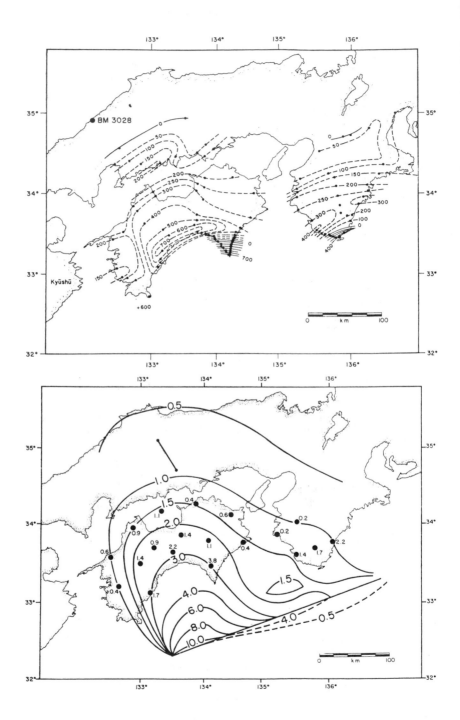

dislocation models which have been fitted to the observations. The analysis of geodetic data to determine coseismic faulting by these means has become standard practice.

The preseismic phase of deformation, by contrast, has proven most elusive in terms of its character and mechanism. Its existence is suggested by fragmentary observations that crustal deformation just prior to a large earthquake is anomalous when compared to the interseismic deformation. This phase belongs to a discussion of earthquake precursory phenomena and it will be taken up in Section 7.2.

In this section we will be concerned primarily with the postseismic and interseismic phases, which predominate in the strain accumulation process. The geodetic observations will be presented first, followed by a discussion of various models that have been employed to explain them.

5.2.1 *Geodetic observations of strain accumulation*

Geodetic measurements have revealed the accumulation of strain from ongoing tectonic processes in many areas. In only two regions, however, are data sufficient to attempt to reconstruct the entire loading cycle – the transform plate boundary encompassing the San Andreas fault in California and along the southern coast of Japan where subduction is occurring beneath the Nankai and Sagami troughs. The observations reveal some overall similarities in the loading cycle for these two cases, while indicating significant differences that arise from the difference in these tectonic environments.

A transform boundary: California Geodetic data bearing on strain accumulation in California can be obtained from triangulation networks that were established in the 1880s and subsequently resurveyed at infrequent intervals and from geodimeter networks that have been regularly surveyed since about 1970. Measurements using very long baseline interferometry (VLBI) and other space-based techniques have been made within the last ten years and are beginning to contribute useful data.

The results of the geodimeter measurements are shown in Figure 5.4 as fault-parallel displacement rates in four profiles across the San Andreas fault. A map indicating the location of these profiles and other localities

---

Fig. 5.3 Deformation produced in a large thrust earthquake: the Nankaido, Japan, earthquake of 1946. Above: observed vertical movements from leveling. Solid and dashed contours indicate uplift and subsidence (in millimeters) relative to mean sea level. Below: horizontal movements towards the Nankai trough (dots, in meters), from triangulation measurements. Contours indicate predicted horizontal movements from a dislocation model in which thrust faulting occurs on a plane dipping 35° landward and outcropping on the inner wall of the Nankai trough (straight line). (From Fitch and Scholz, 1971.)

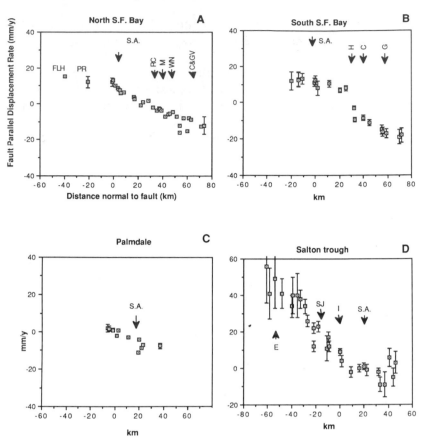

Fig. 5.4 Four profiles across the San Andreas fault showing interseismic fault-parallel displacement rates determined from geodimeter surveys. For locations see Figure 5.5. Faults given by arrows are: SA, San Andreas; RC, Rodgers Creek; M, Maacama; WN, West Napa; C & GV, Cordelia and Green Valley; H, Hayward; C, Calaveras; G, Greenville; I, Imperial; SJ, San Jacinto; E, Elsinore. Also indicated in A are Farallon Lighthouse (FLH), and Point Reyes (PR). Data sources are given in the text.

mentioned in this section is given in Figure 5.5. These profiles are all of deformation well into the interseismic phase: Profiles A and B are in the region of the 1906 rupture; C is in the region ruptured in the earthquake of 1857; and D crosses the San Andreas in the Salton trough, where it is thought to have not ruptured in about 300 yr (Sieh, 1978).

It may be noticed first, by comparison with Figure 5.2 [which is similar to the coseismic displacement field of the 1906 San Francisco earthquake, see Fig. 4.9 in Kasahara (1981)], that the strain accumulation field is much broader than the coseismic strain release. In both profiles near

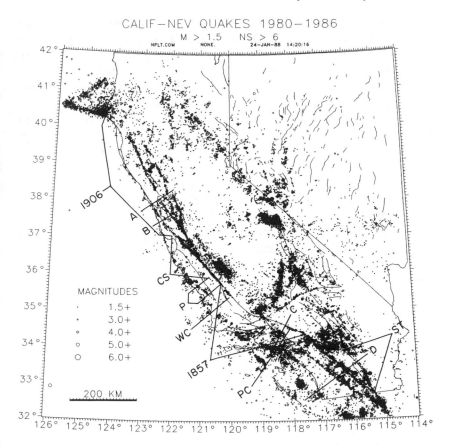

Fig. 5.5 Seismicity and fault map of California, showing the locations mentioned in the text. Geodimeter profiles in Fig. 5.4 are indicated by A–D. Other locations are: 1906, 1857, ruptures of those earthquakes; CS, creeping segment; ST, Salton Trough; WC, Wallace Creek; PC, Pallet Creek; P, Parkfield. (Base map from Hill, Eaton, and Jones, in press.)

San Francisco Bay (A and B) the total deformation rate across the networks is about 35 mm/yr, occurring distributed over a region extending at least 80 km to the northeast of the San Andreas fault. The few measurements on the southwest side of the fault indicate that very little deformation is taking place there. Triangulation measurements over the same region are consistent with these results (Thatcher, 1979). South of profile B the faults continuously slide stably (the so-called creeping section, Fig. 5.5). Because geodimeter measurements there indicate block motion with no strain accumulation, Savage and Burford (1973) were able to there determine a long-term net slip rate of 32 ± 0.5 mm/yr for the system encompassing the San Andreas, Hayward, and Calaveras faults. This agrees with the total strain accumulation rate in the profiles to the north, and with a geological estimate of the Holocene slip rate for the San Andreas fault at Wallace Creek to the south (WC in Fig. 5.5), where these faults have joined (Sieh and Jahns, 1984). Surface creep on the Hayward fault is known to extend into the region of profile B so the offset in this profile of about 7 mm/yr is probably due to slip on that fault. The geologic slip rate of the San Andreas fault on the San Francisco Peninsula is only about 15 mm/yr (Hall, 1984), with no creep, so if the sum adds to 33 mm/yr there must be a significant slip rate on the Calaveras fault, although its rate and the extent to which it is creeping are not known independently. In profile A to the north of San Francisco Bay, there are also a number of active faults to the east of the San Andreas, indicated in Figure 5.4a. Prescott and Yu (1986) have suggested that the width of the deformation zone there also results from either stable slip or strain accumulation on those faults, although there is no independent evidence for this at present.

These measurements, however, evidently do not cross the entire deformation field, because the total plate motion is estimated to be 48 mm/yr (DeMets et al., 1987; formerly thought to be 55 mm/yr, Minster and Jordan, 1978, 1987). Geological and VLBI measurements indicate that only a small part (less than 5 mm/yr) of this discrepancy can be accounted for by extension in the Great Basin to the east, and several schemes have been developed in which the missing deformation takes place on offshore faults (Weldon and Humphreys, 1986; Minster and Jordan, 1987). The San Gregorio fault is the likely candidate in the San Francisco region for this additional slip. It has a geologically estimated slip rate of 7 mm/yr at one of its few coastal exposures (Weber and Lajoie, 1977), but measurements to the Farallon Lighthouse (FLH), shown in Figure 5.4b, do not reflect this deformation. Instead, VLBI measurements to a point about 300 km east of Point Reyes (PR) indicate about 45 mm/yr fault-parallel motion between these stations, suggesting that much of the missing deformation is taking place to the east of the geodimeter networks in an area in which no active faults are known (Kroger et al., 1987). The resolution of this discrepancy and the mechanism by which the deformation field is so broad remains an important problem of San Andreas tectonics.

Far to the south, on the other hand, the deformation across the Salton trough, shown in profile D, indicates a rate of about 55 mm/yr. There, too, it is distributed across a wide zone that contains several subparallel active faults. According to the interpretation of those data by Savage et al. (1979), about 47 mm/yr slip occurs on the Imperial fault, which is partitioned to the north as 25 mm/yr on the San Andreas and 17 mm/yr on the San Jacinto fault, with a smaller amount of slip occurring on the Elsinore fault. Geological data, however, indicate that the San Jacinto

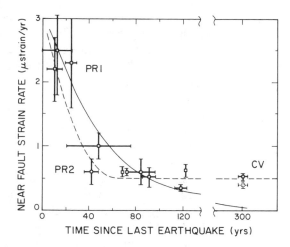

Fig. 5.6 A composite plot of near-fault shear strain rate as a function of time since the last earthquake in the region of measurement. PR1 and PR2 are Point Reyes data, CV is Coachella Valley (Salton Trough): dashed point is original, solid point is renormalized by a factor of $\frac{33}{25}$. Notice the scale break between the point CV and the other points. The curves represent two extreme interpretations. (Data from Thatcher, 1983.)

fault is moving at a slower rate, about 10 mm/yr (Sharp, 1981). With the exception of the Imperial fault, most of this slip occurs beneath the schizosphere, with the shallower portions of the faults locked.

In the Palmdale region (profile C) the San Andreas fault strikes 20° more northwesterly than to the north or south. Thus, the strainfield is different, being almost uniaxial north–south compression (Savage et al., 1981). The tectonics in this region are not purely transcurrent, there being considerable crustal shortening to the south in the Transverse Ranges. The fault-parallel shear strain rate is somewhat smaller in this region than the others, a fact that has been viewed with some significance in modeling, but for the reasons just given, this region may not be entirely comparable with the others.

In order to study the temporal development of these strainfields during the seismic cycle, we show in Figure 5.6 a composite plot of near-fault strain rates measured in various places and at various times since the last earthquake in the region of the measurement. This diagram clearly brings out the postseismic phase of deformation and shows how it merges with the interseismic phase. The plot follows that of Thatcher (1983), with one modification. The point for Coachella Valley (CV) is for a part of the San Andreas fault that is now thought, based on geological and geodetic data, to be moving at a long-term rate of only about 25 mm/yr (Savage et al., 1981; Weldon and Sieh, 1985). If this is correct, then in order to compare it with the other points, which are on sections of the fault moving at about 33 mm/yr, it needs to be multiplied by a factor of $\frac{33}{25}$, as shown by the solid symbol. The dashed symbol is the original rate as shown by Thatcher.

Two extreme positions in interpreting these data are shown by the curves. The solid curve is that given by Thatcher. It emphasizes an inferred long-term nature of the postseismic transient and a steady decrease in the interseismic rate. It is motivated by some modeling results, to be discussed in the next section. The dashed curve is a contrary interpretation, emphasizing the short-term nature of the transient and the lack of evidence for a change in the interseismic rate. Although the data clearly show that high rates of deformation occurred in the 10–20 yr following the 1906 earthquake, the different interpretations of its duration depend on only two points, the high rate for the first Point Reyes epoch (PR1, 1930–38) and the normal rate for the second (PR2, 1938–61). Interpretations of the interseismic behavior depend almost entirely on how point CV is treated, which is why this was brought up earlier.

These data, which indicate a postseismic relaxation transient followed by interseismic deformation at a nearly steady rate, show that the crustal deformation cycle is viscoelastic in character. We defer the question as to what viscous elements may account for this behavior until a discussion of the modeling results. We now turn to a very different tectonic environment, where a similar pattern emerges.

Subduction zones: Japan   A law enacted by the Japanese Diet shortly after the 1891 Nobi earthquake initiated a scientific program for the study of earthquakes and their attendant hazards. One result was that geodetic measurements for the purpose of detecting crustal deformation were made over much of Japan, starting in the 1890s and being repeated at regular intervals since then. These data sets, from both triangulation and leveling, constitute the largest body of data regarding the crustal deformation cycle. During this period, two large, mainly underthrust earthquakes ruptured the subduction plate boundaries off the southern coast of Japan: the Nankaido earthquake of 1946 and the Kanto earthquake of 1923. Some aspects of the Nankaido earthquake were discussed previously in Section 4.4.3 and its coseismic deformation field is shown in Figure 5.3. It is possible to piece together a general idea of the crustal deformation cycle in a subduction environment from these examples.

The Nankaido coseismic deformation, modeled by Fitch and Scholz (1971) and Ando (1975), was produced by thrusting on a fault dipping landward from the Nankai trough with a projected surface trace given by

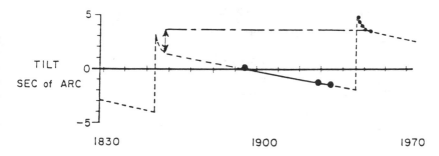

Fig. 5.7 History of tilt at Muroto Point, Shikoku, and its relation with the Nankaido earthquakes of 1854 and 1946. Arrows indicate inferred amount of permanent deformation. (From Fitch and Scholz, 1971.)

the straight line in the lower part of Figure 5.3. This produced uplift on the most seaward promontories, a broad zone of subsidence farther inland, and a general horizontal extension towards the Nankai trough. Geodetic leveling repeated prior to the earthquake, from ~ 1890 to ~ 1930, indicated a generally opposite pattern of interseismic vertical movements. From this it was possible to envision a form of elastic rebound theory in which subduction occurring against a locked interface drags down the overlying plate edge, which then rebounds during earthquakes such as that of 1946. The temporal development of the cycle can be illustrated by tilt at Muroto Point, shown in Figure 5.7. The previous Nankaido earthquake was in 1854, with the same uplift at Moroto as in 1946. In this figure the extrapolation is made by assuming the two earthquakes were identical. A postseismic phase is clearly evident, and in addition a permanent deformation is indicated. This latter feature, which is not expected from a simple elastic rebound theory, is shown both by a comparison of the extrapolated interseismic and postseismic phases with the coseismic vertical movements and by flights of uplifted marine terraces that are found along the coastal promontories.

Although the interseismic deformation is roughly opposite the coseismic, in detail the patterns are quite different. Thatcher (1984a) has shown that they continually evolve during the cycle. His findings are summarized in Figure 5.8 as sketches of profiles of vertical movements across Shikoku and Honshu for various periods. The coseismic movements were followed by rapid postseismic deformation that was concentrated in the most trenchward regions, overlying the coseismic fault plane. This phase died out by about 1950, but the deformation continued to evolve spatially, moving farther inland and developing towards longer wavelengths with time. The decay time for the coastal postseismic phase was only a year or two, whereas the postearthquake transient phase observed farther inland persisted for several decades. Thatcher con-

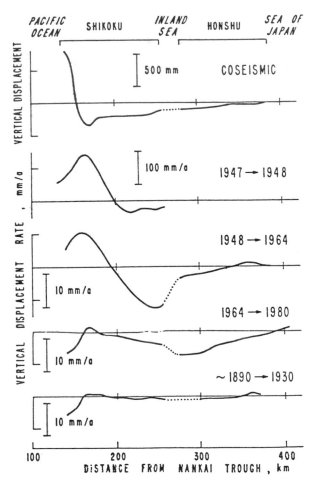

Fig. 5.8 Sketches of profiles of vertical movements across Shikoku and Honshu showing the temporal development of deformation through much of the Nankaido earthquake cycle. (From Thatcher, 1984a.)

cluded that two postseismic phases, with different mechanisms, were present. Again, if the postseismic deformation is added to the interseismic movements (extrapolated over the recurrence interval), then subtracted from the coseismic movements, a significant residual is found. This permanent deformation roughly agrees with the height and distribution of the uplifted terraces and also predicts a broad area of subsidence in the general region of the Seto Inland Sea.

Studies of the deformation following the Kanto earthquake have revealed similar patterns (e.g., Ando, 1974; Scholz and Kato, 1978; Thatcher and Fujita, 1984). An example is given in Figure 5.9, where the

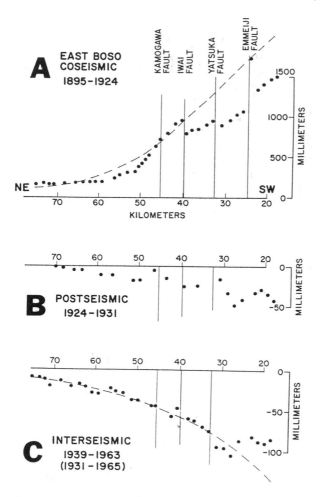

Fig. 5.9 Vertical motions in the Boso Peninsula during coseismic, postseismic, and interseismic periods of the 1923 Kanto earthquake. The origin of the distance scale is at the surface trace of the fault in Sagami Bay. Vertical lines indicate the locations of mapped active faults, which produced offsets in the coseismic profile. The postseismic profile is the residual after the interpolated interseismic deformation has been removed. (From Scholz and Kato, 1978.)

different phases can be seen to have different spatial patterns. Notice also that there they are disturbed by block motions on a series of imbricate faults on the overriding wedge. In the Kanto region the short-term postseismic phase is very evident, but the longer-term phase is not clearly displayed, perhaps because of the more limited geographical extent of the survey area.

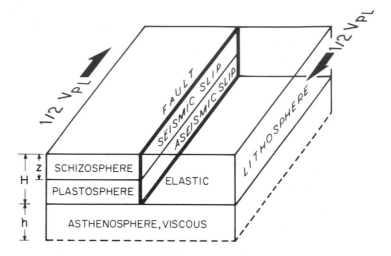

Fig. 5.10 Generalized model of faulting in a lithosphere coupled to an asthenosphere. Coseismic slip in the lithosphere extends to depth Z, below which slip occurs aseismically, so Z is the depth of the schizosphere–plastosphere boundary. The lithospheric thickness is H, below which is a viscous asthenosphere of thickness h.

5.2.2   *Models of the deformation cycle*

The data just reviewed show that Reid's simple concept, sketched in Figure 5.1, is incorrect in two major aspects. First, the strain accumulation field is not the inverse of the coseismic strain release field but has a generally broader spatial extent. Second, the presence of a postseismic phase indicates that the lithospheric response is viscoelastic. There have been two basic approaches to explaining these features: by aseismic slip on the fault beneath the schizosphere, introduced by Fitch and Scholz (1971); and by coupling to a viscous asthenosphere, first proposed by Elsasser (1969) and analyzed by Nur and Mavko (1974).

A generalized model that contains both elements is shown in Figure 5.10. An elastic lithosphere of thickness H ruptures coseismically to a depth Z, below which it may slip aseismically. This corresponds to our division of the lithosphere into schizosphere and plastosphere. Below the lithosphere is a viscous asthenosphere characterized by a time constant, which in the case of a half-space is $\tau = 2\eta/\mu$, where η is its Newtonian viscosity and μ its rigidity. This model has been studied for the case of a vertical fault by Savage and Prescott (1978), who attempted to separate the effects of deep slip from asthenospheric relaxation. They showed that surface measurements cannot distinguish between these two effects because the surface deformation field produced by asthenospheric relaxation can always be duplicated by some reasonable distribution of slip at depth. To return to our remarks in Chapter 3 concerning faulting in the plastosphere, for example, it is to be expected that this region will slip continuously during the interseismic period. If it is velocity strengthening, it should also exhibit a transient period of accelerated slip in the postseismic period (see Sec. 6.4.2). With suitable slip distributions, these two effects can explain both the postseismic and interseismic deformation phases. For asthenospheric relaxation to be important, two conditions must be met. The coseismic slip depth Z must be comparable to the lithospheric thickness H, and τ must be a small fraction of the cycle recurrence time T.

A considerable amount of work has been done by a number of authors on variations of the model shown in Figure 5.10. We review, as an example, the model of Li and Rice (1987), which has the advantage, from an explanatory point of view, that the near-fault motions are not prescribed kinematically. Their model is for a vertical strike-slip case, in which the lithosphere is driven by a remote plate velocity v_{pl}. The deep fault is assumed to slip at constant stress, which, because only differences are considered, means that it is slipping freely. The fault is underlain by an asthenosphere as described above.

The results of one of their models is shown in Figure 5.11, where the surface shear strain rate is plotted as a function of distance from the fault for various fractions of the recurrence time. At short times after the earthquake, strain is concentrated close to the fault and occurs at a high rate, and there is a region of reversed shear farther from the fault. At longer times, as stress diffuses outwards in the asthenosphere, the strain field broadens and the near-fault rates decrease. Li and Rice fit their models to the data of Figure 5.4a and c, and to Thatcher's curve in Figure 5.6, with reasonable agreement. They concluded, with $v_{pl} = 35$ mm/yr and $T = 160$ yr, that $H = 20$–30 km and $\tau = 10$–16 yr.

The case for asthenospheric relaxation in the Li and Rice result depends largely on evidence for a long-term fall in the interseismic rate that, from our comment concerning point CV in Figure 5.6, is not strongly demanded by the data. Furthermore, the contributions of the other active faults to the observations of Figure 5.4a, suggested by Prescott and Yu, and which are certainly present in profiles 5.4b and d, have not been considered, nor has the discrepancy over the supposed 48 mm/yr total plate motion rate. So, although these results are instructive, they are not conclusive. Further geodetic data, particularly as concerns the temporal broadening of the strain field, need to be obtained.

The cycle at subduction zones has been modeled with an elastic dislocation model by Savage (1983) and with asthenospheric relaxation by Thatcher and Rundle (1979). Both models were compared with data from Japan that neither fit particularly well. Another attempt was made by Thatcher and Rundle (1984), using a more flexible model of the type

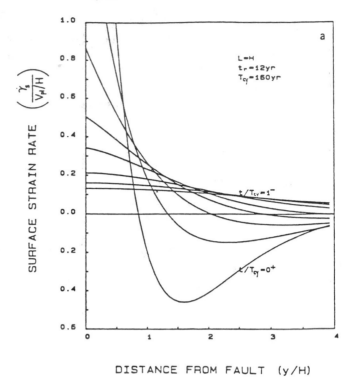

Fig. 5.11 Surface shear strain rate as a function of distance from the fault for selected times during the seismic cycle (cycle time = T_{cy}). At short times strain is concentrated near the fault, then gradually diffuses away. $Z = H$, $\tau = 12$ yr in the case shown. (From Li and Rice, 1987.)

shown in Figure 5.10 and with the additional data of Thatcher (1984a). They concluded that the short-term postseismic phase observed there must be due to deep slip, but that the long-term phase is more likely a result of stress diffusing away in a viscous asthenosphere, in the manner suggested by Elsasser (1969), although their models still did not fit the data very precisely. None of these models explains the permanent deformation.

As pointed out by Turcotte, Liu, and Kulhawy (1984), the viscoelastic relaxation models require an intracrustal asthenosphere at a depth of 20–30 km. This might be rationalized in terms of a strength model such as shown in Figure 3.18, but otherwise has no independent confirmation. From the time constant of the postseismic deformation, the viscosity of this asthenosphere must be in the range 10^{18}–10^{19} Pa-s, one or two orders of magnitude less than viscosity of the (mantle) asthenosphere found from postglacial uplift studies (Peltier, 1981). However these are not truly comparable. As Walcott (1973) points out, postglacial uplift requires volume flow in the asthenosphere, which is controlled by a diffusion constant

$$D = \rho g h^3 / 12\eta \qquad (5.1)$$

where h is the thickness of the asthenospheric channel. In this case, the relaxation time depends on the scale of the region and only D can be determined, so there is a trade-off between h and η. For shear deformation of the asthenosphere, such as in the strike-slip case, however, no volume flow occurs and τ is proportional to $H\eta/h\mu$ (Li and Rice, 1987). The two could be compatible with a narrow low-viscosity asthenosphere channel. This is not strictly the case for subduction zones, where some volume flow must also be required in asthenosphere relaxation models.

A different sort of model of the seismic cycle was explored by Tse and Rice (1986) that yields a picture much richer in detail than the models explored above and that illustrates many features previously discussed. They studied the slip distribution during repeated loading cycles on a fault that obeys the friction constitutive law, Equation 2.27, in which the parameters were taken from laboratory data. The frictional velocity parameters $a - b$ that they assumed are shown in Figure 2.22: For their assumed temperature gradient this produces a transition to stable slip at 11 km (see Secs. 2.3.3 and 3.4.1).

The results of one of their models are shown in Figure 5.12. The fault is driven at a remote plate velocity $v_{pl} = 35$ mm/yr and the cycle time is 92.9 yr. In the case shown, the critical slip distance is 40 mm and a 30% dynamic overshoot is assumed. As loading begins, slip occurs below the stability transition, gradually penetrating upward into the velocity-weakening region above 11 km. Slip is pinned in the central part of the unstable zone but some slip occurs near the surface. At 300 days before the instability, nucleation starts in the central zone, resulting in a small slip patch between curves A and B. During the coseismic slip, B to B', the stress drop in the coseismic region transfers stress to the stable region below. This causes the coseismic slip to penetrate into the region of velocity strengthening to 13–15 km depth. This is followed by rapid postseismic slip, B' to F, at depths from 10–20 km over the next nine years. Steady deep slip then resumes and the cycle is repeated. Stuart (1988) has adapted this model to the subduction zone case and was able to fit the Nankaido uplift data over the whole seismic cycle as well as Thatcher and Rundle were, except for the late postseismic stage, where the Thatcher–Rundle model does somewhat better.

The details of the model depend upon the parameters assumed in the friction law. They are particularly sensitive to the critical slip distance $\mathscr{L}$, because it determines the nucleation zone size and the slip within it (see Eq. 2.31). With $\mathscr{L} > 190$ mm they found that the fault slid steadily at v_{pl} at all depths and that for $80 < \mathscr{L} < 160$ mm the fault slipped

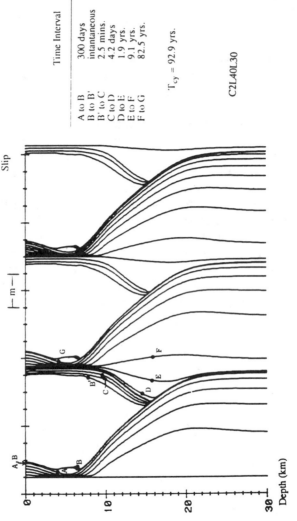

	Time Interval
A to B	300 days
B to B'	intantaneous
B' to C	2.5 mins.
C to D	4.2 days
D to E	1.9 yrs.
E to F	9.1 yrs.
F to G	82.5 yrs.

$T_{cy} = 92.9$ yrs.

C2L40L30

Fig. 5.12 History of slip on a strike-slip fault through several cycles. In this case, the fault is assumed to obey a rate–state-variable friction law at all depths and no asthenosphere is included. The parameters are chosen so that the lower stability transition is at about 11 km. (From Tse and Rice, 1986.)

episodically within the zone shown to be locked in the figure. For values of $\mathscr{L}$ less than 80 mm, the fault showed stick-slip behavior. If **a** − **b** was assumed to be less negative (less velocity weakening) in the region near the surface, it was observed that nucleation extended to the surface and the amount of shallow afterslip was increased. In general, the smaller the value of $\mathscr{L}$ and the more negative the value of **a** − **b** in the upper region, the more unstable is the behavior and the smaller are the preseismic and shallow postseismic slips compared to the coseismic slip (see Sec. 2.4 for an independent estimate of $\mathscr{L}$).

These results show that many of the features observed in the seismic cycle and upon which we have remarked can be predicted by a model that assumes nothing more than a fault on which the frictional behavior is the same as observed in laboratory friction measurements. The value of $\mathscr{L}$ employed to predict geophysically realistic behavior is consistent with that calculated independently from contact theory, as described in Section 2.4, and the **a** − **b** values used were determined experimentally. The model is entirely frictional and no additional assumptions regarding a brittle–plastic transition or asthenospheric relaxation were required to produce these features: This suggests that although those factors may be present, they need not be essential to the process. It is particularly relevant to the subject of this section that this model predicts accelerated postseismic slip in the upper plastosphere superimposed upon long-term steady slip throughout it, features which, though perhaps not explaining all of the postseismic and interseismic deformation phases, are probably basic to an explanation.

5.3   THE EARTHQUAKE CYCLE

This discussion of the tectonic loading and relaxation process has emphasized its cyclic nature but so far has not addressed the question as to whether it is periodic, which is a question of great practical importance. This involves the question as to how sensitive to initial conditions are the instability and subsequent rupture growth. At the outset, the answer to this can be investigated only through observation, the principal findings of which are described in this section.

5.3.1   *Earthquake recurrence*

Earthquake recurrence, in the sense used here, refers to the time between subsequent rupturing on a given segment of fault, hence to the period of the loading cycle. Sometimes, the term also is used to indicate the mean time between earthquakes within a specified region in which many active faults may be present. The latter is a statistical concept useful in

Table 5.1. *Nankaido earthquake characteristics*

	1707	1854	1946	Ratio of last cycle to previous cycle
Rupture Length, km	500	300	300	1.7
Uplift, Muroto Point, m	1.8	1.2	1.15	1.5
Recurrence Interval, yr[a]		147	92	1.6

[a] Refers to rupture of AB block, Figure 5.15. Data from Ando (1975).

earthquake hazard analysis but which is conceptually different than our interest here. The data that can be used to study earthquake recurrence are quite limited because recurrence times are generally large – a hundred years or more. Recording seismometers have been in existence for only a little more than a century, and in many tectonically active regions reliable historical records cannot be obtained for periods earlier than about a hundred years before that. In many remote areas, no historical records are available. As a result, for many areas of the world, one complete seismic cycle has not been documented yet, and for others only one is known, which does not carry any information about periodicity. We are left with just a few case studies.

Nankaido earthquakes   This region, located adjacent to the ancient Japanese capital of Kyoto, has an unusually long and well-documented history. The lateral extent and the amount of coastal uplift is known for the last three events – those of 1707, 1854, and 1946 – and this information is listed in Table 5.1. (The distribution of uplift along the coast is known for 1854 and 1946, but for 1707 is known for only one place, Muroto Point, so our discussion centers on that locality.) These data were used by Shimazaki and Nakata (1980) to test several models of earthquake recurrence. The models they used are shown in Figure 5.13. On the left (a) is shown a simple interpretation of Reid's concept: Successive earthquakes, each with the same stress drop, occur when the stress reaches a critical value. This model leads to perfect periodicity.

In the second model (b) they assume that each earthquake occurs at a critical stress level but that stress drop (slip) varies. This is called time predictable because the time to the next earthquake (but not its size) can be predicted from a knowledge of the slip in the previous one. The third model (c) makes the opposite assertion that earthquakes start at variable stress states but the stress falls to a constant base level. This model is called slip predictable because the slip in the next earthquake can be

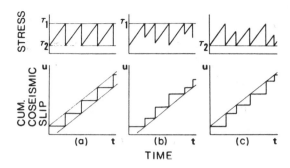

Fig. 5.13 Simple earthquake recurrence models: (a) Reid's perfectly periodic model; (b) time-predictable model; (c) size-predictable model. The time-predictable model is motivated by the observation of the Nankaido earthquakes, but the interpretation in terms of stress is based on a *W*-model. An alternative is discussed in Sec. 4.5. (From Shimazaki and Nakata, 1980.)

predicted from the time since the previous one, but its time cannot be predicted.

The test was made by comparing uplift at Muroto Point for the last three earthquakes with the time between them (Fig. 5.14a). The observations agree quite well with the time-predictable model, as can be seen also by comparing the last two entries in the last column of Table 5.1. At face value the greater uplift in 1707 than in the later events contradicts the periodic "Reid" model as well as some other recurrence models that will be described later.

Ando (1975) compiled an earthquake history of this same plate boundary going back more than 1,200 yr. His results, based on felt reports and tsunamis along this coast, are shown in Figure 5.15. He concluded that the boundary could be divided into four segments, which fail individually or together in large earthquakes. Noting that the 1946 and 1854 events seemed to be the same size, and from comparing the first two entries of the last column of Table 5.1 that the slip in 1707 was greater in proportion to its greater length, we can use a constant slip–length ratio rule to estimate the slips throughout Ando's history. The result for sector AB is shown in Figure 5.14b. This shows that the time-predictable model works back to 1605 but does not agree at all well with the data for earlier times, nor do the other models. This does not seem to be a result of our crude way of estimating slips; the recurrence times prior to 1605 are about double that afterwards. The most likely explanation is that earthquakes are missing from the earlier record. Japan was in a state of civil war for about 600 years during this period, so it is quite likely that the earlier historical record is more fragmentary.

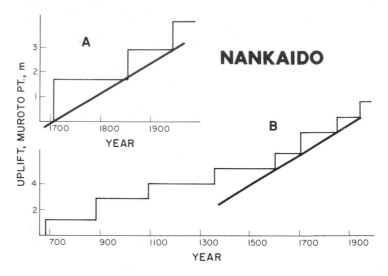

Fig. 5.14 History of uplift at Muroto Point in response to Nankaido earthquakes, compared with the time-predictable model: (a) data since 1707, when measurements are available; (b) an extrapolation further back in time, using Ando's history (Fig. 5.15) and a scaling law to determine uplift. Bold line is interpretation with the time-predictable model.

The Nankaido earthquakes after 1605 are consistent with the time-predictable model; earlier data are incomplete and cannot test the models. This example does demonstrate two germane facts, however: On any one stretch of fault, subsequent large earthquakes do not necessarily have the same slip or length.

Parkfield, California   The village of Parkfield gives its name to a short segment, 20–30 km in length, of the San Andreas fault in central California (Fig. 5.5). This segment lies between the continuously slipping creeping section of the fault to the north and the northern end of the rupture zone of the great earthquake of 1857, which ruptured about 350 km of the fault with an average slip of about 3.5 m (see Fig. 5.23). The Parkfield segment ruptures in earthquakes that, being rather short, have small slips and consequently short recurrence times. This segment is known to have ruptured six times in the past 130 years (Fig. 5.16). The last three earthquakes – those of 1966, 1934, and 1922 – were well-recorded instrumentally; although the 1902 earthquake was recorded also, the location and size of it and of the earlier ones were estimated from felt reports and observations of ground breakage. The 1966 earthquake was described in Section 4.4.1.

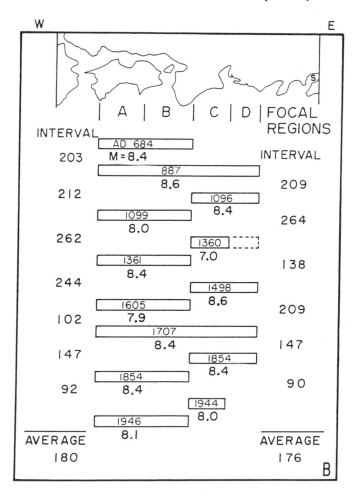

Fig. 5.15 History of earthquakes along the Nankai trough. The region is divided into four rupture zones (A–D). The boxes indicate rupture lengths and dates of earthquakes, and the intervals between earthquakes are also indicated. [From Yonekura (1975), based on Ando (1975) and other sources.]

At first glance the series looks periodic, with a recurrence time of 22 yr. There is a serious discrepancy, however, with the 1934 event occurring 10 yr early, according to that scheme. Bakun and McEvilly (1984) reexamined the seismograms for the last three events and found them to be nearly identical and with the same moment (within 25%). The 1934 and 1966 events were evidently in the same place, had about the same ground breakage and even were each preceded 17 min by $M_L = 5.1$

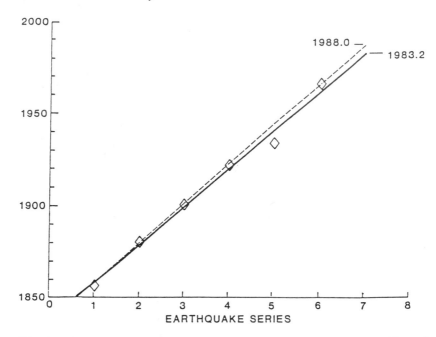

Fig. 5.16 History of Parkfield earthquakes. Dashed line is a linear fit without the 1934 earthquake, solid line includes it. (From Bakun and McEvilly, 1984.)

foreshocks. The events before 1922 are not so well determined in either size or location.

Parkfield is in a stability transition region that presently consists of an irregularly shaped locked patch that protrudes into a zone that stably slides at velocities that increase to the north (Fig. 5.17). This complicates things because the recurrence models refer to total moment or energy storage and release, whereas the coseismic slip of the Parkfield earthquakes accounts for only about half the moment release in that segment (Harris and Segall, 1987). If there is any variation in the relative fraction of aseismic to coseismic slip, it will alter the recurrence times. The partition between the two is not so well determined; even with an abundance of geodetic data since 1966, Segall and Harris (1987) could estimate the strain energy recharge time only to within the broad limits of 14–25 yr.

Bakun and McEvilly (1984) considered various possible explanations for the 1922 to 1966 earthquake series, shown in Figure 5.18. In (a) a constant rate of aseismic slip is assumed in the interseismic period (this agrees with neither recurrence model); in (b) creep is assumed nil from 1922–34, producing agreement with the time-predictable model; in (c) slip was assumed smaller in 1922 because of an assumed larger fault

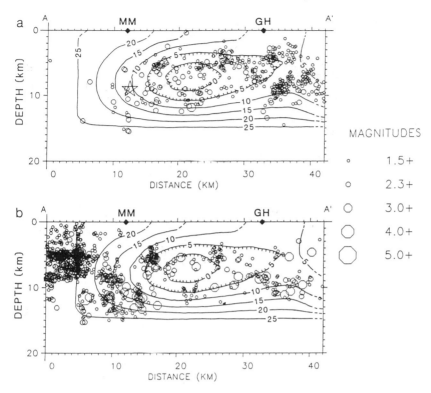

Fig. 5.17 Plan view of the fault at Parkfield showing the interseismic slip distribution compared with (a) aftershocks of the 1966 earthquakes (mainshock, star); (b) seismicity between 1969 and 1985. Hatchures indicate locked region; MM is Middle Mountain; GH is Gold Hill, just north of the jog shown in Figure 3.33. (From Harris and Segall, 1987.) Contours in mm/yr.

area; in (d) the strength or loading rate is assumed variable. There was, however, a significant difference in the initial conditions for the 1934 event than for those in 1922 or 1966 – it was preceded by 55 hr by an $M = 5.0$ foreshock 3 km to the northwest and this may have triggered the 1934 mainshock (Bakun and Lindh, 1985). The correct answer remains equivocal. However, by any reckoning the next earthquake in the series is imminent. The region has consequently been instrumented heavily so, shortly, we should have considerably more information to bring to bear on this problem.

This example brings up another problem, the contribution of aseismic slip in relieving accumulated strain. In general this is not known, but may range from nil in the case of "locked" fault segments where no creep is observed, to nearly 100% as in the creeping segment of the San

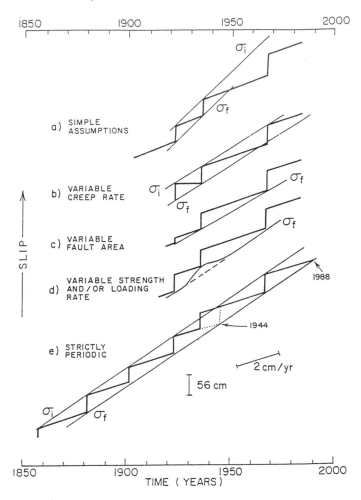

Fig. 5.18 Several different attempts to explain the aperiodicity of the three most recent Parkfield earthquakes. (From Bakun and McEvilly, 1984.)

Andreas fault. This creates obvious problems in earthquake recurrence estimation. For example, in another part of the San Andreas system where creep is known to occur, the Imperial fault, several earthquake cycles indicate that seismic moment release falls short of the total expected from geodetic and geologic estimates by a factor of 2 to 3 (Anderson and Bodin, 1987). This problem is particularly common in subduction zones. Sykes and Quittmeyer (1981) found that earthquake recurrence in subduction zones was consistent with the time-predictable model, but only if the ratio of seismic to total slip was assumed to be

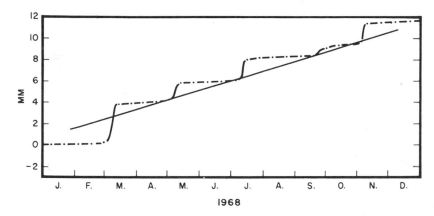

Fig. 5.19 A creep record from the San Andreas fault south of Hollister, California, during 1968. (Data courtesy of R. D. Nason.)

in the range 30–90%. This topic will be further explored in Sections 6.2 and 6.4.

Evidence from aseismic slip   In central California there is a so-called creeping segment of the fault that slips stably. This region and its mechanics are described in detail in Sections 6.4.1 and 6.4.3. Fault slip in this region often occurs in a self-sustaining oscillatory mode known as episodic creep. Study of it is relevant to the present subject in indicating whether or not slip is initiated at a critical frictional stress.

An example of a creep record from this region is shown in Figure 5.19. It shows behavior that appears periodic or in accordance with the time-predictable model. A time-predictable model was used to predict successfully such a creep episode late in the afterslip of the 1966 Parkfield earthquake (Scholz, Wyss, and Smith, 1969). All creep records are not as regular as that shown in Figure 5.19, however. Because the measurements are made with short-baseline instruments stretched over the fault, they often may reflect near-surface behavior that may be sensitive to rainfall or other local perturbations.

A more significant observation was made at the site where fault creep was first discovered, the Cienega Winery (CW). Fault creep has been occurring at a steady rate of 11 mm/yr at that site at least since 1948. In 1961 a moderate earthquake occurred nearby and produced a surface offset at CW. The creep record for this period is shown in Figure 5.20. Creep ceased following the earthquake and did not resume until the strain had reaccumulated sufficiently to restore the stress to its previous level, as shown by the merging of the creep record into its previous rate. This clearly indicates a critical friction level for this section of fault.

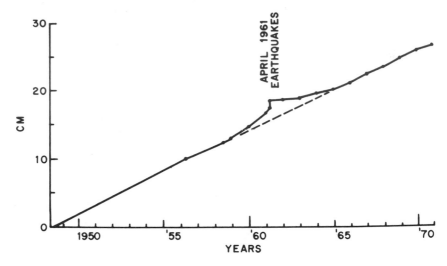

Fig. 5.20 Creep record for Cienega Winery, 1948–72, showing the effects of the 1961 earthquake. For a fuller discussion of this and related effects of coupling between earthquakes and creep, see Burford (1988). (Data courtesy of R. D. Nason.)

Notice also the preseismic acceleration of creep: This pattern is predicted by nucleation models, discussed in Section 7.3.1.

5.3.2   *Geological observations of recurrence times*

In view of the limited information that can be gleaned from instrumental and historical data, the geological record provides the most abundant source of information on recurrence of past earthquakes. Geological studies of active faults can provide rates of slip averaged over the Holocene and also the approximate dates and amounts of slip in prehistoric earthquakes. From this data average recurrence intervals can be obtained.

San Andreas fault   An example of this type of study is that done at Wallace Creek in central California by Sieh and Jahns (1984) (WC, Fig. 5.5). Wallace Creek, an ephemeral stream, flows southwest from the Temblor Range and crosses the San Andreas fault, where it has been repeatedly offset. A photograph of the site is shown in Figure 5.21. There one can see the modern channel (offset 120 m), an older abandoned channel (offset 380 m), and a yet older one indicated by the white arrow (offset 475 m). Also seen are five gullies (A–E) that have been offset smaller amounts by recent earthquakes.

Fig. 5.21 Photograph of the San Andreas fault at Wallace Creek, showing the various offset features. View is from the southwest. (Photo by R. E. Wallace, courtesy of the U.S. Geological Survey.)

The geological history of this site, as reconstructed by Sieh and Jahns, is shown in Figure 5.22. In the earliest time recorded (a) an aggrading alluvial fan progressively buried small scarps along the fault. (b) Presently recognizable offsets were developed after 13,250 yr b.p., when Wallace Creek began entrenching and younger fans began developing on gullies downstream from the fault. (c) Wallace Creek began to be offset and its first channel was abandoned about 10,000 b.p. (d) After an additional 250 m of offset, the second channel was abandoned 3,700 b.p., when (e) the formation of the present channel was begun, which is now (f) offset 130 m.

From this they deduced a long-term slip rate for the fault of 32 ± 3 mm/yr, which, as mentioned before, agrees with the rate determined geodetically to the north. This site was last ruptured in the earthquake of

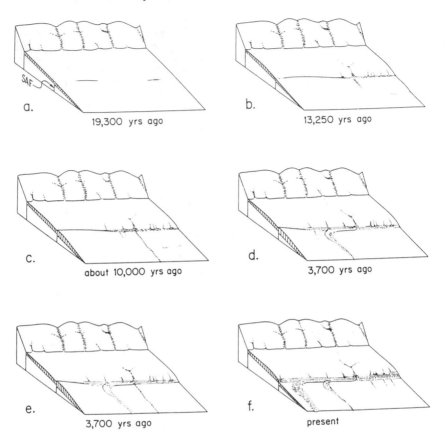

Fig. 5.22 Geological history of Wallace Creek, according to Sieh and Jahns (1984). See text for details.

1857, when it slipped 9.5 m. From offsets of the gullies they were able to recognize four earlier earthquakes at the site with similar slips, although they did not determine their dates. By dividing the slips by the long-term rate, they deduced recurrence intervals of 240–450 yr for these events.

Sieh (1984) made a similar study at Pallet Creek (PC, Fig. 5.5), which is farther south on the section of the fault ruptured in 1857. There he recognized 11 earlier earthquakes, each of which had slips comparable to the 2 m in 1857. Recurrence intervals averaged between 145–200 yr, but with variations between about 50 and 250 yr. Compared to Wallace Creek, earthquakes are more frequent at Pallet Creek, but their individual slips are less. The long-term rate at Pallet Creek was found to be only 9 mm/yr, which disagrees with Wallace Creek, with other sites to the south, and with the geodetic data in the nearby Palmdale net (Fig. 5.4c). This difference is explained by distributed slip at the site, not measured within the cross-fault excavations.

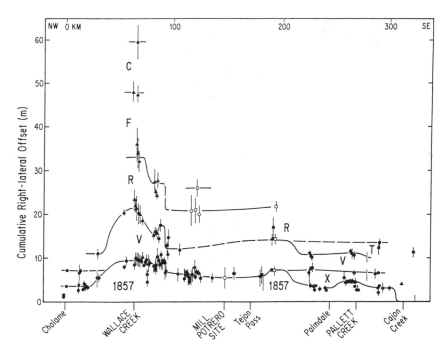

Fig. 5.23 Slip distribution in the 1857 Fort Tejon, California, earthquake and several prehistoric earthquakes in the same place. (From Sieh and Jahns, 1984.)

A number of earlier offsets from these and other sites can be correlated roughly and interpreted as having occurred in individual prehistoric earthquakes, as shown in Figure 5.23. It appears that the immediate predecessor to the 1857 earthquake also slipped at both Wallace Creek and Pallet Creek. What is noticeable from this figure, and in general from the data gathered at these two sites, is that the slip in a given earthquake seems to be a characteristic of the site, an observation to which some significance will be attached later.

Wasatch fault   This fault is the easternmost and one of the most prominent normal faults in the Great Basin structural province. It is an intraplate fault, bounding the Wasatch Mountains just to the east of Salt Lake City, Utah (Figure 5.24). It has a total length of 370 km, composed of a number of distinct segments. Trenches were excavated across this fault zone at a number of sites to obtain paleoseismic data, reported by Swan, Schwartz, and Cluff (1980) and by Schwartz and Coppersmith (1984).

The Holocene vertical slip rate of the fault was found to be about 1.3 mm/yr along the central segments and to decrease to very small values at the ends. This is expected due to the bounded nature of the fault, as in

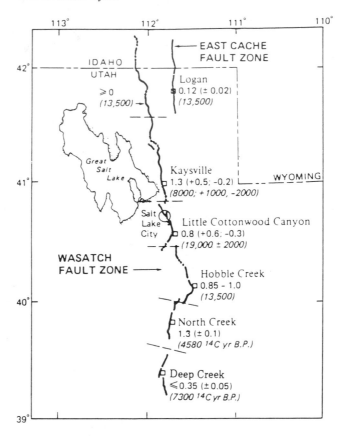

Fig. 5.24 Map of the Wasatch and East Cache fault zones, Utah. For each segment the slip rate is given in mm/yr and the age of displaced datum (in italics) in years before present. (From Schwartz, Hanson, and Swan, 1983.)

the example shown in Figure 3.10. Individual earthquake slips had values that clustered around 2 m, and their average recurrence interval was about 2,000 yr. Earthquakes at different sites did not correlate in time, which led Schwartz and Coppersmith to propose that the fault is broken into segments that rupture independently, as shown in Figure 5.24. This behavior, in which the fault is composed of segments that rupture independently and that have slips characteristic of that segment, they called the characteristic earthquake model. Whether this is a general behavior of faults is debatable and will be dealt with in the next section.

Subduction zones   These are not amenable to direct examination, but as mentioned in the previous section, coseismic movements often produce permanent uplifts that result in flights of uplifted marine terraces (Fig. 5.25). Uplifted shorelines were produced by the 1923

Fig. 5.25 Uplifted beach terraces, Turakirae Head, ESE of Wellington, New Zealand. A flight of five uplifted terraces may be seen here, the most recent of which was uplifted in the earthquake of 1855. The terrace of the last interglacial period may be seen in the distance. Turakirae is at the intersection of the Wairarapa fault with Cook Strait. The Wairarapa fault is largely dextral, with a vertical component that is recorded by these terraces and that has produced the uplift of the Rimutaka range to its west. The 1855 earthquake was described by Lyell (1868) from accounts of eyewitnesses who traveled to London. An uplift of about 2.5 m occurred at Muka-muka, near Turakirae, and the Wairarapa fault was ruptured to a distance of at least 90 km inland. An examination of offset features on the fault shows a dextral offset in 1855 of about 12 m over a distance of at least 128 km (Grapes and Wellman, 1989). Vertical motion was minor in the north and increased to a maximum near Turakirae. (Photograph by Lloyd Homer, New Zealand Geological Survey.)

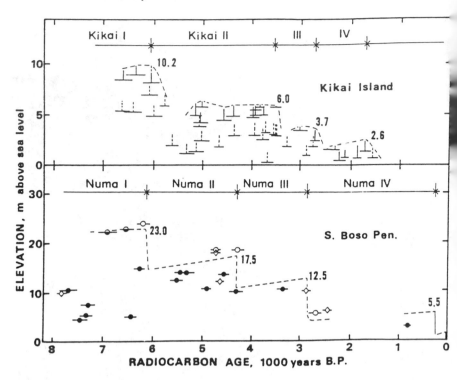

Fig. 5.26 Age and uplift of two flights of marine terraces in Japan: (a) Kinkai-jima, in the northern Ryukyu islands; (b) Boso Peninsula, SSE of Tokyo (same place as the leveling data, Fig. 5.9). (From Shimazaki and Nakata, 1980.)

Kanto earthquake, discussed earlier, and a study of uplifted terraces in the same region was made by Matsuda et al. (1978). On the Boso Peninsula (site of the leveling line, Fig. 5.9) several terraces older than 1923 are prominent. Matsuda et al. identified the next older one with the Genroko earthquake of 1703. The uplift distribution in 1703 was quite different than in 1923, indicating that the earlier earthquake was much larger and extended farther along the Sagami trough to the southeast. The elevation pattern of the older Numa terraces is different again, so that neither earthquake is typical (or "characteristic") of the overall deformation in the region.

The uplift and age of the Boso terraces is shown in Figure 5.26, together with that from Kikaijima, one of the northern Ryukyu islands. If each of these is produced by a single earthquake, they indicate individual uplifts of 7–8 m and recurrence intervals of 1,000–2,000 yr. The strain recharge time for the Kanto earthquake is estimated to be ~ 200 yr (Scholz and Kato, 1978) and historic records indicate that the recurrence time for Genroko-type earthquakes is about 800 yr (Matsuda et al., 1978). It seems likely that more earthquakes occur on this plate boundary than are indicated by individual terraces. One possibility is that coseismic uplifts interfere with eustatic rises and falls in sea level so that only a few of them get preserved. Another possibility is that slip on imbricate strands, such as indicated in the leveling data of Figure 5.9, accentuate just a few of the terraces.

Somewhat more enigmatic are the uplifted terraces on Kikaijima and nearby Kodakarajima in the northern Ryukyu Islands (Nakata, Takahashi, and Koba, 1978; Ota et al., 1978). No large interplate earthquakes are known historically for this region (e.g., McCann et al., 1979). The Ryukyu arc is extensional, at least in its southern portion, and so it is thought to be seismically decoupled (Ruff and Kanamori, 1983), a view that conflicts with the presence of these terraces (see also Taylor et al. (1987), and Sec. 6.3).

One of the best-studied flights of uplifted terraces is on Middleton Island, offshore of the region of the 1964 Alaskan earthquake (Plafker and Rubin, 1978). There, six terraces have been recognized and dated, which yields, if each is identified with an individual earthquake, recurrence times of 500–1,350 yr. However, since Middleton Island was a site of imbricate faulting in the 1964 earthquake, this may be another case, as at Boso Peninsula, where only those events that produce imbricate faulting are recorded by terrace uplift.

Empirical earthquake recurrence models   On the basis of their studies, Sieh (1981) and Schwartz and Coppersmith (1984) proposed several models describing the patterns in which slip may reoccur on faults. These are shown in Figure 5.27. In the *variable slip model*, both the amount of slip in a given place and the length of rupture may vary from earthquake to earthquake, but the net long-term slip is uniform along the fault (or variable, as the case may be). In the *uniform slip model*, the latter two conditions hold, but the slip at a given point is the same in each earthquake. Finally, in the *characteristic earthquake model*, the fault ruptures in a series of characteristic earthquakes, each identical, with the same slip distribution and length. In this last model, if the slip in a given earthquake is observed to vary along the fault, then the long-term slip rate must also vary accordingly. These latter two parts of Figure 5.27 are drawn to be reminiscent of the slip in the 1857 earthquake, and may be compared with Figure 5.23.

The studies of both the San Andreas and Wasatch faults indicated that slip in individual events was, approximately, characteristic of a site, so Sieh and Schwartz and Coppersmith concerned themselves primarily with debating the last two of these models. Sieh favored the uniform slip model, because on a long continuous fault, such as the San Andreas, the variations of slip rate demanded by the characteristic earthquake model

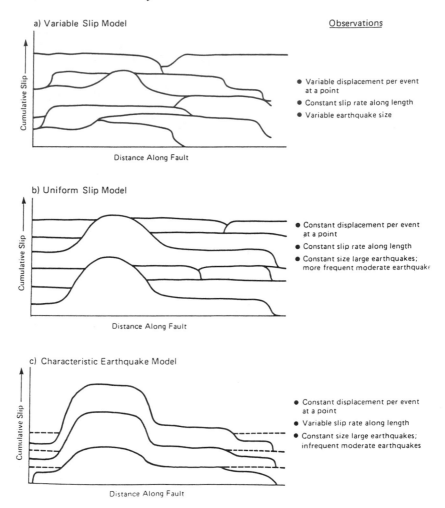

Fig. 5.27 Diagrammatic representation of three hypothetical models of earthquake recurrence. They are drawn to be reminiscent of the slip distribution of the 1857 Fort Tejon earthquake, shown in Figure 5.23. (From Schwartz and Coppersmith, 1984.)

would produce very large along-strike deformation, which is not observed. He also noted that the smaller recurrence times at Pallet Creek than at Wallace Creek indicated that some earthquakes ruptured the former place but did not reach the latter.

Schwartz and Coppersmith took the opposite view. They argued that the smaller slip rate at Pallet Creek agreed with the slip rate variation predicted by the characteristic earthquake model. They were also swayed by the observed earthquake size distribution of the type shown in Figure 4.15a, which they claimed was a consequence of the characteristic earthquake model. However, as discussed in Section 4.3.2, this distribution arises from the lack of self-similarity between large and small earthquakes and has nothing in particular to do with the characteristic earthquake model. Indeed, if a long enough fault is chosen, a power-law size distribution is observed, as shown in Figure 4.15b.

There are observations in a few other cases that earthquakes on the same fault reach can have different lengths. The Nankaido case, 1707 versus 1846 or 1946, has been discussed before, as have the Imperial Valley, 1979 versus 1940, and Borah Peak, 1983 versus 6,000 yr b.p. (Secs. 4.4.1 and 4.5.1). It is interesting to note that although the 1979 Imperial Valley earthquake ruptured only the northern half of what had broken in 1940, the part that broke in both earthquakes slipped about the same amount both times, in accordance with the uniform slip model. On the other hand, the Nankaido case clearly demonstrates that slip in a given place may vary significantly between earthquakes, suggesting the variable slip model. Another case of that type is an earthquake series off the Colombia–Equador coast (Kanamori and McNally, 1982). The subduction zone interface there ruptured in a 500-km-long earthquake in 1906, and the same section was subsequently ruptured in its entirety by three shorter earthquakes in 1942, 1958, and 1979. The sum of the moments of the three shorter earthquakes was less than that of the longer one by a factor of five, indicating that the slip in each of the shorter events was much smaller than in the longer one in the same place, just as observed at Nankaido (see Table 5.1).

In summary, models like the characteristic earthquake model and uniform slip model represent end-member cases, with one perhaps applying to some situations better than to others. Neither can be expected to apply to all cases or times. For a fault of finite length like the Wasatch, where the slip rate varies strongly along strike and which is strongly segmented by discontinuities, the characteristic earthquake model probably works better than the others, although it cannot hold indefinitely or else infinite strains accumulate in the segment boundaries. For a plate boundary like the San Andreas, however, the requirement of uniform slip rate is probably stronger. The fault is on the whole more continuous, which probably will lead to one of the other models being better. As noted in Section 3.2.2, the geometric segmentation of faults appears to depend on net slip, and this also may result in a spectrum in the dynamic rupture segmentation characteristics.

Left unstated here are the physical laws governing recurrence. Because of its importance to the overall problem, we will return to this point with a more theoretical discussion in Section 5.4.

5.3.3   *Recurrence estimation with insufficient data*

If one has, as at Parkfield, a long history of prior earthquake occurrences or a geodetic record of the strain accumulated since the last one, one can with some degree of confidence predict the approximate time of the next earthquake, either by the time-predictable model or by induction. However, such is seldom the case. The more common situation is when the earthquake history encompasses only one or a fraction of a prior cycle and when the strain accumulation must be estimated from, say, plate tectonic rates. In this case one can usually make only qualitative statements concerning the most likely places in which large earthquakes are to be expected in the future, with little or no specification as to their occurrence time. Nonetheless, such statements can be quite useful.

A tenet of plate tectonics is that the rates of plate motion must be steady over geologic time periods and must be continuous along plate boundaries. If it is assumed further that a significant portion of this motion must be released seismically, it then follows that segments of plate boundaries that have not ruptured for the longest time are those most likely to rupture in the near future. These places are called *seismic gaps*. As an example, consider the Aleutian Arc, first studied systematically by Sykes (1971). This region is shown in Figure 5.28, which shows rupture zones for past large earthquakes, determined by aftershock distributions, and gives the temporal history of those and preinstrumental earthquakes. Notice that through time the rupture zones tend to fill in the entire plate boundary, with little or no overlap. The first observation affirms the assumption made earlier, and the second demonstrates the efficiency with which a previously relaxed region stops ruptures, as noted in Section 4.5.1.

The seismic gaps can be recognized readily and are indicated in the figure. The figure shown is from Sykes et al. (1981); since the original figure was made by Sykes (1971), several of the gaps have been filled in by large earthquakes: the Sitka gap and part of the Yakataga gap. During the earlier study, the far western Aleutians, where the arc is almost parallel to the slip vector and becomes at least partly transcurrent in its nature, was not known to have been ruptured by a large earthquake in historic times. It was therefore impossible to determine whether it had the potential to produce a large earthquake, because it may slip aseismically. Subsequently, it was found to have been the site of several large earthquakes in the nineteenth century and its status could be clarified. This example demonstrates that one needs either positive evidence for a previous large earthquake or negative evidence for fault creep before identifying a given fault segment as a seismic gap.

A map of the Circumpacific region showing seismic gaps on the main plate boundaries as of 1989 is given in Figure 5.29. In this map different

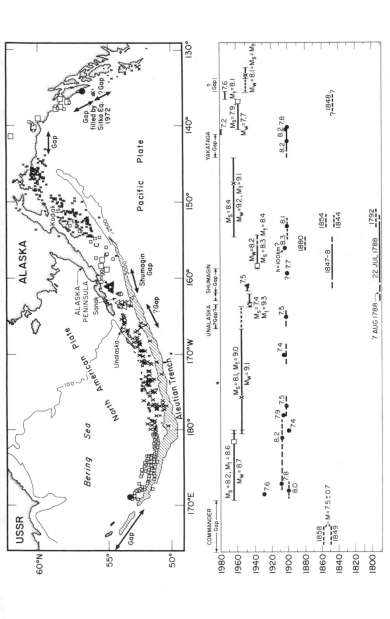

Fig. 5.28 Earthquake history of the Aleutian arc. In the map at the top, aftershocks of the most recent series of large earthquakes are shown with different symbols. Notice the abutting feature of their rupture zones so defined. At the bottom is a space–time diagram in which the rupture lengths and times of rupture are indicated for as far into the past as historical records permit. (From Sykes et al., 1981.)

symbols are used to designate sections of plate boundaries with different "seismic potential." This term is used to classify gaps according to the imminency of the earthquakes expected to fill them. In general, plate boundary segments that are known to be seismic but have not ruptured for the longest time have the highest seismic potential. Regions for which there is evidence for aseismic slip have low seismic potential. The meaning of this term should not be confused with the implications of the slip-predictable model: One may presume instead that the size of the future earthquake will be dictated by the size of gap that it fills. The more information one has regarding the slip, dimensions, and times of prior earthquakes and the long-term slip or loading rates, the more definite one can be in making long-term predictions of this type. Sykes and Nishenko (1984), for example, estimated the size, expected recurrence times, and probabilities for major earthquakes along the San Andreas fault in California by an extension of this method with the use of additional data.

The application of this approach to intraplate regions is not so clear, because the stationarity of rates in space and time that is its essential assumption may not hold. In the Great Basin, for example, Wallace (1981, 1987) has shown that a group of faults in one region may be active for a few thousand years, and then the activity shifts to another region (Fig. 5.30). He notes that the activity in central Nevada during the current century is anomalously high and may not be typical of the long-term deformation rates. Thus, the continuity in deformation required for plate boundaries is not so clearly required, at least on the short term. Seismic gaps identified in such regions, such as the one discussed by Savage and Lisowski (1984), lack the definition and sense of immediacy of those on plate boundaries. Ambraseys (1970) has made the same observations about earthquake activity in the Middle East: Activity there is observed to migrate from one block boundary to another, concentrating in one zone for several centuries with long quiescent periods in between.

5.3.4 *Seismicity changes during the loading cycle*

So far in this section we have concentrated our discussion on the principal earthquake that relaxes the loading cycle. Within the loading region of the principal shock there are likely to be many smaller faults that also produce earthquakes. Secondary seismic activity on these faults might be expected to rise and fall with the loading cycle. This pattern was noticed first by Fedotov (1965) for earthquakes in the Kamchatka–Kurile region. The average recurrence interval in that region is about 140 yr, and for the first 40–60% of that interval, following decay of the aftershocks, the rupture zone, and its immediate surroundings, was

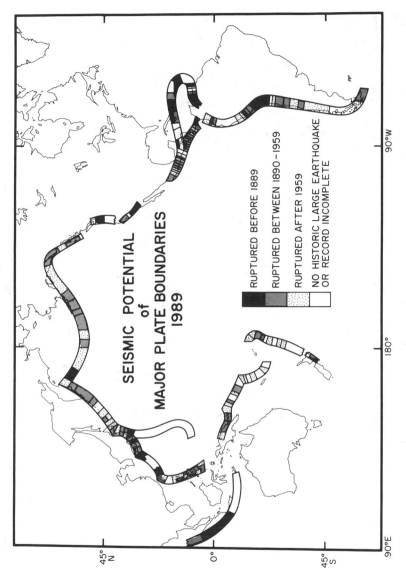

Fig. 5.29 Seismic gaps and seismic potential for the Circumpacific region. (Data courtesy of S. Nishenko, U. S. Geological Survey).

263

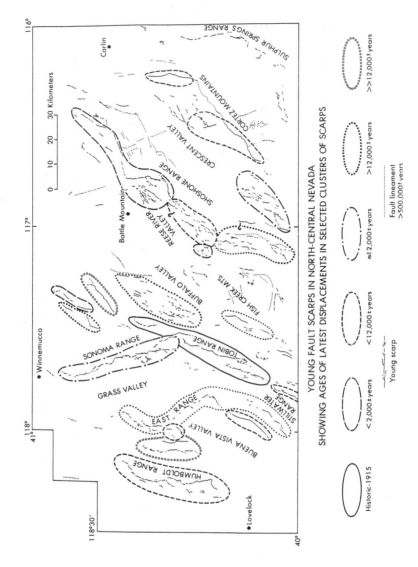

Fig. 5.30 Young fault scarps in north-central Nevada showing ages of latest displacements in selected clusters of scarps. (From Wallace, 1981.)

marked by a striking quiescence. Activity was then observed to build up, to a more or less steady "background" rate, during the later part of the cycle.

Mogi (1979) further documented this cycle in the seismicity for earthquakes in Japan and elsewhere. He also pointed out a number of cases where, particularly near the end of the cycle, the activity concentrated around the periphery of the rupture zone, defining a quiescent region in the center and forming what he called a doughnut pattern. In the part of southwest Japan that fronts the Nankai trough, for example, intraplate seismicity in inland regions appears to increase in activity in the 50 yr prior to the principal underthrusting earthquakes on the plate boundary (Shimazaki, 1976; Seno, 1979). This activity is particularly concentrated behind the lateral edges of the rupture zones (Mogi, 1981). Mogi suggested that this increase in activity reflects the overall increase in regional stress as the loading cycle progresses. From this point of view, the doughnut pattern arises because of the stress concentration around the rupture periphery. In many subduction zones there is no clear increase in intraplate seismicity in the arc, so that the doughnut pattern, if it should be called that, is manifested only by increased seismicity near the ends of the rupture zone in the last few decades prior to the principal rupture (Perez, 1983).

Seismicity changes related to the loading cycle may also occur on the subducted plate. As in the case mentioned of the Rat Island earthquake (Sec. 4.4.2), normal faulting earthquakes are sometimes observed in the outer rise just after large underthrust events in the adjacent trench. However, compressional (thrust) earthquakes are sometimes observed in the same location late in the loading cycle, suggesting a very dramatic change in the stress regime there (Christensen and Ruff, 1983). This change, however, has not been observed over the entire loading cycle for a given arc segment, so we do not have a direct test that such a marked change in state of stress occurs. It has also been noted, in studies of the Mexican subduction zone, that large normal faulting earthquakes often occur there in the subducted plate downdip of the plate interface several years before the interface-rupturing earthquake (McNally and Gonzalez-Ruis, 1986; Dmowska et al., 1988). Dmowska et al. developed a model to explain this phenomenon. In their model, the subducted plate is pulled by its negative buoyancy so as to increase downdip tension in the slab as the loading cycle progresses. This is then released when the interface slips in a large thrust event, and the opposite effect occurs on the outer rise.

An example of such behavior is given in Figure 5.31, for the region of the 1906 California earthquake. The region was typified by a high rate of activity in the 50 yr prior to 1906, and in particular, this period was marked by several large earthquakes on the Hayward fault on the eastern

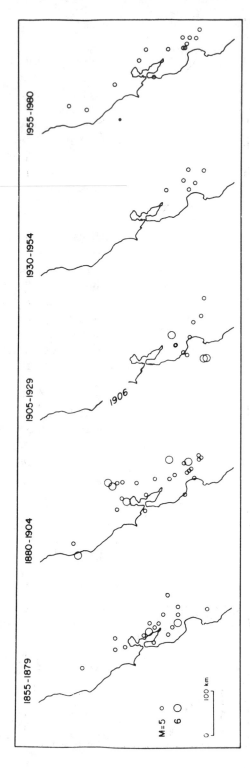

Fig. 5.31 Earthquake history of the San Francisco Bay region in 25-yr periods from 1855–1980. (From Ellsworth et al, 1981.)

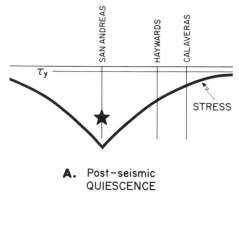

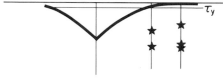

Fig. 5.32 A simple model to explain the pattern shown in Figure 5.31. The stress drop from the 1906 San Francisco earthquake reduces the stress on the nearby Hayward and Calaveras faults, and they become quiescent. Later in the cycle, as the stress is restored, they become active again, but the San Andreas is still quiet. (From Scholz, 1988.)

side of San Francisco Bay (Fig. 5.5) (Tocher, 1959). As can be seen in Figure 5.31, 1906 was followed by a 50-yr period of almost total quiet over the entire region. Starting with an earthquake on the San Francisco Peninsula in 1957 a new period of activity has begun. This has been marked by a renewal of activity on the East Bay faults, notably by the 1979 Coyote Lake ($M = 5.7$) and 1984 Morgan Hill ($M = 6.1$) earthquakes on the Hayward and Calaveras faults, and an earthquake on the Livermore fault ($M = 5.9$), but the San Andreas fault has continued to be inactive.

A simple model that can account for this behavior is shown in Figure 5.32. The stress drop in 1906 also reduced the stress on the adjacent Hayward and Calaveras faults, rendering them and the surrounding regions inactive. As the stress reaccumulates, it first reaches a critical level on the outer faults, so that they become reactivated before the San Andreas does. A similar conceptualization can be used for other situations, such as subduction zones.

EARTHQUAKE DATES

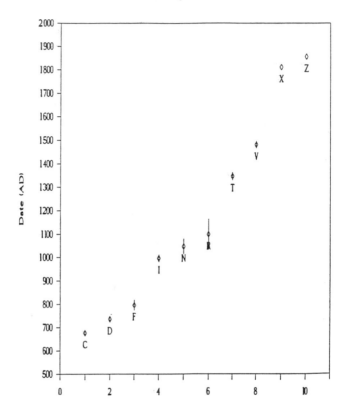

Fig. 5.33 Dates of the last 10 earthquakes to have ruptured the San Andreas fault at Pallet Creek, obtained from trenching studies. Error bars indicate uncertainties in the radiocarbon dates. (From Sieh, Stuiver, and Brillinger, 1989.)

5.3.5   *The question of earthquake periodicity*

In the foregoing, several specific cases have been pointed out in which earthquake recurrence departs markedly from perfect periodicity. Some mechanisms that may be responsible for deviations from periodicity have been discussed, and a more detailed analysis of possible physical reasons for aperiodicity follows in the next section. Here we sum up with a discussion of two extensive sets of data that bear on this issue, one geological and the other primarily historical.

In Figure 5.33 are shown the dates of the last 10 large earthquakes at Pallet Creek (Sieh, Stuiver, and Brillinger, 1989). This data set is unusual in that the radiocarbon dates have sufficiently small errors that periodic-

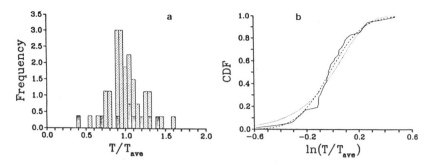

Fig. 5.34 Recurrence-time data from historic data in which more than three recurrences of earthquakes on a given plate boundary segment are known. (a) Recurrence time T normalized to the average recurrency time T_{ave} for that segment. (b) Fits of various probability distribution functions to the data: dashed, lognormal, and dotted, Weibull functions, respectively. (From Nishenko and Buland, 1987.)

ity may be discussed. It can be seen that although the mean recurrence time can be well determined at 131 ± 1 yr, the sequence is aperiodic: Deviations from this mean are significant. The earthquakes appear to occur in four clusters in which two or three earthquakes occur at short intervals followed by much longer quiet periods. It is interesting that a single earthquake at Wallace Creek, where the slip in individual events is typically larger than at Pallet Creek, corresponds in time to each of these clusters. This suggests that each cluster at Pallet Creek may be produced by overlapping earthquakes of different size and lateral extent.

Nishenko and Buland (1987) have collected data from a number of plate boundaries where three or more earthquakes have ruptured the same zone. They normalized each recurrence time T to the mean for that region, T_{ave}, and displayed the result as a histogram of T/T_{ave} (Fig. 5.34a). If this behavior were perfectly periodic, the histogram would be a spike at T_{ave}. This is not what is seen, but mean recurrence time is well defined and the standard deviation is a function of that mean, so the behavior is quasiperiodic. Nishenko and Buland showed that this data was well described with a lognormal distribution (Fig. 5.34b).

There are no errors in the time measurements for the historical data, so that the lack of periodicity shown in Figure 5.34 cannot be attributed to scatter in the data. Although earthquake recurrence is not periodic, a recurrence probability function can be well determined. In Section 7.4.3 we will show how this result can be used to analyze seismic hazard. We now return to a discussion of possible physical reasons for this lack of periodicity.

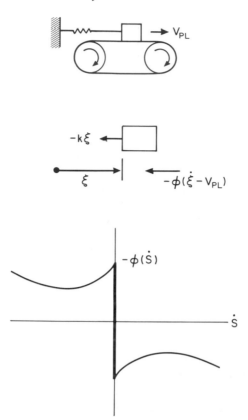

Fig. 5.35 A simple mass–spring–slider model of earthquake recurrence:
From the top, a schematic of the system – the mass rides on a conveyor
belt and is restrained by the spring; the forces in the system and a
schematic diagram of the friction force function $\Phi(\dot{s})$ that will produce
an instability.

5.4   EARTHQUAKE RECURRENCE MODELS

To examine the question of earthquake recurrence we need to look more
closely at the dynamics of the entire system. To do that we will consider
a simple model of the system to be that shown in Figure 5.35a, in which
a mass-slider restrained by a spring rides on a conveyor belt that moves
at a constant velocity v_{pl}. The velocity of slip between the mass and belt
is

$$\dot{s} = \dot{\xi} - v_{\text{pl}} \tag{5.2}$$

where ξ is the departure from the unstretched position of the spring.
Consider the friction force to be a function of slip velocity $\dot{s}$, velocity

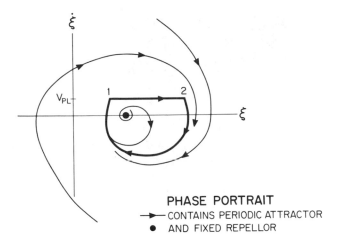

PHASE PORTRAIT
→ CONTAINS PERIODIC ATTRACTOR
● AND FIXED REPELLOR

Fig. 5.36 Phase portrait of the dynamic system. Thick orbit is a periodic attractor, closed circle a fixed repellor.

weakening at low velocities and velocity strengthening at high velocities, as discussed in Chapter 2. The friction force $-\Phi(\dot{\delta})$ is shown schematically in Figure 5.35. It rises to some critical value as the spring is extended at the rate v_{pl} and then slip begins, meeting at first a reduced resistance that leads to an instability if k is small enough, as discussed before. The equation of motion of the block is

$$m\ddot{\xi} + \Phi\left(\dot{\xi} - v_{pl}\right) + k\xi = 0 \qquad (5.3)$$

which is a reparameterized and more explicit form of Equation 2.33.

This system is a classical one in nonlinear dynamics. It was first introduced by Rayleigh (1877) to analyze the action of a bow on a stringed instrument, and its solution is often used as an example in standard texts on the subject (e.g., Stoker, 1950). The phase portrait of the system is shown in Figure 5.36. It contains a periodic attractor, shown as the dark curve, within which is a focal point repellor, so that any initial state on the phase plane either inside or outside the attractor will converge onto it, as shown by the light curves. Once on the attractor, it will cycle clockwise. From point 1 to point 2 is the loading portion, moving at v_{pl}. Point 2 is the critical friction where slip begins, and the loop from 2 back to 1, where locking reoccurs, depends only on the form of $\Phi(\dot{\delta})$ and the spring stiffness k. The stress drop, from state 2 to state 1, is constant. This is a typical example of a limit cycle, and it is strictly periodic. An analysis of this system using a single-state-variable friction law, of the form of Equation 2.27, has been given by Rice and Tse (1986).

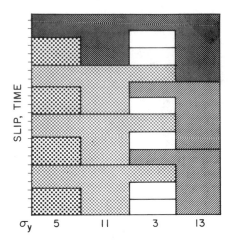

Fig. 5.37 Results of a complete cycle of a simple four-block coupled model.

The observations reviewed in the previous section do not seem to support this simple model. This should not be too surprising, however. This model system has only one degree of freedom and would be expected to apply only if the friction and loading rate is uniform over the fault. For an extended fault, this is usually not the case: We have already cited many reasons for expecting that fault friction is highly variable. In this case, the critical stress is reached only at one point in the quasistatic loading process; the rest of the fault is ruptured by dynamic triggering as described in Section 4.2.3.

We can make another simple model incorporating this latter feature. Consider a fault composed of segments in which each segment has a different value of σ_y. Let $\sigma_f = \gamma\sigma_y$, and say that a segment denoted by a single prime ruptures when $\sigma_1' = \sigma_y'$ or is triggered by rupture of an adjacent double-primed segment when

$$S' = \left(\frac{\sigma_y' - \sigma_1'}{\sigma_1'' - \sigma_f''} \right) < 1 \qquad (5.4)$$

where the critical S value, chosen to be 1, indicates a moderate triggering sensitivity. Assume, as an example, a four-segment fault with σ_y values 5, 11, 3, and 13, consecutively along the fault, and $\gamma = 0.8$. Start with $\sigma_1 = \sigma_f$ in all segments and load them at the same rate. Assume no overshoot, that is, $\sigma_2 = \sigma_f$.

The result is shown in Figure 5.37. Five different-sized earthquakes are produced, and a complete cycle of the model, as shown, from one initial state to the same initial state, is a great deal longer than the recurrence time of a single event. Notice that the stronger segments act as barriers most of the time, but when they rupture they act as asperities and trigger the adjoining segments. Now compare the results of this model with those of Figure 5.27. The long-term slip is the same in all segments, and the slip in each segment tends to have a characteristic value, but this is not the characteristic earthquake model because earthquakes of different length are produced. It is close to the uniform slip model, but not quite the same. Notice that the slip and recurrence time in the segment 5 alternates as 5, 4. So the slip in this segment is not constant or periodic. This segment behaves, to a certain extent, like the variable slip model, but with only slight departures from its characteristic slip value. Notice also that the segments 11 and 3 act somewhat like Wallace Creek and Pallet Creek, with one large slip event in the former segment corresponding to three smaller ones at the latter, one of these being simultaneous with the large event.

The geological data discussed in the previous section seem consistent with this model, but are not definitive one way or the other. Slip in each event is not determined better than about 20%. The radiocarbon dating method cannot establish synchronism in rupture, so that the measurement precision is not sufficient to test the model. However it is demonstrated that even a very simple heterogeneous fault model can produce a fairly rich behavior that on the whole is aperiodic and may be aperiodic even within a given homogeneous segment. Yet each segment can obey the dynamics of Equation 5.3, with mild coupling between them. A physical model of this sort has been explored experimentally and numerically by Burridge and Knopoff (1967) and by Cao and Aki (1984). A more thorough study of the stability of a two-block model was made by Nussbaum and Ruina (1987).

We will now decide if this model can explain the substantial variation in slip and recurrence time demonstrated for the Nankaido earthquakes. This does not seem possible because in all three cases listed in Table 5.1 the adjacent CD region also ruptured, so changes in triggering conditions did not occur. This model, in which a region can be triggered prematurely to produce a smaller slip event, should follow the predictions of the slip-predictable model. Yet the Nankaido events seem to obey the time-predictable model. In any case it does not seem likely that very great departures from periodicity can be produced by this mechanism, because that would require strong triggering [at much higher critical values of S (Eq. 5.4)]. The robustness of most seismic gaps in surviving throughout much of their cycle indicates that they are not so easily triggered.

If we consider that the Nankaido case demonstrates slip variability of a greater degree than may be explained by the above, we may then ask the question: Are either of the two proposed variable slip models, time-predictable and slip-predictable, consistent with the dynamic system of Figure 5.35? The simple answer, if these models are interpreted in terms of the loading histories as shown in Figure 5.13, is that they are not. This is because this dynamic system is characterized by two fixed critical stresses, at states 1 and 2, whereas these models require variation in one or the other of these. However, the data in Table 5.1 suggest another possibility. Notice that the ratio of the length of the 1707 event to the later ones is the same as that of the corresponding uplifts at Muroto Point and of the recurrence times. This indicates that the slip is proportional to fault length, which seems to be a general property of large earthquakes, as noted in Section 4.3.2. If we now interpret this with an *L*-model, discussed in that section, we infer that these earthquakes had variable slip but the same stress drop. This interpretation makes the time-predictable model consistent with the dynamic system with which we started. The variation of slip arises from the variations of fault length. This is the equivalent of changing spring constants for different relaxations (which rescales the abscissa of Fig. 5.36). Two fixed critical stresses are still present. The slip-predictable model, on the other hand, cannot be so rationalized to be compatible with this dynamic system, except to the limited extent allowed by dynamic triggering.

Also, we may consider the effect of variable loading rates, v_{pl}, due to the viscoelastic nature of the loading cycle, as discussed by Thatcher (1984b). The effect of nonlinearity in the loading cycle discussed by him can be incorporated in the dynamic model by using a soft (nonlinear) spring. This distorts the shape of the attractor but does not change its periodic nature. This effect therefore makes extrapolation of geodetic data to determine the recurrence time more difficult, but otherwise does not change the basic behavior. Irregular transients in crustal deformation may also prematurely trigger earthquakes (Thatcher, 1982), but we don't consider these a general mechanism for the aperiodicity that seems to occur.

A more drastic possibility exists. For nonlinear equations such as 5.3 it is generally impossible to prove that the solution obtained is unique and that it characterizes the phase space for all choices of parameters. In many cases chaotic behavior occurs over some range of parameters, and this is both aperiodic and unpredictable. For example, in the case of the van der Pol equation, which is similar to Equation 5.3 but with a different characteristic, it is known that forced periodic oscillations will produce chaotic behavior (Hubermann and Crutchfield, 1979). More to the point, Gu et al. (1984), in their study of the system of Figure 5.35 with a two-state-variable friction law, observed chaotic behavior near the critical stiffness (Sec. 2.3.3). This shows that, depending on the exact form of Φ, aperiodic behavior is possible over some part of the phase space for this system. Because we do not actually know the correct form of Φ as it applies to the real case, this leaves open the possibility that earthquake recurrence is truly unpredictable in the strict sense.

Chaotic behavior is characterized by a strange attractor in the phase plane, which unlike the periodic attractor of Figure 5.36, follows a different orbit each cycle. The orbits of strange attractors depart only gradually from one cycle to the next, so that if one knows the initial conditions one may be able to predict a single cycle reasonably well. However, the spatial heterogeneity of faults introduces many degrees of freedom, so that considerable uncertainty in the outcome should always be expected. Put a simpler way, if one has a good knowledge of the past rupturing history of a section of fault, one may be able to estimate the approximate time of the next rupture, but still not know how far, once initiated, that rupture will propagate.

A general dynamic model of spatially extended dissipative systems of this type has been developed by Bak, Tang, and Wiesenfeld (1988), which predicts some of the characteristics of earthquakes. Their model has the property that it evolves to a self-organized critical state that is characterized spatially by a power-law size distribution of clusters and temporally by $1/f$ noise. A simple visualization of this is of a pile of sand, in which the self-organized critical state is the maximum angle of repose, the size distribution of landslides is a power law, and the sand flow off the pile is $1/f$ noise. This is the only dynamic model that predicts, without a built-in assumption, the earthquake size distribution given in Equation 4.31. This model suggests that the whole schizosphere is in a self-organized critical state, which may be why, as pointed out in Section 6.5.4, it is everywhere near failure. It also predicts, as noted above, that the size of a given event cannot be predicted because it depends in a very sensitive way on the initial conditions.

The seismic cycle, whether it is represented by the simple system shown in Figure 5.35 or by a more complex system, must satisfy an energy balance, so even if aperiodic it must be characterized by a mean recurrence time, as the observations indicate. Because the strength of the fault, though spatially variable, must also be a well-defined quantity, it is also reasonable that the variance from the mean is a well-defined function of the mean, because the mean recurrence time is a measure of the stressing rate. These concepts were well-phrased by Gilbert (1909), who said:

There is a class of natural and artificial rhythms in which energy gradually passes into the potential form as internal stress and strain and is thus stored until a resistance of fixed amount is overcome, when a catastrophic discharge of energy takes place. The supply of energy being continuous and uniform, the discharges recur with regular intervals. ... If the stresses of the earthquake district affected only homogeneous rock and were always relieved by slipping on the same fault plane, the cycle of events would be regular; but with complexity of structure and multiplicity of alternative points of collapse, all superficial indication of rhythm is lost.

Although these conditions prevent us from making accurate long-term earthquake predictions, they do provide a firm basis for earthquake hazard analysis, which will be discussed in Section 7.4. Before we go on to that problem, however, we need to discuss the variable roles earthquakes play in different tectonic environments, which is the topic of the next chapter.

6 Seismotectonics

We now discuss the role of earthquakes in a variety of tectonic settings and, in particular, the relative role of seismic and aseismic faulting. The stability of faulting, which has been considered only for continental fault zones, is examined for oceanic faults and subduction zones. We also review what lessons may be learned from induced seismicity.

6.1 INTRODUCTION

In the foregoing discussion, the general principles that govern the mechanics of earthquakes and faulting were described and various illustrative examples were given. We have yet to place these phenomena within the perspective of the total tectonic process of which they form a part. Since Lyell's day it has been acknowledged that earthquakes play a role in active tectonics, but the amount of the total deformation produced by this mechanism, and the role of earthquakes in different tectonic environments is still a moot point. Seismicity often is used to deduce the tectonics of a region, so, logically, we may seek to know how much is revealed and how much remains hidden by seismology. The study of earthquakes as a tectonic component has become known as seismotectonics, which is the topic of this chapter.

It has long been known that earthquake activity is a symptom, if not the agent, of active tectonics. A global map of seismicity, as shown in Figure 6.1, reveals most of the actively deforming regions of the earth. The major plate boundaries can be traced out easily, as well as the diffuse bands of seismicity associated with the distributed deformation zones such as the Alpine–Himalayan belt and the Great Basin in the western United States. The existence of seismicity thus can be used to deduce the tectonic activity of a region, particularly in submarine regions and other areas that are otherwise inaccessible. It was the recognition of a continuous belt of seismicity, for example, together with a few profiles across the North Atlantic, that allowed Ewing and Heezen (1956) to predict the existence of a worldwide system of midocean rifts. Similar

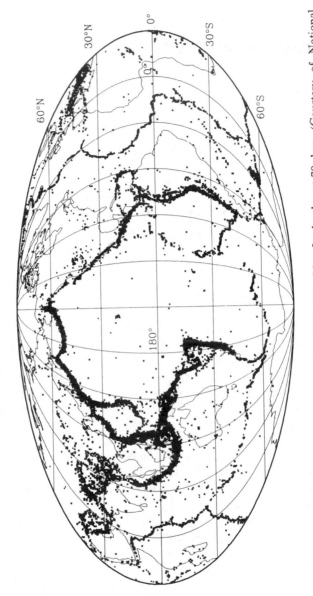

Fig. 6.1 Map of global shallow seismicity, 1963–88, $M > 5$, depth < 70 km. (Courtesy of National Earthquake Information Center, U.S. Geological Survey.)

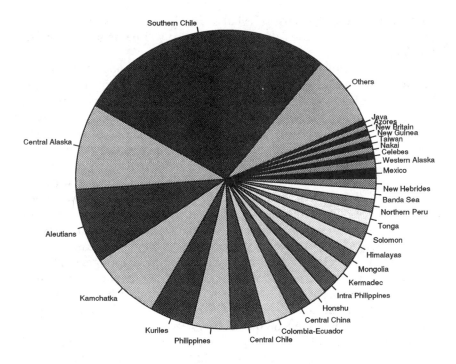

Seismic moment released (1904-1986)

Fig. 6.2 A pie diagram illustrating the distribution of global seismic moment release, 1904–86. (Courtesy of J. Pacheco.)

maps at smaller scale, constructed with data from regional and local seismic networks, allow the tectonics to be studied in much finer detail.

In determining the question as to what degree earthquakes are the agents rather than the symptoms of deformation, however, such seismicity maps may be misleading. Even when they distinguish, by different symbols, earthquakes of different magnitude, they do not give a very adequate indication of the different sizes of the earthquakes. This point is illustrated with the pie diagram in Figure 6.2, which gives the distribution of seismic moment release worldwide for the period 1904–86. It is seen that moment release, even among the major plate boundaries, is distributed very unevenly. About one fourth of the total moment was released in a single earthquake: the great Chile earthquake of 1960 that ruptured about 800 km of the subduction zone interface at the Peru–Chile trench. This is not surprising, given the size distribution as described by Equation 4.31. This power-law distribution must be convergent (i.e., sum to finite moment) so it follows that most of the moment must be contained in the few largest events. The same is true for the size distribution in any region; the moment release will be dominated by the few largest events.

Further examination of Figure 6.2 shows that about 85% of the moment release occurs at subduction zones. More than 95% is produced by shallow, plate-boundary earthquakes, the remainder being distributed between intraplate events and deep and intermediate focus earthquakes. Volcanic earthquakes also produce insignificant amounts of moment release. In the examination of the seismotectonics of any region, one needs to keep an awareness of the relative moment released in different earthquakes and of that with respect to the moment release expected from the total deformation of the region. The common use of magnitude often obscures these differences because it is a logarithmic measure of moment and because, unless moment magnitude is used, it saturates for the earthquakes that contribute the most to the moment release.

Earthquake focal mechanisms may of course be used to great effect in deducing the tectonics of a region. Although, as mentioned in Section 4.1, the double-couple radiation pattern for earthquakes had been known since the 1920s, it wasn't until the Worldwide Standardized Seismic Network was established in the early 1960s, which provided data of sufficient quality, that the technique came to fore. Two papers published in that era deserve special credit for establishing the basis for this type of analysis. The first is Sykes's (1967) use of focal mechanisms to prove Wilson's (1965) transform fault hypothesis. The second is the paper of Isacks, Oliver, and Sykes (1968), which used focal mechanisms to confirm the kinematics of plate tectonics and to deduce the mechanics of subducted slabs. That paper embodied the main techniques of qualitative seismotectonic analysis that we take up in the next section.

6.2 SEISMOTECTONIC ANALYSIS

6.2.1 *Qualitative analysis*

Earthquakes that result from sudden slip along a fault produce a quadrantal radiation pattern that can be determined by a variety of seismological techniques, and which is called a focal mechanism (or fault plane) solution (Fig. 6.3). The four lobes are divided by two planes, one of which has the orientation of the fault plane and the other, called the auxiliary plane, has the slip vector as its normal. This provides the geometrical orientation of the seismic moment tensor (Eq. 4.25). The principal axes of the moment tensor bisect the dilatational and compressional lobes and are called the P and T axes. These are also the principal

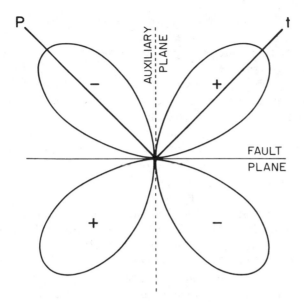

Fig. 6.3 The geometry of the double-couple earthquake fault plane solution. Compressional and dilational first motions of P waves are indicated by + and −. Fault slip is right-lateral in this example.

axes of the stress-drop tensor, hence their names refer to the direction of compression and tension, respectively.

Two types of ambiguities are present in the interpretation of focal mechanisms for tectonic analysis. The distinction between the fault plane and the auxiliary plane cannot be made from the focal mechanism itself but must be based on a judicious comparison with the local geology. This usually is sufficient to resolve the question, but one needs to keep the alternative in mind. For example, in the case of the strike-slip earthquakes in the South Iceland seismic zone (Figs. 3.7, 3.8, and 4.28), if one had not been aware of the north–south-striking surface ruptures, one might have selected the auxiliary planes in the focal mechanisms as indicating left-lateral motion along an east–west transform fault connecting the spreading ridge that extends up the Reykjanes Peninsula with the eastern rift that ends at Hekla.

The other uncertainty arises in attempting to relate the P and T axes with the greatest and least principal compressive stress directions. This would be correct only if the stress-drop tensor is the same as the stress tensor, which is unlikely ever to be the case, because the stress-drop tensor contains only shear stresses. The P and T axes are at 45° to the fault plane, which is not the correct direction for Coulomb failure. However, because the fault is a preexisting plane of weakness, the ambiguity is greater than that. The problem is very similar to that in estimating stress directions from the orientation of active faults, discussed in Section 3.1.1.

There are two ways of interpreting focal mechanism data. A kinematic interpretation may be made in which the slip planes and vectors yield a tectonic interpretation consistent with geologic and other data. In other situations a dynamic interpretation, in which the focal mechanisms indicate consistent regional stress directions, may be preferable. These two types of interpretation may not be consistent with one another. For example, at the intersection of a transform fault with a spreading ridge, the strike-slip focal mechanisms on the former are kinematically consistent with normal faulting on the latter, but the two types of focal mechanisms have quite different P and T axes.

The paper by Isacks et al. (1968) illustrates both methods of analysis. They interpreted shallow interplate earthquakes kinematically as determining the local relative plate-motion directions. In contrast, they interpreted focal mechanisms for intermediate- and deep-focus Wadati–Benioff zone earthquakes dynamically, suggesting that they indicate downdip compression or extension in the subducting slab. The most appropriate analysis is usually obvious in the context of the application, but the distinction between the two needs to be kept in mind.

6.2.2 *Quantitative analysis*

So far we have considered only geometrical information. If we include the magnitude of the seismic moments in the analysis, we can make a quantitative estimate of the contribution of seismic faulting to the deformation. Consider the simple case of a single fault of area A on which N earthquakes of known moment occur within time T. Then the seismic slip rate of the fault is

$$\dot{u} = \frac{1}{\mu A T} \sum_{k=1}^{N} M_0^{(k)} \tag{6.1}$$

(Brune, 1968). If, on the other hand, we consider some volume V in which many faults are active, we may calculate the strain rates in the volume according to the formula of Kostrov (1974),

$$\dot{\varepsilon}_{ij} = \frac{1}{2\mu V T} \sum_{k=1}^{N} M_{0ij}^{(k)} \tag{6.2}$$

where $\dot{\varepsilon}_{ij}$ are the rates of tensor (irrotational) strain components defined in the normal way,

$$\varepsilon_{ij} = \frac{1}{2}\left(\frac{\partial u_i}{\partial x_j} + \frac{\partial u_j}{\partial x_i}\right)$$

Molnar (1983), based on the assumption that one usually can distinguish the fault plane from the auxiliary plane, defined an asymmetric
moment tensor

$$M_{0ij}^* = \mu \overline{\Delta u_i} n_j \tag{6.3}$$

Then the rate of change of the total strain or deformation tensor,
$e_{ij} = \partial u_i / \partial x_j$, may be calculated according to the formula,

$$\dot{e}_{ij} = \frac{1}{\mu VT} \sum_{k=1}^{N} M_{0ij}^{*(k)} \tag{6.4}$$

Note that $e_{ij} = \varepsilon_{ij} + \omega_{ij}$ (ω_{ij} is the rotation tensor), so it includes both
strains and rotations.

However, because M_0^* includes rigid body rotation, which cannot be
determined from the seismic moment tensor, this has to be assumed
arbitrarily, which limits the applicability of this formula (Jackson and
McKenzie, 1988). The latter authors show that Equations 6.1 and 6.2 are
compatible, and give several instructive examples of their application.

In order to use these formulae to calculate deformation rates that may
be compared sensibly with long-term geologic rates, the time period T
for which the data are complete must be longer than the recurrence time
for the faults in the region. If not, the method may yield an underestima-
tion. The catalogue need not be complete for small events, because, as
remarked before, most of the moment is carried in the few largest
earthquakes. This introduces the possibility of error in either direction
because the largest events during a sample interval may be either larger
or smaller than that characteristic of a typical period of the same
duration.

Brune (1968) used Equation 6.1 to compare the seismic slip rate of the
Imperial and several other faults with that estimated from geodetic data.
Davies and Brune (1971) used it to compare slip rates on the major plate
boundaries with those predicted from sea-floor spreading data. They
obtained reasonable agreement, using about 50 yr of data, within what
was then the error in the determinations: about a factor of 2–3. In many
areas they did not have an independent estimate of the depth of faulting
and so could use that as an adjustable parameter. In some cases, such as
oceanic transform faults, they found that the seismic moment release rate
was considerably smaller than that inferred from sea-floor spreading
rates, if they assumed what seemed a reasonable depth of active faulting
(this problem is reanalyzed in Sec. 6.3.3).

There has not been a more recent global survey of this type, but it is
recognized now that some sections of plate boundaries are deficient in
their seismic moment release as compared with plate tectonic estimates.

This may be a permanent condition and is quantified with the seismic coupling coefficient χ, defined as

$$\chi = \frac{\dot{M}_0^s}{\dot{M}_0^g} \tag{6.5}$$

where $\dot{M}_0^s$ is the seismic moment release rate and $\dot{M}_0^g$ is the moment rate calculated from geologically measured fault slip rates or, in the case of plate boundaries, the slip rate calculated from plate motions. This is an important measure of the degree of instability of faults, and hence of their rheological properties, and will be discussed at some length in later sections.

Deformation rates in intraplate Japan were calculated by Wesnousky, Scholz, and Shimazaki (1982) with Equation 6.2, using 400 yr of seismic data. Although this is shorter than the recurrence intervals of individual faults there, they were able to show that this time period was long enough to provide a reasonably stable moment release rate for the region as a whole. They then calculated the deformation rates over the late Quaternary, by using the geologically measured slip rates of active faults, with the same formula. This yields an estimate of deformation due to all faulting, both seismic and aseismic, and over a much longer period. The two estimates were found to agree within the uncertainties, about a factor of 2. This agreement is reasonable, considering that fault creep has not been observed for any intraplate fault in Japan, and it also serves as an important test of the method. The results showed that northeast and central Honshu are shortening, in permanent deformation, at about 5 mm/a (millimeters per annum) in an east–west direction. Southwest Honshu is also experiencing east–west shortening, but at a rate that decreases to about 0.5 mm/a at the western end of the island. This is in response to east–west convergence along the Japan trench at about 97 mm/a, so the permanent deformation is about 5% of the convergence rate.

When estimates made from moment sums can be compared to rates from sea-floor spreading and other sources, a quantitative measure of seismic coupling can be made. Peterson and Seno (1984) showed, by this method, that variation in the seismic coupling coefficient occurs among subduction zones. In the study of Jackson and McKenzie (1988), it was found that northeast Iran, the North Anatolian fault zone, and the Aegean Sea were almost totally coupled seismically, but in the Zagros of southern Iran, the Caucasus, and the Hellenic Trench, as little as 10% of the deformation was being accommodated seismically. Mechanical bases for such variations in coupling will be discussed in Section 6.4. First we describe the various roles seismicity plays in different tectonic settings.

6.3   COMPARATIVE SEISMOTECTONICS

It is axiomatic that the role of earthquake activity in the tectonic process varies widely with tectonic environment. Here we attempt a brief review of these differences, paying particular attention to aspects that may be illustrative of some of the physical principles that have been discussed earlier. No attempt is made at completeness; indeed, a thorough treatment would require a treatise of its own.

6.3.1   *Transcurrent systems: California and New Zealand*

The Alpine fault system of New Zealand is a continental right-lateral strike-slip plate bounding system that has often been compared with the San Andreas system of California (Richter, 1958; Hatherton, 1968; Scholz, 1977; Yeats and Berryman, 1987). Whereas the San Andreas system connects two spreading ridges, in the Gulf of California to the south and the Gorda ridge to the north, the Alpine fault connects two subduction zones of opposite polarity, marked by the Hikurangi trench in the north and the Puysegur trench in the south. Both fault systems have similar total offsets and rates of motion. The net offset of the San Andreas fault is in the order of 400 km, which has occurred since the Pacific and North American plates came into contact at about 29 Ma (Hill and Diblee, 1953; Atwater, 1970). The Alpine fault has a net slip of 480 km, as shown by the offset of a Mesozoic subduction complex from Otago–Fiordland in the south to Nelson–Buller in the north (Wellman, 1955; see also Fig. 3.21). Although it was thought earlier that much of this displacement may have taken place in the late Mesozoic (Suggate, 1963; Grindley, 1975), it has been suggested more recently that most of it may have occurred since the middle Tertiary (Christoffel, 1971; Norris, Carter, and Turnbull, 1978; Cutten, 1979; Stock and Molnar, 1987). Current rates of motion, estimated from sea-floor spreading data, are 55–46 mm/yr for the San Andreas (Minster and Jordan, 1987; DeMets et al., 1987) and 46 mm/yr for the Alpine fault (Chase, 1978).

The southern part of the San Andreas system is compared with the Alpine system in Figure 6.4. There, the New Zealand map has been rotated 90° so that the slip directions of the two systems more closely agree in the figure. The fault systems run from upper left to lower right in the figures. In the upper left in both, the systems are composed of a single master fault. The plate slip vectors are oblique to these faults, in a compressive sense. In central California this obliquity is only about 6°, so that the convergence may be accommodated by minor structures, such as folds with axes subparallel to the fault (e.g., Mount and Suppe, 1987). In New Zealand, however, this obliquity is about 30°, resulting in major

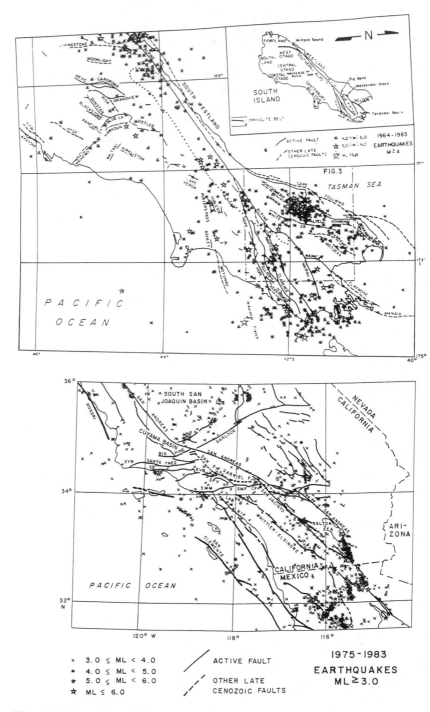

Fig. 6.4 Maps of active faults and seismicity for Southern California (bottom) and the South Island of New Zealand (top). The maps have been rotated so that the slip vectors of relative plate motion nearly are aligned in the two localities. (From Yeats and Berryman, 1987.)

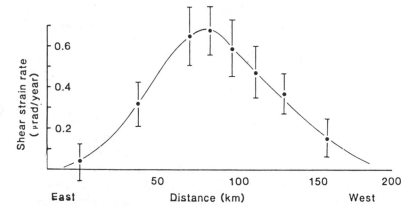

Fig. 6.5 Distribution of shear strain in a transect normal to the Alpine fault in the northern part of the South Island of New Zealand. (From Bibby, Haines, and Walcott, 1986.)

continental convergence and producing the uplift of the Southern Alps along a great escarpment at the Alpine fault (Scholz et al., 1973). Late Quaternary uplift rates along the Southern Alps may be as high as 10 mm/yr (Adams, 1980a).

In the central parts of the diagrams, the faults undergo a major compressive fault-bend jog, and a series of strike-slip faults, more nearly aligned with the plate slip vector, splay from them, so that the plate margin zone broadens to several hundred kilometers. Uplift, due to folding and reverse faulting in the Transverse Ranges, is associated with this bend in California. In New Zealand, where the most pronounced part of the bend is shorter, uplift occurs to the northeast in the Spenser Mountains, and there is a broad zone of reverse faulting in the Nelson–Buller region to the west. Several large earthquakes, notably that of Murchison in 1929 and Inangahua in 1968, have occurred on these reverse faults in historic times, but they have typical recurrence times of 10^4 yr, as compared to 400–500 yr more typical of the strike-slip faults. Although recurrence times for the reverse faults in the Transverse Ranges are not as well known, they are also probably much longer than on the strike-slip faults of the San Andreas system (Wesnousky, 1986).

The rate of shear-strain accumulation, determined from geodetic measurements across the northern fourth of the South Island is shown in Figure 6.5 (Bibby, Haines, and Walcott, 1986). These rates, when integrated, agree reasonably with the plate tectonic estimates, but do not show any obvious correlation with the faults, but rather a maximum in the central part of the island, tapering off in either direction. Holocene slip rates for the Marlborough faults are 5, 6, 6, and 20 mm/yr for the

Wairau, Awatere, Clarence, and Hope faults, respectively, which ac-counts for about 80% of the plate motion (Yeats and Berryman, 1987; Vanigen and Beanland, in press). Farther south, where the Alpine fault constitutes the entire plate margin, there may be a larger discrepancy between the fault slip rates and the plate motion estimates. In the southern part of the Alpine fault, Wellman and Wilson (1964) obtained a rate of 25 mm/yr for the Alpine fault, and Lensen (1975) reported a rate of only 10 mm/yr for the central part. This led Walcott (1978) to suggest that 25 to 75% of the plate motion takes place by penetrative aseismic deformation, perhaps within the Alpine schists. More recent estimates from the southern part of the Alpine fault, however, give rates of 27–41 mm/yr (Berryman et al., in press). The uncertainty in rates is mainly in the ages of the offset features, so further work may resolve this discrep-ancy.

As discussed briefly in Section 5.2.1, a similar discrepancy for the San Andreas system has been much studied. For the region shown in Figure 6.4, the San Andreas is slipping at about 25 mm/yr (Weldon and Sieh, 1985), and the San Jacinto at 10 mm/yr (Sharp, 1981). The other onshore faults contribute little to the deformation, so Weldon and Humphreys (1986) have suggested that about 20 mm/yr slip takes place on offshore faults.

During the 130 yr of complete historic record in New Zealand, the Alpine fault system has been relatively inactive compared with the San Andreas system. There have been only two large strike-slip earthquakes on the Alpine system: one of 1848, variously attributed to the Wairau or Awatere fault, and the earthquake of 1888 that ruptured about 50 km of the Hope fault. There have been no earthquakes on the Alpine fault proper during this time. Allis, Henley, and Carman (1979) have sug-gested that, as a result of rapid uplift in the Southern Alps, geotherms are unusually elevated on the Alpine fault, perhaps rendering it aseismic. However, no fault creep has been recognized on the Alpine fault, and Adams (1980b), Hull and Berryman (1986), and Berryman et al. (in press) have reported evidence for discrete slip events from prehistoric earthquakes. This would lead one to conclude that the Alpine fault constitutes a seismic gap of high potential (Nishenko and McCann, 1981), and that the different seismic expressions of the Alpine and San Andreas faults during historic times reflects their being in different stages of the seismic cycle.

6.3.2   *Subduction zone seismicity*

Much of the discussion so far has concerned crustal fault zones, al-though, as mentioned earlier, the bulk of the global seismic moment release occurs at subduction zones, which are much different in structure and seismic behavior. The major seismic moment release at subduction zones occurs at the frictional interface between the subducting plate and the overriding plate, which is not strictly a fault (in the sense that its two sides were not formed by brittle fracture and then modified by continued shear). Instead, as the schematic diagram in Figure 6.6 shows, one side, which consists of sediment-covered sea floor, is being introduced continually into the system. Clearly, the fate of this sediment – to what degree it is subducted and the extent to which subducted material is consolidated and metamorphosed – will have a great deal to do with the frictional behavior of the interface. Furthermore, because of the nearby presence of the subducted slab, major plate-driving forces are in close proximity to the subduction zone (Forsyth and Uyeda, 1975), so we cannot assume that they are remote and the stresses on the interface are given by some sort of standard "Andersonian" state. The variability of these factors may be important causes of why subduction zones have a much greater range in seismic behavior than do crustal fault zones.

Ignoring such variations for the moment, let us return to Figure 6.6 to discuss the typical seismicity of subduction zones. The principal moment release occurs by thrusting along the plate interface during great underthrusting earthquakes. The ruptures of such earthquakes apparently do not propagate all the way to the surface at the trench, however, but terminate within the accretionary wedge (Byrne, Davis, and Sykes, 1988). This is a sedimentary wedge of variable width that has been built up, for the most part, by the scraping of sediments off the subducting sea floor. The downdip termination of rupture in these earthquakes occurs at depths that vary between 40–60 km (J. Pacheco, unpublished work, 1989). Hence this marks the base of the schizosphere in the sense defined for crustal fault zones. Because this is considerably deeper than in crustal fault zones and because the dip of the interface is often quite shallow, the downdip width of rupture in subduction zone earthquakes may be very large, often in excess of 200 km. This factor, along with their great total length, explains why subduction zones account for the major portion of the world's seismic moment release. The along-strike length of individual earthquakes is proportionately large, so that these earthquakes are by far the largest, both in rupture area and moment.

Minor seismicity occurs in both upper and lower plates. In the upper plate it is confined to a region defined by the *aseismic front* (Yoshii, 1979), a sharp cutoff in seismicity some distance in front of the volcanic arc, and by the beginning of the accretionary wedge, which is also aseismic (Engdahl, 1977; Chen, Frohlich, and Latham, 1982; Byrne et al., 1988). These upper-plate earthquakes typically indicate compression normal to the arc, but this may vary substantially in different arcs, as will be discussed later. The lower plate is characterized by normal faulting earthquakes close to its axis of maximum bending, from which

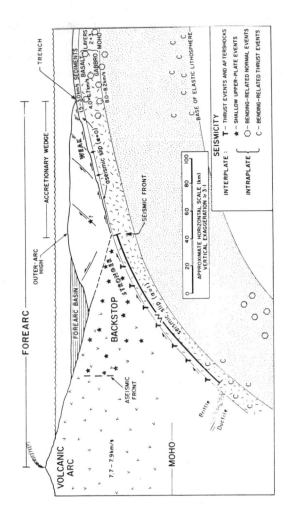

Fig. 6.6 Schematic cross section of a subduction zone, illustrating many of their characteristic features. (From Byrne, Davis, and Sykes, 1988.)

they result. This typically occurs at the outer trench wall and well back from it, but in some cases it may persist farther towards the arc.

At greater depth the lower-plate earthquakes merge into the Wadati–Benioff zone which may extend to depths as great as 650 km. These intermediate- and deep-focus earthquakes reflect internal stresses in the downgoing slab, and although they exhibit double-couple focal mechanisms and therefore may be treated with the usual seismotectonic analysis, they probably do not arise from the type of frictional instability which is the main topic here, owing to the very high pressures at those depths, but by another mechanism, outlined in Section 6.3.6.

In developing the rheological model of crustal fault zones in Chapter 3 we had a large body of research to build upon. There is much less available for understanding the mechanics of subduction zones. Whereas much was learned about crustal fault zones by the study of exhumed ancient faults, very little has been written about structures that may correspond to deeply eroded subduction shear zones. In the author's view the most likely structure to be such a fossil subduction zone is that typified by the Median Tectonic Line (MTL) of Japan (Fig. 6.7). By this we do not mean the westernmost 300 km of this structure, which is presently an active transcurrent fault known by the same name, but the profound contact between the Ryoke and Sanbagawa metamorphic terrains that runs for over 1000 km through southern Japan. The Ryoke zone is a high T/P (temperature-to-pressure ratio) metamorphic belt with grade increasing toward the MTL. Like the main central thrust of the Himalayas it contains abundant S-type (possibly anatectic) granites and reaches sillimanite metamorphism adjacent to the MTL. The Sanbagawa, in contrast, is a low T/P belt, of the blueschist series, and it also increases in grade as the MTL is approached.

The structure of these rocks has been studied in the Kanto Mountains by Toriumi (1982). He found that they constitute a shear zone, with strain and temperature increasing dramatically toward the MTL, where the rocks have been thoroughly mylonitized (Fig. 6.8). Therefore the contact is a shear zone, and these rocks have been sheared together in situ. This constitutes a remarkable thermal contrast, which can be explained by the asymmetry of shear heating during subduction: The upper plate has undergone shear heating resulting from thousands of kilometers of subduction, whereas the lower plate has just entered the system (Scholz, 1980). A profile of the thermal structure of the Sanbagawa belt during its metamorphism is shown in Figure 6.9. The thermal path shown is what would be expected from subduction, and the sharp thermal front on the "hanging wall" of this structure is due to the shear heating in the MTL and contact with the Ryoke belt. This interpretation of these classic paired metamorphic belts differs greatly from that of Dewey and Bird (1970), who misidentified this situation as being the

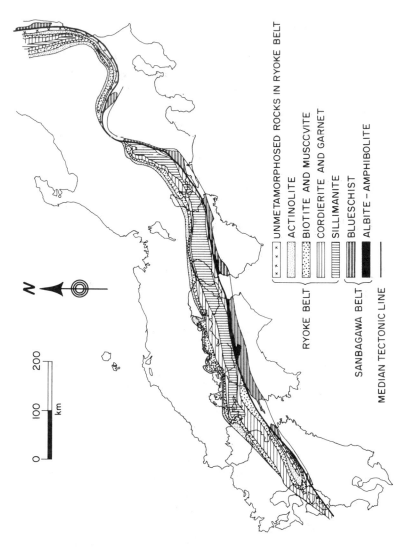

Fig. 6.7 The Median Tectonic Line of Japan and the metamorphic belts associated with it. (From Scholz, 1980.)

RYOKE BELT
- ⟨ UNMETAMORPHOSED ROCKS IN RYOKE BELT
- ⟨ ACTINOLITE
- ⟨ BIOTITE AND MUSCCVITE
- ⟨ CORDIERITE AND GARNET
- ⟨ SILLIMANITE

SANBAGAWA BELT
- ⟨ BLUESCHIST
- ⟨ ALBITE – AMPHIBOLITE

—— MEDIAN TECTONIC LINE

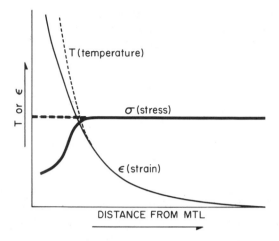

Fig. 6.8 Temperature, strain, and stress (inferred from grain-size pale-opiezometer) across the Sanbagawa belt approaching the Median Tectonic Line. Section is in the Kanto Mountains. (After Toriumi, 1982.)

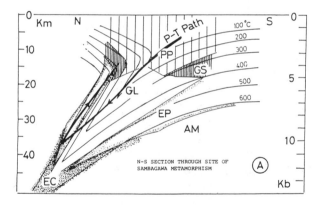

Fig. 6.9 Schematic cross section of the Sanbagawa belt, showing the *P–T* path and inferred thermal structure. Abbreviations are: PP, prehnite–pumpellyite; GL, glaucophane (blueschist); GS, greenschist; EP, epidote amphibole; EC, eclogite. (From Grapes, 1987.)

same as the Mesozoic metamorphism in California, where a small metamorphic aureole around the Sierra Nevada outcrops hundreds of kilometers arcwards from the blueschists of the Franciscan terrain. Similar, but less well studied structures as the MTL occur in Hokkaido (the Hidaka–Kamuikotan belt) and in the South Island of New Zealand, which also has been called the Median Tectonic Line (Landis and Coombs, 1967).

These observations lead me to think that, at least in many cases, the shear interface is within the lithified, subducted sediments, and it is the mechanical and thermal state of these sediments that determine the seismic behavior of subduction zones. The aseismicity of the accretionary wedge may be explained readily by an upper stability transition because of the presence there of unconsolidated, velocity-strengthening, weak sediments (as in the explanation of the upper stability transition for crustal faults, Sec. 3.4.1; Marone and Scholz, 1988; Byrne et al., 1988). The structure of the accretionary wedge itself may be used to show it is a zone of very high pore pressures and weak basal friction (Davis, Dahlen, and Suppe, 1983). The great depth of the lower friction stability transition probably cannot be explained by lithology, however. If the interface is composed of lithified siliceous sediments, they are not likely to be frictionally unstable at higher temperatures than granite, about 300°C. This suggests that the greater transition depth results from a much lower thermal gradient than in crustal shear zones. Certainly this is not unexpected because of the cooling effect of the subducted slab, but heat flow measurements in the forearc are not reliable enough to test this hypothesis directly (e.g., Watanabe, Langseth, and Anderson, 1977).

However, one can use the sediments themselves as a geothermometer. The stability fields of the various metamorphic assemblages are shown in Figure 6.10. Metamorphism in crustal fault zones follows the usual Barrovian trend through the greenschist and amphibolite fields. Subduction zone sediments, such as the Sanbagawa, however, follow the low T/P blueschist trend. This has a geothermal gradient about one-third the former, indicating by comparison that a temperature of 300°C will occur at about 45 km in subduction zones, which is about where the lower stability transition is observed, judging from the lower cutoff in interface (thrust) seismicity. This also provides independent justification for Stuart's (1988) modeling result: In order to adapt the frictional model of Tse and Rice (1986) to fit the Nankaido geodetic data, he had to assume that the lower stability transition was at about 40 km.

Whereas the above comments may explain the general characteristics of subduction zones, there is, as mentioned before, a great deal of variability between them. This has been discussed at some length by, among others, Uyeda (1982), who has described as the end members the "Chilean" type and the "Mariana" type, as shown in Figure 6.11. The Chilean type, highly compressional, has a pronounced outer rise, abundant calc-alkaline volcanism in the arc, a shallow-dipping Wadati–Benioff zone, and, most importantly for our subject, great earthquakes along the interface. In contrast, the Mariana type is extensional, has back-arc spreading, little or no outer rise, few andesites in the arc, and no great earthquakes.

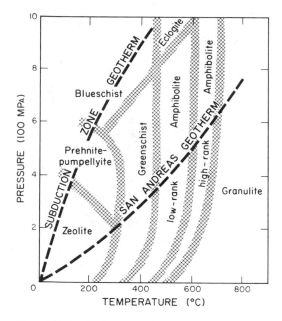

Fig. 6.10 *P–T* diagram showing the fields of metamorphic facies. A typical continental geotherm and metamorphic path is shown, as well as one following the high *P/T* path typical of subduction zones. (Following Ernst, 1975.)

Based on the 80-yr seismic record, it appears that subduction zones differ greatly in their seismic coupling coefficient χ, as estimated by the maximum earthquake size (Ruff and Kanamori, 1980) or by summing moments (Peterson and Seno, 1984). Thus χ ranges from essentially nil in the case of the Mariana arc to nearly one in the Chilean arc. In this respect subduction zones seem to differ greatly from major crustal faults in continental terrains, for which fault creep (aseismic slip) appears to be, with a few important exceptions, a rare phenomenon, and which therefore have nearly total seismic coupling. We defer an analysis of the possible causes of this behavior until later, following a more thorough discussion of aseismic slip (Sec. 6.4.3).

Subduction zones and their earthquakes are in general much larger than crustal shear zones, but their rupture characteristics show many of the same features. Because of their greater size and rupture duration, the gross interior features of subduction zone earthquakes can be evaluated from long-period waves recorded at teleseismic distances. A number of examples are shown in Figure 6.12. Two crustal earthquakes, Guatemala and Tangshan, are included for comparison. North is up in the figure, each event is drawn at the same scale, and for the subduction events the

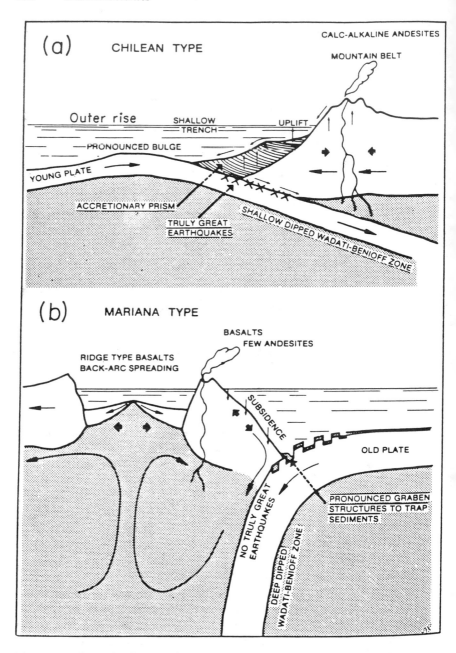

Fig. 6.11 Schematic diagram showing the two end members of subduction zone types: (a) Chilean type, highly compressive; (b) Mariana type, extensional. (After Uyeda, 1982.)

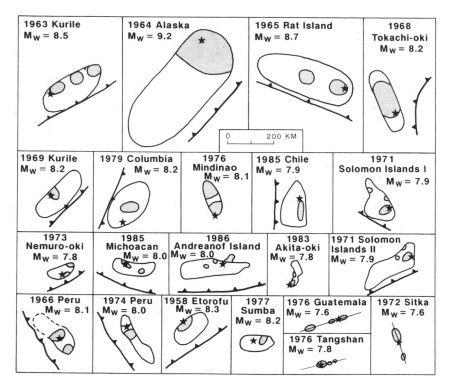

Fig. 6.12 Rupture zones and zones of concentrated moment release for a selection of large earthquakes. Rupture zone is indicated by the enclosed region, epicenter by a star, and zones of concentrated moment release by shading. Most are subduction zone events, in which the trench is indicated by a hatched curve. All figures are to the same scale, and north is up. (Courtesy of W. Thatcher, unpublished work, 1989.)

position of the trench is indicated, hatchures on the overriding side. Zones of concentrated moment release within each rupture are indicated by shading, and the epicenter is given by a star. Many of these same earthquakes have been analyzed in finer detail by Kikuchi and Fukao (1987) who show that the high moment release zones can be subdivided further, with local areas of stress drop an order of magnitude greater than the mean stress drop for the entire rupture. Heterogeneity in rupture, which we have discussed at length with regard to crustal fault zones, is thus also a feature of subduction zone events. Notice that the high moment release zones are often small fractions of the rupture zones (although this will depend, to an extent, on how moment release is contoured). Also, the high moment release zones tend to be on the far side from the trench, that is, at the deep end of the rupture, and there is a tendency for the epicenter to be located within or near such a zone.

Although these are not universal features, they do correspond to some of the common properties of crustal faulting already discussed: The first reflecting the tendency for rupture to initiate near the base of the fault and for stress drop to be higher there, both resulting from the increase in stress and strength with depth, and the second, the extent to which high stress drop favors dynamic triggering (e.g., Sec. 5.4). Nearly unilateral rupture propagation is also common, which probably reflects the tendency for subduction earthquakes to initiate and terminate, like crustal earthquakes, at structural irregularities that can be recognized from bathymetric features transverse to the arc (Mogi, 1969; Kelleher et al., 1974).

Except for a very few cases, we have observations of, at the most, only a single rupture event at most sites along subduction zones, so we cannot know whether or not the zones of high moment release are ephemeral or permanent features (see Sec. 4.5.2). However, Lay, Kanamori, and Ruff (1982) have noted regional variations in the rupture characteristics of subduction zone earthquakes. Specifically, arcs with low seismic coupling tend to have earthquakes with small rupture zones and the zones of high moment release are small isolated patches within those regions. In contrast, arcs with high coupling have very large earthquake rupture zones within which the zones of high moment release are also very large. Whereas these statements constitute a truism, Lay et al. went on to suggest that the zones of high moment release reflect the presence of "asperities," which they interpret as permanent, mechanically distinct features of subduction zone interfaces. In their view, then, seismic coupling is determined by the fraction of the interface that is comprized of asperities. This leaves unexplained the mechanical basis for such features, which we take up in Section 6.4.3.

6.3.3 Oceanic earthquakes

Although some subduction zones are fully oceanic, in the sense that the upper plate as well as the lower is oceanic lithosphere, for the most part we have been discussing earthquakes rupturing continental crust. In that case we have been guided by the rheological model developed in Section 3.4.1, which was based on a quartz-rich composition. The oceanic lithosphere, in contrast, should obey an olivine-rich rheology, which, according to its flow law, will retain its strength to considerably higher temperatures.

The main oceanic plate boundaries are spreading ridges and transform faults. The ridges are seismically active, but this activity is secondary to the spreading process. Seismic activity in the crestal area is confined mainly to swarms of low-magnitude events that probably are associated with magmatic intrusion (e.g., Sykes, 1970a). Normal faulting events occur on the flanks close to the crest, driven by isostatic uplift of the depressed central rift valley. These events seldom exceed $M_s = 6.5$, and their summed moments show that they contribute no more than 10–20% to the spreading (Solomon, Huang, and Meinke, 1988). Moment centroid depths collected by Solomon et al. show that the schizosphere is 5–10 km thick in the vicinity of the median valley and decreases in thickness with increasing spreading rate. However, most spreading must occur aseismically, probably by dike intrusion along the central volcanic zone. Sea-floor spreading is primarily a thermal, rather than mechanical process, and seismicity associated with it is of secondary significance. In Iceland, for example, rifting episodes are accompanied by intense swarm activity, but with events that seldom exceed $M = 4.5$. The only large-magnitude earthquakes in Iceland are the strike-slip events in the south Iceland seismic zone.

As Davies and Brune (1971) noted, the oceanic transform faults, taken as a whole, also seem deficient in their seismic moment release rate, as compared to their slip rates. The exceptions are several of the longer transforms on the slow-spreading Mid-Atlantic Ridge. Brune (1968) found that for the Romanche Fracture Zone (FZ) in the South Atlantic, the seismic moment release rate agreed with the spreading rate if seismic faulting was assumed to extend to about 6-km depth. Kanamori and Stewart (1976), assuming a faulting depth of 10 km, showed that the Gibbs FZ in the North Atlantic was fully seismically coupled. These assumed depths of faulting are poorly constrained, but a microearthquake survey with ocean bottom seismometers on the Rivera FZ showed that microearthquakes there do not occur deeper than 10 km, so it is unlikely that faulting can extend much deeper (Prothero and Reid, 1982). Bergman and Solomon (1988) found that moment centroid depths for earthquakes on transform faults in the North Atlantic lie in the range 7–10 km, so their maximum faulting depth cannot be greater than twice that.

Burr and Solomon (1978) calculated the depth of faulting that would be consistent with full coupling for a large number of oceanic transform faults. They obtained 5–15 km for slow-moving, long transforms in the Atlantic, and 1–2 km for short, fast-moving transforms in the Pacific. They also observed that the largest earthquakes tended to occur near the midpoints of the long, slow-spreading faults. In another survey, Kawasaki et al. (1985) calculated χ with an assumed active depth of 10 km. They found χ to be near 1 for the Gibbs and Romanche transforms, but much smaller for the other cases they studied, which were fast-moving cases from the Pacific. Although there was considerable scatter in their data, they showed that χ was negatively correlated with spreading rate and positively correlated with $\sqrt{t_{mp}}$, where t_{mp} is the age of the crust at the midpoint of the transform. All of these observations strongly suggest that the seismic coupling, or the thickness of the seismogenic region, is in this case controlled by temperature.

Using a simple plate-cooling model, Bergman and Solomon found that the depth of seismic faulting on the North Atlantic transforms is limited to temperatures below 600–900°C. These temperatures are not very well constrained because, if the seismic moment release is weighted to greater depths, then the maximum depth of faulting, and hence temperature, would be less. On the other hand, if shear heating is included in the models, the limiting temperature would be higher (Chen, 1988), and if nonconductive heat loss, such as from water circulation, is important, it would be lower. Within these uncertainties, the observations are consistent with the seismicity being limited by the brittle–plastic transition in an olivine-rich oceanic lithosphere strength model, of the type illustrated in Figure 3.18.

The limiting depth for seismic faulting for the Gibbs transform ($t_{mp} = 22$ Myr), from Bergman and Solomon, is 10–20 km and 8–10 km for the 15°20′ transform, also in the North Atlantic. These depths are consistent with full coupling on the Gibbs and Romanche transforms, according to the observations described above. On the other hand, the faster-spreading cases discussed by Kawasaki et al. and Burr and Solomon have $t_{mp} \approx 2$–4 Myr, so their average seismically active thicknesses, predicted from the cooling model, would be very small. Even if they were fully coupled within that thickness, they would appear to have very low values of χ if these were calculated based on an assumed schizospheric thickness of 10 km, as done by Kawasaki et al. Control of the seismic thickness by an isotherm would explain the $\sqrt{t_{mp}}$ dependence of χ they found. So, within the uncertainty, these data can be interpreted with a model in which a fully coupled schizosphere overlies an aseismically slipping plastosphere with the schizospheric boundary defined approximately by a 600–800°C isotherm. We do not need to invoke a model in which a low χ value arises because both aseismic and seismic slip initiate on the *same part of the fault*, which would fundamentally violate the friction constitutive laws of the type given in Chapter 2. However, the simple plate-cooling model does not work well in the vicinity of the ridge crest, where it would predict a vanishing lithospheric thickness. The depths of microearthquakes obtained in ocean bottom seismometer studies do not shallow in the predicted way as the ridge crest is approached, where they still occur at depths of 8–10 km beneath the sea-floor [Trehu and Solomon (1983), Toomey et al. (1985), and Cessaro and Hussong (1986)].

There are also intraplate earthquakes within the ocean basins, and these have been the subject of considerable research because of the

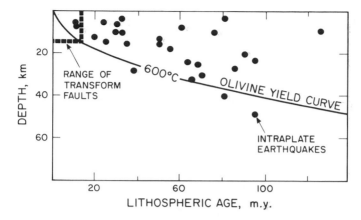

Fig. 6.13 Depth of oceanic intraplate earthquakes as a function of lithospheric age. The 600°C isotherm, from a simple plate cooling model, is taken as an approximate yield curve for an olivine-rich rheology. This approximately limits the earthquake depths. (Earthquake data from Wiens and Stein, 1983.)

possibility that they may reflect the plate-driving forces (cf. Sykes and Sbar, 1973). They can occur to greater depth than in continental crust, and their maximum depth increases with age, approximately following a 600°C isotherm, according to the same plate-cooling model used above, as shown in Figure 6.13 (Chen and Molnar, 1983; Wiens and Stein, 1983; Bergman, 1986). This result is consistent with the interpretation of the transform fault earthquakes.

Although there are insufficient friction data available for ultrabasic rocks to prove that their frictional instability field corresponds with the "brittle–ductile transition" calculated from this simple type of model, these results indicate that to first order the correct result is obtained by this method. This is probably for the same reasons as given in Section 3.4.1 in the case of the granitic model.

The number of oceanic intraplate earthquakes and their cumulative moment release decrease with the age of the lithosphere. This disagrees with expected "ridge push" stresses, generated by the gravitational potential of ridge topography, which should increase with age (Wiens and Stein, 1983). Bergman (1986) argues that the majority of them occur in young sea floor (< 35 Myr) in response to thermoelastic stresses. The older sea floor is largely free of them because the older lithosphere is stronger than the driving forces, except in a few exceptional places where the lithosphere is either unusually weak or there are stress concentrations. These include several sites of concentrated activity in the Indian Ocean, near the Ninetyeast and Chagos–Laccadive Ridges, where new plate boundaries may be forming (Sykes, 1970b; Wiens et al., 1986), and near Hawaii, a hot-spot locality.

6.3.4   *Continental extensional regimes*

Although sea-floor spreading is largely aseismic in character, extension in continental crust is accompanied by a substantial degree of active faulting, accompanied by large earthquakes. Continental extensional regimes are quite varied, ranging from rifts to broad extensional regions such as the Great Basin in the western United States, to incipient zones of back-arc spreading as in the Taupo volcanic zone of New Zealand. A recent collection of papers on the tectonics of these regions is given in Coward, Dewey, and Hancock (1987).

Large normal faulting earthquakes, breaking the entire schizosphere, occur in these zones. They typically nucleate at depths of 6–15 km on faults dipping between 30–60° (Jackson, 1987). Reflection profiles in the Great Basin show that normal faults there become listric below the base of the schizosphere (Smith and Bruhn, 1984): Large earthquakes rupture only the steeply dipping planar sections of these faults above this depth (Doser, 1988), and there is no evidence of seismic activity on low-angle extensional detachment faults. A major problem, discussed by Jackson, is that it is not possible to produce very large extension solely with high-angle faults, but the activity of such faults is not compatible with active low-angle faults. A solution has been offered by Buck (1988). He hypothesizes that active normal faults become shallow dipping with progressive slip, as a result of isostatic adjustment and flexure. When the dip of the fault becomes too shallow, it is abandoned and new high-angle normal faults are formed. In this model then, the detachments in extensional terranes are not active, which agrees with the seismic observations and also avoids the mechanical problem mentioned in Section 3.1.

Well-defined rifts, like those of East Africa, are clearly accompanied by major active faulting, as testified by the great escarpments of the normal faults that bound the rift valleys. In the northern part of the African rift system, in Ethiopia and Kenya, the formation of the rift was preceded by a broad thermal doming, so that the rift valley itself can be considered the surficial response to an underlying thermal process (e.g., Baker and Wohlenberg, 1971). However, in central and southern Africa the rifts are structurally controlled, following zones of weakness along the old Pan-African greenstone belts and avoiding the Precambrian cratonal blocks (Vail, 1967; Versfelt and Rosendahl, 1989). Far to the south the tip of the rift is propagating by active normal faulting that has not been preceded by doming (Scholz, Koczynski, and Hutchins, 1976).

The African rifts are opening too slowly to assess, within the limited instrumental period, the quantitative role of seismic activity in the process. That the major bounding faults are seismically active is attested by the $M = 7.1$ earthquake of 1928 in the Subukia Valley, Kenya (described in Richter, 1958). That earthquake produced about 30 km of primary surface faulting, with a maximum slip of about 3.5 m, along the Laikipia Escarpment of the Eastern (Gregory) Rift.

In contrast, the Great Basin (Basin and Range) province of the western United States is a very broad (up to 400 km) region within which active normal faulting occurs on many subparallel range fronts. However, as remarked earlier (Sec. 5.3.3, Fig. 5.30) the activity seems to migrate between subregions within the province, so that the geological deformation is much more broadly distributed than is indicated by the most recent seismicity. The range-bounding faults in the Basin and Range may be quite long, up to several hundred kilometers, and have throws of kilometers, judging from the uplifts of the ranges. Their topographic expression is very different from the normal faults in the African rifts. The latter bound downdropped grabens and show little evidence for uplift on the footwall side. In contrast, the prominent topography of the Basin and Range province is produced by the uplift of ranges on the footwall side of such faults. This may represent a different stage of development, or some fundamental difference in the tectonics. An example is given in Figure 6.14a, which shows a section of the scarp of the 1983 Borah Peak earthquake (see Sec. 4.4.1). Behind it one can see Borah Peak, the highest mountain in Idaho at 3,890 m.

It has been argued that the Basin and Range province tectonically represents back-arc spreading within an ensialic environment, now in a mature stage of development (Scholz, Barazangi, and Sbar, 1971). By comparison the Taupo (Central Volcanic) Zone of the North Island of New Zealand is thought to be a much more recently activated feature of this type (Karig, 1970). The Taupo Zone is currently much hotter and more volcanically active than the Basin and Range, and the style of extensional tectonics is very different. Although there are numerous normal faults, they are short, a few tens of kilometers at most in length, and have little topographic expression. Nonetheless, large normal faulting earthquakes do occur. An example is shown in Figure 6.14b, which shows the main surface rupture of the Edgecumbe earthquake of 1987 ($M_s = 6.6$). In the photograph a rift is shown in a monoclinal flexure that was typical of the surface expression of this earthquake. The fault had been unrecognized previously as it was marked only by an older monclinal warp, about 1 m high, that can be seen to the right in the photograph. Comparison with Figure 6.14a dramatizes the difference in the tectonics of the two regions. In spite of this difference, the Taupo

a

Fig. 6.14 Fault scarps produced by normal faulting earthquakes. (a) Borah Peak, Idaho, 1983. View of the scarp of the 1983 earthquake, looking east, with Borah Peak in the background. (Photograph by R. E. Wallace, courtesy of U.S. Geological Survey.) (b) Edgecumbe, New Zealand, 1987. View looking north, Mt. Edgecumbe is in the background. This is a rift on the top of an earthquake-produced monocline. The swale of the monocline, which was evident before the earthquake, may be seen to the right of the rift. (Photograph by Quentin Christie, courtesy of NZ Soil Bureau, D.S.I.R., New Zealand.)

Zone seems to be fully coupled seismically. There have been six earthquakes of size comparable to the Edgecumbe event in the last 120 yr, and this can account for the 7 mm/yr extension rate for this region, indicating that the behavior of the upper crust there is entirely brittle to a depth of 10–15 km (Grapes, Sissons, and Wellman, 1987).

6.3.5   *Intraplate earthquakes*

Although something like 95% of the global seismic moment release is produced by plate boundary earthquakes, there are significant numbers of earthquakes that occur well away from plate boundaries. These intraplate earthquakes are important because they greatly expand the

b

Fig. 6.14 *(cont.)*

Table 6.1. *Classification of tectonic earthquakes*

Type	Slip rate mm/yr	Recurrence time yr
I. Interplate	$v > 10$	~ 100
II. Intraplate, plate boundary related	$0.1 \leq v \leq 10$	$10^2 - 10^4$
III. Intraplate, midplate	$v < 0.1$	$> 10^4$

region of possible seismic hazard from the proximity of plate boundaries. Their role in tectonics is poorly understood, both from the viewpoint of the origin of the forces that generate them and what sort of structures localize them.

One way to distinguish between interplate and intraplate earthquakes is based on the slip rate of their faults and hence their recurrence time, as explained in Table 6.1. Intraplate earthquakes are classified into two types. Type II occur in broad zones near and tectonically related to plate boundaries or in diffuse plate boundaries. Examples are earthquakes of the Basin and Range province of western North America, which very

Table 6.2. *Parameters of selected strike-slip earthquakes*

Earthquake	M_s	M_0 10^{20} N-m	L km	W km	$\overline{\Delta u}$ m	$\Delta\sigma$ MPa	$\dot{u}$ mm/yr
Borrego Mtn. 1968 (C)	6.7	0.1	40	13	0.57	1.3	10
Coyote Lake 1979 (C)	5.7	0.0035	25	8	0.06	0.2	10
Imperial Valley 1979 (C)	6.5	0.06	42	10	0.47	1.4	35
Izu 1930 (J)	7.2	0.25	22	12	3.14	7.8	3.1
Izu-hanto-oki 1974 (J)	6.5	0.059	20	11	0.91	2.5	0.9
Morgan Hill 1984 (C)	6.1	0.02	30	10	0.20	0.6	10
Parkfield 1966 (C)	6.0	0.014	30	13	0.12	0.3	35
Tango 1927 (J)	7.6	0.46	35	13	3.29	7.6	0.8
Tottori 1943 (J)	7.4	0.36	33	13	2.80	6.5	0.4

Data from Kanamori and Allen (1986). (C): California; (J): Japan. Slip rate of California faults from geodetic data; for Japanese faults, calculated from $\overline{\Delta u}/T$, where T is estimated recurrence time. Stress drop calculated as $\Delta\sigma = \mu(\overline{\Delta u}/W)$.

broadly may be considered to be part of the Pacific–North America plate boundary (Atwater, 1970), or inland earthquakes in Japan, which are tectonically a part of the compressional Pacific–Eurasian plate margin. In contrast, Type III occur in midplate regions and seem to be unrelated to plate boundaries. This classification is of course somewhat artificial, because there is a continuous spectrum of earthquake types.

An important reason for this classification is that intraplate and interplate earthquakes, so defined, have distinctly different source parameters, as was shown in Figure 4.13. The intraplate earthquakes in that figure are all large Type II events, which systematically have higher stress drops than the interplate earthquakes. Because stress drop depends on rupture geometry we cannot compare this parameter easily for all the earthquakes in Figure 4.13. For this comparison, rupture parameters for a subset of strike-slip earthquakes, from intraplate Japan and the San Andreas fault system, but all with a similar L/W ratio, are given in Table 6.2.

Table 6.2 indicates that stress drops for the Japanese earthquakes average about a factor of 9 higher than the California cases. If we adopt a frictional model in which stress drop is proportional to total stress, this indicates that the faults in Japan are considerably stronger than the San Andreas.

These comparisons are based on mean values of static parameters. Somerville et al. (1987) compared the source parameters of eastern and western North American earthquakes by determining t_r and M_0 obtained by waveform modeling. They concluded that both sets of earthquakes had similar characteristics, consistent with constant values of stress drop of about 10 MPa, but with much greater scatter in stress drop in the western data set. However, most of the western cases they studied were Type II intraplate events, according to the scheme of Table 6.1, and so the comparison was mainly between Type II and Type III intraplate earthquakes, which do not appear to differ in stress drop. Dynamic measurements of this type emphasize the maximum values of stress drop. If we compare again the 1979 Imperial Valley and 1983 Borah Peak earthquakes (Sec. 4.4.1), there are localized regions within the former where the stress drop is as high as the mean value of the latter, even though the mean stress drop of the Imperial Valley earthquake is much less.

If we consider the entire class of faults, as shown in Figure 4.13, the obvious differences between these two classes of earthquakes are that the geological slip rates for the intraplate faults are several orders of magnitude slower than the plate boundaries, their earthquake recurrence time proportionally longer, and their total slip much less. If the difference in stress drop is strictly a rate effect, the data in Table 6.2 indicate that it is about 4 MPa per decade in slip rate. A velocity-weakening friction law (Eq. 2.27) predicts that slower-moving faults will be stronger, and have larger stress drops, but as Cao and Aki (1986) have shown, the laboratory measured values of $a - b$ are much too small to account for the observed differences (even though, by a different interpretation of the data, they estimate the effect as four times less than reported here). Calculations from experimental data by Zhao and Wong (1990) suggest agreement with the magnitude of the effect reported by Cao and Aki, but cannot explain the effect if it is as large as shown here.

There seem to be four alternatives. One, suggested in Scholz, Aviles, and Wesnousky (1986), is that interseismic healing is greatly augmented by chemical processes that are not accounted for in the cited laboratory estimates of $a - b$. The second is the evolutionary model in which fast-moving faults weaken with age, discussed in Section 3.4.3. The third is that the San Andreas fault is characterized by a less negative value of $a - b$, so that it is less unstable and has smaller relative stress drops. The data for the California earthquakes in Table 6.2 suggest that this latter effect may be important. The Parkfield, Coyote Lake, and Morgan Hill earthquakes are all within a weakly coupled, stability transition region of the fault (see Sec. 6.4.1), and have stress drops a factor of seven smaller than the other two San Andreas earthquakes.

The fourth possibility is that the strength of faults decreases with net slip. This is suggested by the data in Figure 3.12, which indicate that faults become smoother with displacement. Because high net slip usually corresponds with high slip rate, the effect of these two parameters cannot usually be separated. The strike-slip faults in the south Iceland seismic zone are an exception. Although they have very little net slip (Sec. 3.2.1) their earthquake recurrence times indicate slip rates of ~ 1 cm/yr, and so they would be classified as Type I, according to the scheme of Table 6.1. However, a large earthquake ($M_s = 5.8$) on one of these faults in 1987 had a mean slip of 3 m on a fault of length 15 km (Bjarnason and Einarsson, in press). This is more in accord with the parameters of the intraplate earthquakes in Figure 4.13, which suggests that net slip may be more important than slip rate in explaining their difference from the interplate events.

The tectonic role of Type II intraplate earthquakes can be understood readily, related as they are to the plate-boundary deformation fields, as described in the Basin and Range and inland Japan cases. The origin of the stresses that produce Type III intraplate earthquakes is more problematic. It has been suggested that they reflect the plate tectonic driving forces, since those must produce a continuous stress field throughout the plates (Sykes and Sbar, 1973). They may be localized at reactivated sites of prior deformation (Sykes, 1978), or at sites of stress concentration resulting from deep structure, such as along the margins of cratons (Wesnousky and Scholz, 1980; Gough, Fordjor, and Bell, 1983). Whatever their underlying cause, their focal mechanisms often indicate consistent stress directions over broad regions, which agree with in situ stress measurements and other indicators of stress direction. This allows the mapping of stress directions on a regional scale, from which one may define distinct *stress provinces* (Zoback and Zoback, 1980). For example, earthquakes in New York and adjacent parts of New England and Canada consistently exhibit NE–SW P-axis directions, even though their focal mechanisms may otherwise vary widely (Sbar and Sykes, 1977).

The faults that produce midplate earthquakes cannot be identified readily, which poses a serious problem for earthquake hazard analysis in such regions (Sec. 7.5). In the eastern United States, for example, only two active faults have been identified: the Meers fault in Oklahoma, recently identified from its morphology (Crone and Luza, 1986), and the fault in the Mississippi embayment responsible for the earthquakes of 1811–12 at New Madrid, Mo. This latter fault was identified on the basis of subsurface structure, microearthquake locations, and the location of liquifaction features (Nuttli, 1973).

The lack of evidence for active faults in midplate regions is probably largely because the tectonic deformation rates are slower than erosion rates, so that traces of activity are obliterated. The 1968 Meckering and 1988 Tennant Creek earthquakes in the Australian shield, for example, appear to have occurred on preexisting faults, but these had no scarps by which they could be recognized as being active (Gordon and Lewis, 1980; Bowman et al., 1988). In other cases it may be that some intraplate earthquakes occur on detachment faults with no surface expression, as has been suggested for the Charleston, South Carolina, earthquake of 1896 (Seeber and Armbruster, 1981).

The widespread occurrence of intraplate earthquakes within continental regions may be taken to mean that much of the continental lithosphere must be at a state of stress near to failure. The common occurrence of seismicity induced by reservoir impoundment leads to the same conclusion (Sec. 6.5.4).

6.3.6 *Mechanism of deep earthquakes*

It was first shown by Wadati (1928) that some earthquakes occur at depths considerably greater than the base of the Earth's crust. These deep-focus earthquakes are now known to occur primarily in the Wadati–Benioff zones within the lithospheric slabs that descend into the mantle at subduction zones. Their properties are reviewed by Frohlich (1989). Their frequency of occurrence decreases approximately exponentially with depth down to 450 km, then increases to a minor peak at about 600 km before ceasing abruptly between 650 and 680 km. Their radiation characteristics indicate similar focal mechanisms as shallow earthquakes – they are dominated by double-couple radiation and their calculated stress drops are not greatly different from shallow events.

The mechanism of these earthquakes has long been a matter of conjecture because the great pressure at these depths would prohibit any brittle or frictional processes from operating. One candidate mechanism is that they result from a dynamically running polymorphic phase transition, first suggested by Bridgman (1945). An observational fact that supports this view is that their moment release rate is proportional to the subduction rate (McGarr, 1977), but the way in which this mechanism can result in double-couple radiation has been a longstanding problem.

One solution, proposed by Kirby (1987) is that a solid–solid phase transition occurs preferentially in a deviatoric stress field, and by virtue of the stress concentration around an implosion site in such a field runs preferentially in a sheet on planes of maximum shear stress, relaxing the shear stresses. The way in which this can occur has been experimentally demonstrated by Green and Burnley (1989). In conducting experiments with the olivine → spinel transition in a germanate analog, they found that the transformation was enhanced by deviatoric stress and that under some conditions it occurred on conjugate planes at an acute angle to σ_1 accompanied by a sudden stress drop and audible report. Microscopic examination of their specimens showed that the spinel was forming in thin lenses normal to the σ_1 direction that were arranged in en echelon arrays in the orientation of conjugate shear faults as in an ordinary brittle fracture experiment, and that the failure of the specimen occurred when the phase transition ran through these arrays. They reasoned that the volume reduction accompanying the phase transition caused it to occur by the formation of Mode $(-I)$ anticracks, which are the same as Mode I cracks except the signs of displacements and stresses are reversed [as in the Fletcher and Pollard (1981) model for stylolite formation]. The anticracks thus form normal to σ_1 and align themselves in en echelon arrays that coalesce to form a shear fault in a way completely analogous (but with different orientation) to the way Mode I cracks form and then coalesce to form a fault in ordinary brittle fracture.

6.4   THE RELATIVE ROLE OF SEISMIC AND ASEISMIC FAULTING

We have been concentrating mainly on dynamic faulting and the resulting earthquakes. However, there are clearly faults that slip aseismically, by stable frictional sliding, and so their tectonic contribution is not measured by seismic activity. Faults that may behave predominantly in this way are those in weakly consolidated sediments and ductile faults of various types, such as decollements. In this context recall that our definitions of schizosphere and plastosphere are lithologically based and do not correspond to some universal isotherm that is independent of the local rock type. Furthermore, the conditions under which seismic rather than aseismic slip occurs depends on the frictional stability transitions, which are known only for a very few rock types but which are certain to be strongly dependent upon lithology. Deformation that takes place in this manner, and by other aseismic means such as folding, is not revealed directly by the seismotectonics. Folding, of course, may be accompanied by faulting, and sometimes by earthquakes, as in the case of the 1983 Coalinga (Stein and King, 1984) and 1987 Whittier Narrows (Hauksson et al., 1988) earthquakes in California. Yeats (1986) has given a review of such fold-related faulting.

We are more concerned here with throughgoing crustal faults that are capable of large displacement and that cut crystalline rock over some extent of their surface. These too are known to slip aseismically under some conditions, as has been noted previously in several places. In this section we describe the known cases in some detail, discuss the possible causes of this behavior, and try to access its quantitative importance.

6.4.1   *Aseismic slip*

Stable aseismic fault slip was first discovered on the San Andreas fault in central California where the fault was found to be progressively offsetting the walls of a winery built upon it (Steinbrugge et al., 1960; Tocher,

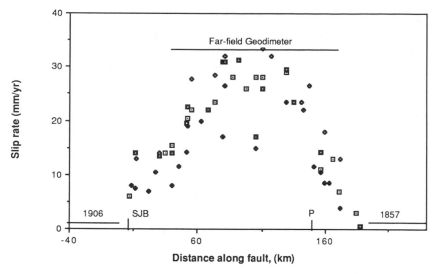

Fig. 6.15 Slip rate of the creeping section of the San Andreas fault in central California, plotted as a function of position along the fault trace. Data types [spans], and sources: closed boxes, alinement arrays [100 m], Burford and Harsh (1980); closed diamonds, creepmeters [10 m], Schulz et al. (1982); open boxes, short-range [1 km] and open diamonds, geodimeter [10 km], Lisowski and Prescott (1981). Far-field geodimeter estimate made over a 50-km span (Savage and Burford, 1973). Notice that the deformation increases somewhat with span, but is mostly constrained within the 100-m alinement arrays. P is Parkfield and SJB is San Juan Bautista. The ends of the 1906 and 1857 ruptures are also shown.

1960). This phenomenon was originally called creep, and although this familiar term is convenient for casual usage, it should not be confused with the type of viscoelastic deformation of the same name that is a common property of solid materials. Subsequent to its discovery, fault creep was traced to the north and south along the contiguous San Andreas fault and on the Hayward–Calaveras fault that branches to the north. The distribution of aseismic slip of this so-called creeping section of the San Andreas fault is shown in Figure 6.15.

Slip at the Cienega Winery, the discovery site, has been at a steady rate of 11 mm/yr since the construction of the building in 1948; the other slip rates shown have been measured for about 15 yr and also have been steady over that period. At most sites the slip is episodic, as shown in Figure 5.19, and although occasional disturbances do occur, such as those produced by local earthquakes (Fig. 5.20), over the long run the slip rates return to their steady values. The slip rates decrease smoothly to zero at the ends of the active segment (Fig. 6.15) and reach a peak at the center, suggesting that the slip extends to as great a depth as the length of the slipping segment. This interpretation is consistent with the result of Tse, Dmowska, and Rice (1985), who modeled it as an end-pinned crack extending through the lithosphere. This agrees with the earlier interpretation (Sec. 5.2.1) that the deformation in the center of this region is essentially by block motion.

The stably slipping part of the San Andreas fault terminates to the north at San Juan Bautista, the southernmost extent of the rupture of the 1906 earthquake, and in the south at the north end of the rupture of 1857. The rates of stable slip of the Hayward and Calaveras faults gradually decrease northwards, and creep has not been observed north of San Francisco Bay. The creeping faults have an unusually high level of small earthquake activity (Fig. 5.5), with very little seismicity off the faults. Thus the seismicity very clearly delineates the faults, which distinguishes them from faults that do not creep, where microearthquake activity is more diffuse and one often has difficulty recognizing active faults on the basis of microearthquakes alone. This high rate of activity is limited to small earthquakes, however, and does not contribute significantly to slip. A moment summation by the author indicates that the seismic moment release rate of the creeping section of the San Andreas accounts for less than 5% of its slip. Exceptions occur at the ends of the creeping segment, where transition zones occur. The transition zone at Parkfield has been described already (Sec. 5.3.1). Several large earthquakes have occurred recently on the northern parts of the Calaveras and Hayward faults (1979 Coyote Lake, 1984 Morgan Hill), and the Hayward fault is known to have ruptured in two large earthquakes in the last century, most notably one in 1868.

The behavior of this region, in which the faults have been directly observed to exhibit steady-state aseismic slip at their tectonically driven rates, is unique among crustal faults. Extensive study in other parts of California has revealed clear evidence of steady aseismic slip in only one other area, the Salton trough. There it is at about an order of magnitude less than the strain accumulation rate (Louie et al., 1985). Later we will see that creep in this region may be restricted to the sedimentary cover, which is unusually thick there. Searches for fault creep also have been made in Japan and New Zealand, with negative results (T. Matsuda, personal communication, 1985; K. Berryman, personal communication, 1988). It therefore seems to be a rare phenomenon. Decollements, where slip occurs aseismically on weak ductile materials like salt, may on the other hand dominate the tectonics of some areas, for example, the Zagros Mountains of southern Iran (Ambrayseys, 1975; Jackson and McKenzie, 1988) and the Salt Range of Pakistan (Seeber, Armbruster, and Quittmeyer, 1981; Jaume and Lillie, 1988).

In Sections 2.3.3 and 3.4.1 we described a situation that may be considered normal, in which a zone of unstable slip occurs over some depth range, limited on top and bottom by upper and lower stability transitions. The creeping section is stable at all depths, so the two stability transitions must merge there, causing the unstable region to vanish. There are three plausible mechanisms by which this might occur:

1. The fault may be lined with an intrinsically more ductile, velocity-strengthening material, so that the lower transition (T_1 in Fig. 3.19) is elevated to T_4.
2. The fault may be lined with unconsolidated or clay-rich gouge to an unusual depth, depressing the upper transition T_4 to T_1.
3. Normal stresses may be unusually low and be below the critical value for instability, as defined in Equation 2.29, at all depths.

The second mechanism may involve low normal stress too, because the degree of consolidation will be controlled by the normal stress.

All of these possibilities have been suggested on geological grounds. Allen (1968) pointed out that serpentine was found first in the fault zone at Parkfield and that it was common throughout the creeping zone. It is known from experiment that the presence of serpentine, even in small quantities, may stabilize sliding (Sec. 2.3.1). Irwin and Barnes (1975), on the other hand, noted that in the creeping section one side of the fault is formed of Franciscan formation rocks overlain by the Great Valley sequence. Where the Great Valley sequence rocks are absent, the fault is locked. They suggested that this creates unusually high pore pressures in the fault zone, as a result of metamorphic fluids being exuded from the Franciscan formation and capped by the impermeable rocks of the Great Valley sequence. In this interpretation, overpressurization would result in low effective normal stresses on this section of fault and thus narrow, and perhaps eliminate, the unstable field.

As described in Section 3.2.2, there is some evidence that the fault zone in the creeping section is characterized by a several-kilometer-wide zone of low-velocity and low-density materials that extends through the seismogenic thickness; a feature that is not recognized on other, locked, sections of the fault. This suggests that the fault zone in the creeping section may be filled with poorly consolidated material throughout the seismogenic thickness, and that this velocity-strengthening material stabilizes the slip, as in the mechanism suggested by Marone and Scholz (1988) to explain the upper stability transition. The lack of consolidation of this material may be explained in turn by high pore pressures, as postulated by Irwin and Barnes. According to this hypothesis, the faults then would be unusually weak in this region, which also would explain the lack of off-fault seismicity.

Episodic slip, typically seen in this area, has been found by both experiment and theory to occur very close to the stability boundary (Sec. 2.3.3). Within the creeping zone, there are also small regions that slip repeatedly in moderate earthquakes and therefore indicate localized unstable patches (Wesson and Ellsworth, 1973). These may be in local irregularities where the normal stress is higher than elsewhere, perhaps because the fault is inclined in a more compressional orientation with respect to the slip vector (Bilham and Williams, 1985). Both seismic and aseismic slip can occur in the transition regions at either end of the creeping segment. A more refined model of this region will be introduced later, in conjunction with a summary of seismic coupling.

6.4.2   *Afterslip*

It has been observed often that surface slip as measured following an earthquake continues to increase for the next year or so, at a diminishing rate with time. This was noticed first in California following the 1966 Parkfield earthquake (Smith and Wyss, 1968) and subsequently was measured for the 1968 Borrego Mountain and 1979 Imperial Valley earthquakes (Burford, 1972; Harsh, 1982). Many earlier examples from Japan are collected in Scholz (1972). This phenomenon, known as *afterslip*, decays in a way that can be fit with

$$u_p(t) = c + d \log t \qquad (6.6)$$

where c and d are constants, u_p is postseismic slip, and t is time measured since the earthquake. Like the steady-state aseismic slip described previously, afterslip often is episodic.

Although its time decay is similar to that of aftershocks, it was shown in the Parkfield case that the cumulative moment of aftershocks is far too small to account for afterslip; it is definitely an aseismic process (Scholz, Wyss, and Smith, 1969). It also varies greatly in its amount proportional to the coseismic slip: At Parkfield the surface slip may have been almost entirely due to afterslip, while at Borrego Mountain, Burford found that it occurred on only one of the two main rupture segments.

It was remarked in Section 4.2.3 that afterslip will be expected as the relaxation of a velocity-strengthening region that has been overdriven by dynamic rupture propagating into it. Consider a spring-slider system with spring stiffness k in which the slider obeys a velocity-strengthening friction law. Assume that the coseismic slip at depth induces a velocity jump in the sediments $(V_c - V^*)$, where V_c is the coseismic slip velocity in the sediments and V^* is a reference velocity taken to be the long-term slip rate. This will induce a stress jump, which is, from Equation 2.28a, and ignoring the transient due to the direct velocity effect, $\tau_c = (a - b)\sigma_n \ln(V_c/V^*)$. If we make the approximation that relaxation occurs at steady-state friction we can obtain an analytical expression for

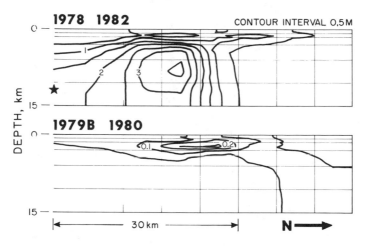

Fig. 6.16 Coseismic and postseismic (top) and immediate postseismic (bottom) slip of the Imperial fault as a result of the 1979 earthquake, obtained by inversion of geodetic data from a dense Mekometer array. The earthquake hypocenter is shown by the star. Compare with the results from accelerometer data in Figure 4.20. (After Crook, 1984.)

afterslip u_p

$$\frac{du_p}{dt} = V^* \exp\left(\frac{\tau_c - u_p k}{(a - b)\sigma_n}\right) \qquad (6.7)$$

which integrates to

$$u_p = \frac{(a - b)\sigma_n}{k} \ln\left(\frac{\alpha k}{(a - b)\sigma_n} t + 1\right) + V^* t \qquad (6.8)$$

where $\alpha = V^* \exp(\tau_c/(a - b)\sigma_n)$. Except at very short times, Equation 6.8 agrees with the numerical integration of Equation 2.27 for a wide range of reasonable choices of parameters.

According to this hypothesis, afterslip should be restricted to the velocity-strengthening regions where the dynamic slip has been impeded. From what has been said before, this should be most prominent where thick sediments cover the fault, and the afterslip should be restricted to the sediments. This was the conclusion reached by Scholz et al. (1969) in the case of the 1966 Parkfield earthquake (Sec. 4.4.1). A more persuasive demonstration can be made for the 1979 Imperial Valley earthquake. There, a dense geodetic network was resurveyed several times after the earthquake, allowing for an inversion for the postseismic slip distribution (Crook, 1984). The results are shown in Figure 6.16, where it can be seen that the afterslip indeed was restricted to the top several kilometers of

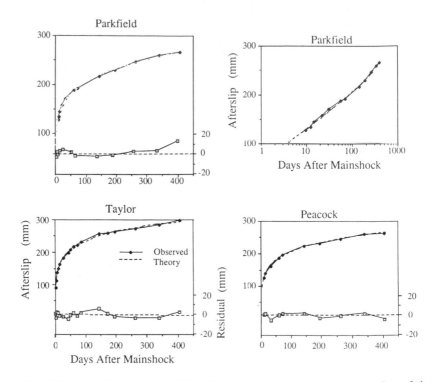

Fig. 6.17 Afterslip of the 1966 Parkfield earthquake plotted as a function of time after the earthquake. The data at three sites – Parkfield, Taylor, and Peacock – have been fit with Equation 6.8 (dashed curve) with the residuals as shown. The Parkfield data are also shown in semilog coordinates for comparison with Equation 6.6.

sediments. Comparison with Figure 4.20 shows that this was the region in which dynamic slip was impeded.

The temporal development of afterslip is shown for three sites at Parkfield in Figure 6.17. These data have been fit with Equation 6.8 and the residuals are shown in the figure. The stiffness k is given by μ/d, where μ is the shear modulus of the sediments and d is their thickness. If we assume that σ_n increases with depth at 15 MPa/km, we may interpret the parameters obtained by fitting the afterslip curves with Equation 6.8. The results of this exercise, for afterslip data for the 1966 Parkfield and 1987 Superstition Hills earthquakes, are that for the range of values of $a - b$ observed in that laboratory, 0.002–0.006, the sediment thickness is predicted to be in the range 1–4 km. Corresponding values of τ_c are 1–2 MPa, which agree reasonably with the estimates of negative stress drop in the sediments shown in Figure 4.21.

It should be remarked, however, that other viscoelastic processes also can result in such relaxation phenomena. McGarr and Green (1975), for example, observed logarithmic relaxation following a rockburst in a deep mine in South Africa, where no sediments were present. In that case a fresh fracture was found, so the recovery may involve subcritical crack growth of the type discussed in Section 1.3.3.

This analysis indicates that the net surface slip, following afterslip, should nearly equal the coseismic slip at depth. However, this analysis does not include distributed deformation within the near-surface sediments, such as discussed by Thatcher and Lisowski (1987), which may lead to surface offset measurements underestimating deep slip.

6.4.3 Seismic coupling

It seems that aseismic slip is a rare phenomenon in faulting in continental crust. Except in areas dominated by active decollement or detachment structures, it usually will not cause a major problem in estimating deformation by summing seismic moments. Faults in oceanic lithosphere also seem to be seismically coupled in the way expected from a standard analysis of an olivine rheology. However, a lack of complete seismic coupling seems to be common in subduction zones, which requires a much fuller discussion.

The great and apparently systematic variation in the size of large earthquakes in different subduction zones was pointed out first by Kanamori (1971). At that time he attributed the difference to variations in the strength of the interface, in which variable weakening was caused by shear heating, and in the extreme cases, by a complete separation of the interfaces. Kelleher et al. (1974) argued that it resulted from differences in interface dip and hence contact area.

The problem was put on a more quantitative footing by Ruff and Kanamori (1980) who used the moment (described in terms of the moment magnitude, M_w) of the largest earthquakes as a measure of seismic coupling, which we have defined more formally in Equation 6.5. This term is suggestively similar to the "strength of coupling," which they defined as the force (per unit length along strike) exerted between the plates at the interface. They then asserted, without further clarification, that "the seismic moment (and therefore M_w)... can be related to the strength of coupling." However, this is not so straightforward; the seismic coupling coefficient, as we have shown, depends on the stability or instability of the frictional constitutive law at the interface, and this may or may not be related to the force exerted across it. We therefore begin by treating their quoted assertion as an unsupported hypothesis, to be tested by the strength of the correlations of M_w with parameters that may be related to the interface forces.

Ruff and Kanamori tried correlating M_w with the depth and lateral extent of the subducting slab, the age of the subducting slab, and the

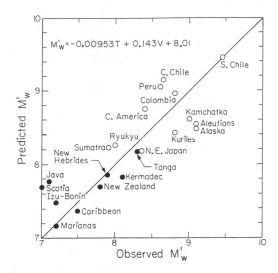

Fig. 6.18 The correlation between M_w' predicted from Equation 6.9 and the observed value for various subduction zones. Solid dots indicate arcs with active back-arc spreading. (From Kanamori, 1986.)

convergence velocity. Individually these correlations were poor: The only acceptable one was found with two independent variables – the convergence velocity v_c and slab age $\mathscr{A}_s$, for which the optimum regression relation is

$$M_w' = -0.00953\mathscr{A}_s + 0.143v_c + 8.01 \qquad (6.9)$$

where $\mathscr{A}_s$ is in Myr, v_c is in cm/yr, and M_w' is a value of M_w equivalent to the total moment release of all earthquakes in a seismic cycle (Kanamori, 1986). This correlation is shown in Figure 6.18.

Peterson and Seno (1984) improved this analysis somewhat by summing the moments and obtaining χ, which they found to vary from 1.57 for south Chile to 0.01 for the Marianas (values greater than 1 can occur because the recurrence time of the largest earthquake observed is larger than the sampling period). They found poor correlations of this with v_c or $\mathscr{A}_s$, although the correlation with age was good on individual subducting plates. They found that χ depends on the sign of absolute motion of the upper plate: It is consistently low for retreating upper plates, variable for advancing upper plates. J. Pacheco (unpublished work, 1989) refined the estimates of χ by properly correcting for the width of the accretionary prism, which should be omitted from the coupled zone width (Sykes and Quittmeyer, 1981), and by determining the lower transition depths directly. This resulted in an increase in the values of χ, but otherwise the findings of Peterson and Seno were confirmed.

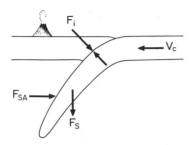

Fig. 6.19 A simple model of the forces on the subduction interface. Although many other factors may be important in individual cases, the horizontal component of the interface normal force, $\mathscr{F}_i$ should increase with convergence velocity v_c, and the vertical component should decrease with slab drag, $\mathscr{F}_s$, which should correlate with $\mathscr{A}_s$, the slab age. This explains the correlation of Equation 6.9 as primarily reflecting variations in $\mathscr{F}_i$. $\mathscr{F}_{sa}$ is a sea-anchor force, proposed by Uyeda and Kanamori (1979).

Correlations with many more parameters were tried by Jarrard (1986a). He concluded that the best correlation with moment release was the one with v_c and $\mathscr{A}_s$ found by Ruff and Kanamori. He also found a reasonably good agreement between M_w' and a qualitative grading of strain class of the upper plate, which strengthens the argument for a connection with interface force. Kanamori (1986) also lists other observations indicating variations in interface force; among them are the presence or absence of an outer rise or back-arc spreading, and compressional earthquakes on the outer rise. Jarrard's strain class, however, was also correlated with the absolute motion of the upper plate and the intermediate dip of the slab: It seems that many of these variables are interrelated in as yet unclear ways.

On balance it appears that seismic coupling correlates positively with convergence rate and negatively with slab age, taken together. A very simple model, sketched in Figure 6.19, shows how these may be related to the force normal to the interface, $\mathscr{F}_i$. If the upper plate responds viscoelastically, on a broad scale, then the horizontal contact force will be proportional to v_c (cf. England and McKenzie, 1982). The vertical contact force also will be inversely dependent on the slab pull force $\mathscr{F}_s$, which is due to the negative bouyancy of the slab and is therefore dominated by $\mathscr{A}_s$ (Richter and McKenzie, 1978). The correlation between v_c, $\mathscr{A}_s$, and M_w' therefore does seem to indicate a correlation between seismic coupling and the interface force. Of course, other forces also may be important locally, such as a sea-anchor force $\mathscr{F}_{sa}$ on the slab (Uyeda and Kanamori, 1979), and other factors, both general and local, may be important in effecting χ, as will be discussed later. But for the principal determinant of seismic coupling we look at the mean normal stress across the interface.

Lay and Kanamori (1981) and Lay et al. (1982) have interpreted seismic coupling with what they call an asperity model. In this model they assume that stresses are distributed heterogeneously on the interface, and that those most highly stressed regions are fully coupled (the asperities) while less-stressed regions normally slip aseismically but may be ruptured in response to rupture of the asperities. The more highly coupled arcs are those with a greater portion of their interfaces composed of asperities. Thus, in Figure 6.12, the zones of concentrated moment release are called asperities and the remaining parts of the rupture zones constitute those aseismically slipping regions whose rupture is triggered by rupture of the asperities. However, we note that in the arcs with the lowest coupling the rupture zones themselves do not cover the interface, so there must be a third category: These are regions that always slip aseismically and do not rupture in response to rupture of adjoining regions.

Examination of ancient subduction shear zones (Sec. 6.3.2) suggests that the interface is lined with sediments, which gradually are consolidated and metamorphosed during subduction. When poorly consolidated, the sediments will be velocity strengthening and stable, but as they become consolidated, $a - b$ will decrease and eventually become negative, and the slip will become unstable. Because consolidation is promoted strongly by normal stress, under geological conditions $a - b$ can be expected to decrease strongly with increasing normal stress, as shown in Figure 6.20 (Marone and Scholz observed in the laboratory that $a - b$ decreased with σ_n for granular materials). Recall now from Section 2.3.3 that there are three possible stability conditions. If $a - b$ is positive, friction is intrinsically stable, stress drop is negative and rupture from an adjoining region will be arrested quickly. If $a - b$ is negative, two possibilities exist: For a given stiffness slip will be intrinsically unstable if the normal stress is higher than the critical value given in Equation 2.32, but if lower it will be conditionally stable, that is, stable unless perturbed by a sudden velocity jump, as by an adjacent rupture, when it can become unstable (see Fig. 2.25). These three possible states are indicated in Figure 6.20, where the stability boundaries are defined at two critical levels of normal stress, σ_{n1}, where $a - b = 0$, and $\sigma_{n2} = k\mathscr{L}(b - a)$.

So there are three possible stability states within the interface: unstable (u), which must be fully coupled seismically; conditionally stable (cs), which may slip stably during the interseismic period but may be triggered by slip initiated in a neighboring u region; and stable (s), which always slips stably. These obviously correspond to the three types of regions within subduction interfaces described before. From Figure

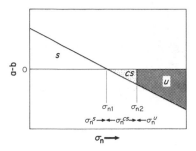

Fig. 6.20 As sediments are subducted, they will be consolidated gradually by the effects of both pressure and temperature. In this conceptual model, the results of Marone and Scholz (1988), which showed that the friction rate parameter $\mathbf{a} - \mathbf{b}$ decreases with normal stress due to compaction, are extrapolated on the assumption that temperature of burial will enhance this trend until they become high enough to cause ductility. This indicates that $(\mathbf{a} - \mathbf{b})$ will change sign with burial, leading to a transition from stable (s) to conditionally stable (cs) to unstable (u) frictional fields. If thermal gradients are about the same in different arcs, the degree to which the different fields are represented in the arc interface will depend on normal stress σ_n proportional to the normal force across the interface $\mathcal{F}_i$.

6.20 we see that these stability fields exist for three ranges of normal stress, $\sigma_n^s \leq \sigma_{n1}$, $\sigma_{n1} \leq \sigma_n^{cs} \leq \sigma_{n2}$, and $\sigma_n^u \geq \sigma_{n2}$. If we invoke equilibrium across the interface and denote the total area of the interface as $A = A^s + A^{cs} + A^u$, where the superscripts refer to the three types of region, we obtain

$$\mathcal{F}_i = \sigma_n^s A^s + \sigma_n^{cs} A^{cs} + \sigma_n^u A^u \qquad (6.10)$$

from which we see that as $\mathcal{F}_i$ increases there will be a corresponding increase in A^{cs} and A^u. We thus are able to interpret the asperity model in terms of the frictional stability relations that we have been investigating, and to account thereby for the dependence of χ on $\mathcal{F}_i$.

A sketch of the resulting interpretation is given in Figure 6.21. Note that the two other features noted with respect to Figure 6.12 are intrinsic to this model; rupture should initiate in the u zones, and u and cs zones should be more prevalent at greater depth.

Kanamori (1986) has also pointed out that there may be an inverse correlation between χ and the extent to which aftershock zones expand in time, as surveyed by Tajima and Kanamori (1985), and mentioned in Section 4.4.2. Certainly if the zone was fully coupled, that is, all u, the entire stressed part of the boundary would rupture dynamically and no such quasistatic expansion would be expected, in accordance with their observations and with most observations of crustal earthquakes on land. Because aftershocks cannot originate within s zones, this observation

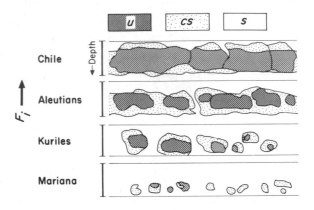

Fig. 6.21 A schematic diagram of the distribution of the different stability fields on the interface of several representative arcs. The difference is interpreted as being due to variations in the interface force $\mathcal{F}_i$.

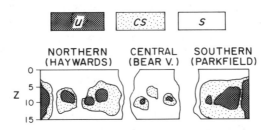

Fig. 6.22 A similar interpretation of the creeping section of the San Andreas fault in central California. Three sections are shown: the northern transitional region; the central region; and the southern transitional region near Parkfield (for a detail of the latter, see Figure 5.17).

seems to require quasistatic rupture expansion in cs regions. This seems plausible, but would require some modeling for further elucidation.

The creeping section of the San Andreas fault also can be interpreted with this basic model, as shown in Figure 6.22. The figure has been broken into three sections to highlight the behavior at the transition zones near the ends on the creeping segment. It has been noticed also that aftershock zones of earthquakes in the creeping segment show considerable expansion in time, similar to that for some subduction zones (Wesson, 1987). This lends additional credence to the suggestion of the previous paragraph.

However much this analysis seems to account for the general trends in seismic behavior of subduction zones, caution should be exercized in using Equation 6.9 to predict the behavior of an individual region, because there are clearly other influential factors besides v_c and $\mathscr{A}_s$. For example, the solid symbols in Figure 6.18 indicate arcs with known or suspected back-arc spreading and the open symbols, arcs with none (see also Fig. 5.29). One therefore might say simply that the extensional arcs are characterized by $M_w' < 8.2$ and the compressional ones, by higher values. This observation agrees with Peterson and Seno's (1984) finding regarding the effect of the absolute upper-plate velocity. It also so happens that all extensional arcs are oceanic and compressional arcs are dominated by ocean–continent contacts (Jarrard, 1986b), so lithology, or the abundance of sediments (Ruff, 1989), also could be a factor.

The northern part of the Philippine Sea plate has subduction with the same polarity on both sides. This kinematically requires that the slabs migrate together, which can be mitigated only by back-arc spreading and trench rollback. Evidently these both happen, but there still must be a sea-anchor force, as shown in Figure 6.19, on the Bonin and Mariana slabs, as suggested by Uyeda and Kanamori (1979). This will have a clear effect on $\mathscr{F}_i$, but it is a local effect, not related, say, to the age of the slab, and not accounted for in the regression shown in Figure 6.18. Similarly, the Tonga and New Hebrides arcs significantly overlap with respect to the Indian–Pacific slip vector, so those slabs must move apart, producing a sea-anchor force of the opposite sign as that in the Marianas.

To give a final example, the Japan trench is observed to become progressively seismically decoupled southwards from Hokkaido to the triple junction with the Sagami trough. Both interacting plates remain the same over this length, and neither the age of the subducting plate nor the convergence rates vary significantly. This effect is completely unexplained by the previous arguments; presumably it occurs because of local forces arising from the migration of the unstable trench–trench–trench triple junction at the intersection with the Sagami trough and Bonin trench.

The conclusion regarding seismic coupling on the Japan trench, cited previously, is based on about 400 yr of historical data (cf. Wesnousky et al., 1984). In most other regions the available seismic record is much smaller, so one has less certainty that a lack of great earthquakes is indicative of the permanent mechanical properties of the interface. A more direct test of the decoupling hypothesis would be geodetic data that indicate a lack of strain accumulation on the upper plate. There is some data of this type for the Hikurangi margin off the North Island of New Zealand. This indicates that compressional strain is accumulating in the southern part of the zone, south of Cape Turnagain, but there is no strain accumulation north of that point, in the section of the upper plate opposite the Taupo volcanic zone (Walcott, 1984, 1987). On the other hand, a lack of strain accumulation has also been reported for the

Shumagin Islands sector of the Aleutian arc (Lisowski et al., 1988), a region which otherwise is thought to be coupled on the basis of its seismic history (Jacob, 1984).

In the model presented, gross differences in the seismic behavior of subduction zones are explained by variations of the mean normal stress across the interface; variations that place parts of the interface into different frictional stability fields. Because earthquakes do not relieve the normal stress and their slip is only about 10^{-5} of their rupture dimensions, one expects the stability field on a given section of interface to be persistent over many seismic cycles. In particular, zones that are within the unstable field, and act as asperities, may result from the subduction of geometrical irregularities or sections of unusually buoyant sea floor, such as seamounts or aseismic ridges (Kelleher and McCann, 1976). This is not to say, however, that variations in moment release observed in the most recent earthquake cycle uniquely define the permanent stability structure of the interface. This latter question is subject to all the caveats of Section 4.5.1.

6.5   INDUCED SEISMICITY

That certain types of human activity, particularly the impoundment of reservoirs, can trigger earthquakes has been known since Carder (1945) noticed an increase of seismicity associated with the filling of Lake Mead on the Arizona–Nevada border. This phenomenon is of interest here because in these cases we have some knowledge of the perturbations involved, allowing further study of some aspects of the earthquake mechanism not afforded by naturally occurring cases. Some discussion already has been made of this, in regard to the Rangely experiment (Sec. 2.4). Here we briefly expand upon several additional points that can be reached by study of this phenomenon. No attempt, however, will be made to provide a thorough review of this subject, which can be found elsewhere (Gupta and Rastogi, 1976; Simpson, 1986; O'Reilly and Rastogi, 1986).

6.5.1   *Some examples*

The most conclusive cases of reservoir-induced seismicity come from those instances where seismic monitoring was initiated prior to impoundment and there was a substantial increase of seismicity upon the first filling. In many other cases, however, a clear onset of felt earthquakes upon impoundment is sufficient to establish a correlation between the two. There are now many known cases, and they seem to fall into

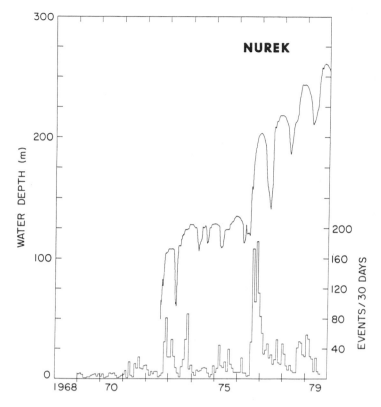

Fig. 6.23 An example of reservoir-induced seismicity with a rapid response: the Nurek reservoir, Tadjikistan, USSR. (From Simpson et al., 1988.)

two broad catagories as far as the response of the crust is concerned (Simpson, Leith, and Scholz, 1988). In many cases there is a rapid response, in which there is an increase in seismicity almost immediately upon reservoir filling. In other cases there is a delayed response, in which the main seismic activity does not occur until some years after the reservoir has been filled and the water level maintained at a stable height. These different responses help us evaluate the several different mechanisms that may be responsible for this phenomenon.

A good example of the rapid response type is given by the Nurek reservoir in Tadjikistan, USSR, shown in Figure 6.23 (Simpson and Negmatullaev, 1981). In this case monitoring began prior to filling, and a pronounced earthquake swarm was observed immediately upon the first filling of the reservoir to the 100-m level. This activity died down and then increased again the next year when the water level was raised to 120 m. Seismicity fell again to a low level, continuing with mild fluctuations

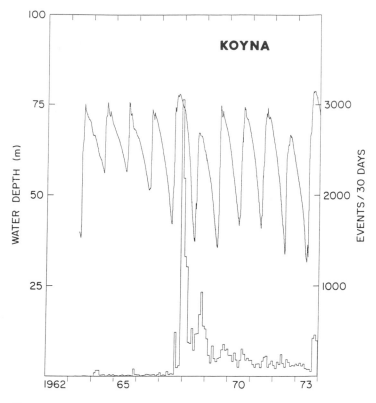

Fig. 6.24 An example of induced seismicity with delayed response: Koyna reservoir, India. (From Simpson et al., 1988.)

until the second stage of filling to 200 m when a greater burst of activity occurred. Other cases of rapid response of this type are at Monticello (South Carolina), Manic-3 (Quebec), Kariba (Zimbabwe), Kremasta (Greece), and Talbingo (Australia).

In Figure 6.24 a case of delayed response, the Koyna dam in the Deccan Traps of western India, is shown. Filling of the reservoir began in 1962 and shortly afterward small earthquakes were noticed. However, large earthquakes did not occur until late 1967, when magnitude 5.5 and 6.2 events occurred. The largest earthquake, which caused considerable damage and loss of life, was a strike-slip event with a normal depth ($\sim$ 5 km), located some 10 km downstream from the dam. It occurred just after the water level had reached its maximum value. In another delayed response case, at Oroville, California, an earthquake ($M = 5.7$) occurred seven years after impoundment, just after a period of record low water level. At Aswan, Egypt, an $M = 5.3$ earthquake occurred in late 1981, some years after the latest stage of filling had begun in 1975.

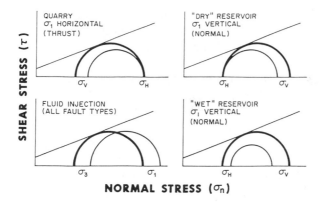

Fig. 6.25 Mohr circle representation of several direct effects of reservoir loading and quarry unloading on the strength of rock underneath. Light circle is before and heavy circle after loading. (From Simpson, 1986.)

Cases of rapid response tend to produce swarms of small earthquakes, located at shallow depths in the immediate vicinity of or just below the reservoir. In the delayed response cases, the earthquakes tend to be larger, at greater depth, and are often at some distance ($\sim$ 10 km) from the deep part of the reservoir.

6.5.2 *Mechanisms of reservoir-induced seismicity*

Direct effects An elementary understanding of the mechanism can be gained by considering the reservoir to produce a static increase in load and pore pressure equivalent to the water depth (Snow, 1972). Thus a 100-m-deep reservoir would produce an increase in vertical stress and pore pressure of 1 MPa immediately under the reservoir. Several illustrative examples are shown in Figure 6.25, where the effect on the Mohr circle with respect to a Coulomb failure envelope is shown. Cases labeled "dry" and "wet" indicate whether or not the reservoir fluid can permeate the rock beneath the reservoir and increase the pore pressure there by diffusion.

The effect of the load will increase or decrease the radius of the Mohr circle, depending on the tectonic environment, and the effect of increasing pore pressure will move the Mohr circle towards the origin. Thus an increase in pore pressure will always favor induced seismicity, whereas the effect of the load will be to trigger earthquakes for a region of normal faulting, inhibit them for thrust faulting, and have no effect for strike-slip faulting. In quarrying operations the vertical load is reduced, which in a thrust faulting environment can trigger earthquakes, as in a case described by Pomeroy, Simpson, and Sbar (1976). On the other hand, reservoirs impounded in thrust terrains can move the stress state away from the failure condition and result in seismic quiescence, as was observed at Tarbela dam in Pakistan (Jacob et al., 1979).

Coupled poroelastic effect   The above analysis considered equilibrium states, sometime after filling is complete. The effect of the load will of course be immediate, due to an elastic response, whereas it might be guessed at first that pore pressure changes would be delayed by the time required for pressure to diffuse to depth or, in the case of reservoirs surrounded by unsaturated layers, for flow to occur to raise the water table. However, there also will be an immediate increase in pore pressure, due to the load elastically compacting the pore space [see Rice and Cleary, 1976; and for application to this problem, Bell and Nur (1978) and Roeloffs (1988)]. We call this the coupled poroelastic effect.

When fluid flow can be neglected, as in sudden loading, the poroelastic constitutive equation is (Rice and Cleary, 1976)

$$\Delta p_{\text{p}} = -(2\mathscr{G}\mathscr{B}/3)[(1 + \nu_{\text{u}})/(1 - 2\nu_{\text{u}})]\,\Delta\varepsilon_{\text{v}} \qquad (6.11)$$

where Δp_{p} is the change in pore pressure due to a volume strain increment $\Delta\varepsilon_{\text{v}}$; $\mathscr{B}$ is Skempton's coefficient, $\mathscr{G}$ is the shear modulus, and ν_{u} is the undrained Poisson's ratio. $\mathscr{B}$ depends on the geometry of the pore space relative to that of the stress field and takes a value between 0 and 1. We can alternatively write the response to a change in normal stress $\Delta\sigma_{\text{n}}$ as

$$\Delta p_{\text{p}} = (\mathscr{B}/3)[(1 + \nu_{\text{u}})/(1 - \nu_{\text{u}})]\,\Delta\sigma_{\text{n}} \qquad (6.12)$$

Thus, for example, if $\nu_{\text{u}} = \frac{1}{3}$, then $\Delta p_{\text{p}}/\Delta\sigma_{\text{n}} = \frac{2}{3}\mathscr{B}$.

Perhaps the most apt natural example of this effect, for the present purposes, is given by some hot springs in Japan that show a coseismic response to earthquakes. These springs are at Dogo, near Matsuyama, on the north coast of Shikoku some 150 km from the rupture zones of the Nankaido earthquakes (see Fig. 5.3). The springs are artesian, have been in commercial operation for many centuries, and have a long history of being affected by earthquakes, most particularly those on the Nankai trough. In the most recent event, in 1946, the water level in four springs dropped 12 m at the time of the earthquake and recovered to their previous levels at an exponential rate over the next 70–80 days, as shown in Figure 6.26 (Toyoda and Noma, 1952). Very similar behavior was observed in the previous Nankaido earthquakes of 1854, 1707, and 1605.

The affected springs are located at an intersection of NW- and NE-striking faults. Nearby springs, located on the NW-striking fault system, were unaffected by the earthquake (R. Grapes, unpublished

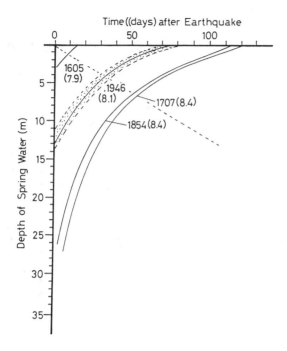

Fig. 6.26 Coseismic response and recovery of water level in the Dogo Hot Springs, Shikoku, to the last four Nankaido earthquakes. Stippled region encloses the data from frequent measurements of four wells in 1946 by Toyoda and Noma (1952). Only the coseismic drop and recovery times are known for the three earlier earthquakes: The curves are sketched in, following the form observed in 1946. (From an unpublished sketch of R. Grapes.)

work, 1988). For a change in uniaxial stress in a cracked solid where the cracks are perpendicular to the stress, $\mathscr{B} \approx 1$, and for parallel cracks, $\mathscr{B} \approx 0$. The Nankaido earthquake stress relief produced a uniaxial tension oriented NW in the vicinity of Matsuyama, and so would cause a reduction in pore pressure in the NE-striking fractures and not in the NW-striking ones, which explains the selective response of the springs. The water level drop at Dogo indicates, from Equation 6.12, a stress drop there of about 0.18 MPa, which is in reasonable agreement with geodetic data. The 80-day recovery time reflects the hydraulic diffusivity of the local natural plumbing system supplying the springs.

In the case of the Dogo hot springs a coseismic stress reduction caused a pore-pressure drop at a considerable distance from the rupture. In the filling of a reservoir, the reservoir load can, by the same mechanism, cause an immediate increase in pore pressure at some depth beneath the reservoir.

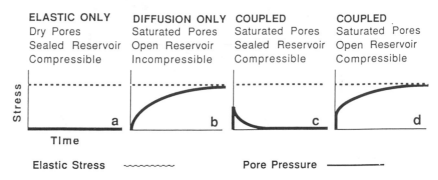

Fig. 6.27 Schematic diagram showing the different effects involved in reservoir-induced seismicity. (From Simpson et al., 1988.)

The coupled poroelastic effect was mentioned briefly earlier regarding the compound Superstition Hills earthquakes, in which a left-lateral strike-slip earthquake triggered delayed rupture on an abutting right-lateral fault orthogonal to the first (Sec. 4.4.3 and Fig. 4.30). The first rupture would reduce the normal stress on the abutting fault, thereby bringing it closer to the failure state. Application of Equation 6.12 shows that it also would reduce the pore pressure by as much as $\frac{2}{3}$ the amount of reduction of normal stress, so that the weakening effect would be slight at first. Subsequently, the pore pressure would be recharged in the way shown in Figure 6.26, bringing the second fault progressively closer to failure. This mechanism can explain the 12-hr time delay between the two earthquakes (Hudnut, Seeber, and Pacheco, 1989).

Interpretation of the two types of induced seismicity   A summary of the different effects is given in Figure 6.27, which indicates changes in stress and pore pressure at some depth beneath a reservoir, filled instantly at the time origin. The elastic and coupled poroelastic effects are instantaneous. Diffusion labeled in the figure refers to diffusion from the reservoir; there is a shorter diffusion time indicated for the relaxation of the coupled effect because in that case it is assumed that the diffusion is from a region of high $\mathscr{B}$ to a nearby region of low $\mathscr{B}$, with a smaller path length than from the reservoir.

Cases of rapid response must be primarily a result of the elastic-coupled effects, and the delayed response to effects dominated by diffusion or flow. These different mechanisms also make some sense with respect to the general characteristics of the seismicity produced in the two types of cases.

Talwani and Acree (1985) have shown a relation between the square of the distance to the seismicity and the delay time after filling, and have used this to obtain an estimate of the hydraulic diffusivity. This has been discussed further by Simpson et al. (1988), who discriminated between the two types but also arrived at a similar value for diffusivity, 10^4–10^5 cm^2/s. This value, which is similar to that obtained from earthquake precursors (Scholz, Sykes, and Aggarwal, 1973), is significantly larger than that expected from the permeability of bulk rock, and implies that the diffusivity must be controlled by large-scale fractures.

6.5.3 *Mining-induced seismicity*

Seismic activity often is caused by mining activity and may constitute a major hazard to mining. Such activity ranges from minor rockbursts or "bumps" to tremors of considerable magnitude. This activity may involve spallation into cavities or from pillars, or collapse, but many mine tremors are physically indistinguishable from natural earthquakes (McGarr, 1984) and it is these that will be discussed here. These tremors exhibit a double-couple focal mechanism and have the same radiative spectrum and stress drop as ordinary earthquakes.

An example is given in Figure 6.28, from a South African gold mine that is being worked at a depth of about 3 km. The activity is concentrated in the immediate vicinity and just in front of the working stope faces. The maximum principal stresses in this mine are vertical, and the mine tremors reflect normal faulting associated with the stress concentration around the stope face, as shown in the cross section. The stope closes down behind the working face, as shown, and it is found that seismic moment release rate is proportional to the rate of stope closure. That is, for some specified time,

$$\sum M_0 = \mu \, \Delta V_s \qquad (6.13)$$

where ΔV_s is the volume of stope closure (McGarr, 1976, 1984).

The absolute stresses in this mine are known, and the stress drops for the mine tremors are found to be a small fraction of them (Spottiswoode and McGarr, 1975; Spottiswoode, 1984). This allows the seismic efficiency (Eq. 4.8) to be calculated, and it is found to be less than 1%. Because mine tremors are indistinguishable from natural earthquakes, there is no reason why these results should not apply to the natural case also. This finding reinforces the statements made in Section 3.4.3 criticizing the common assumption that stress drop is equal to stress.

6.5.4 *Induced seismicity as a stress gauge*

One of the most interesting features of induced seismicity is how widespread it is. It is by no means simply restricted to cases in which reservoirs have been built over what one might otherwise recognize as an

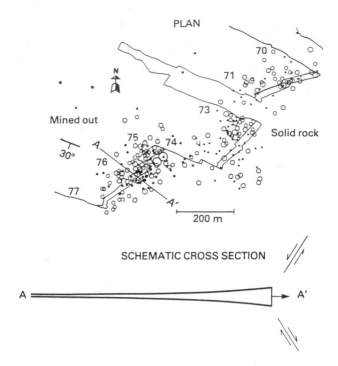

Fig. 6.28 Plan and schematic cross-section views of seismicity associated with the East Rand Proprietary Mines gold mine, at a depth of about 3 km in the Witwatersrand, South Africa. The mine tremors, indicated by open circles, occurred during a 100-day period in 1972; the position of the stope faces at the beginning and end of that period are shown. The strata and stopes dip about 30° toward the SSW. As shown in the cross section, the stopes are slots, 1–2 m high, which close down behind the working face. The mechanisms of the earthquakes are indicated in front of the working face. (From McGarr, 1984.)

active fault. Just among the cases mentioned previously are those far from any naturally occurring seismicity and several, such as the cases in Quebec, India, and Australia, which are in midplate environments. The analysis shows that in the optimal case the effect of reservoir impoundment is to move the state of stress closer to the failure condition by an amount equivalent to an increase of shear stress of only about 1 MPa or less. The obvious conclusion must be that a large part of the earth's lithosphere must be stressed that close to failure, even in regions not undergoing active tectonic deformation. From this point of view, a study of the many cases of reservoirs that do not induce seismicity would be interesting.

The lesson that seems to be learned from this is that the greater portion of the lithosphere is stressed to near its failure strength, consistent with the point made in Section 5.4 that the schizosphere is in a self-organized critical state. The schizosphere can relax that stress only by earthquake stress drops, which typically lie in the range 3–10 MPa, so the change induced by a reservoir is significant with respect to that amount. Natural intraplate earthquakes indicate that strain accumulates in all regions, though at greatly varying rates. It is hardly surprizing that many small regions are within 1 MPa of failure, which may be induced by the filling of a nearby reservoir. It is simply a case of a very rapid artificial loading cycle being imposed on a very slow natural one.

7 Earthquake Prediction and Hazard Analysis

The most important social benefit from earthquake research is the use of that knowledge to reduce the hazard earthquakes pose to mankind. These applications may take several forms, which range from the construction of various kinds of hazard maps that permit the prediction (in a probabilistic sense) of the exposure to future ground shaking to the actual prediction of specific earthquakes. Here the current status of these developing fields is summarized.

7.1 INTRODUCTION

There was a time when the weather belonged to the gods. ... There was a time when the earthquake was equally enveloped in mystery, and was forecast in the enigmatic phrases of the astrologer and oracle; and now that it too has passed from the shadow of the occult to the light of knowledge, the people of the civilized earth – the lay clients of the seismologist – would be glad to know whether the time has yet come for a scientific forecast of the impending tremor.

G. K. Gilbert (1909)

7.1.1 *Historical*

Throughout the ages the prediction of earthquakes, along with wars, pestilences, famines, and floods, has been the occupation of soothsayers and other such self-proclaimed prophets. The skepticism that accompanied the rise of the scientific study of earthquakes has not, however, removed prediction as a serious goal. Lyell, for example, devoted considerable attention to the subject. He was impressed that throughout history, in many disconnected societies, nearly identical traditions could be found that held that earthquakes are preceded by unusual phenomena. Among such traditional precursors are peculiar weather (particularly involving unusual rainfall), uplift and subsidence of the ground, and anomalous behavior of animals. Lyell thought that the universality of such popular myths must indicate some underlying truth.

Most scientists are now more apt to believe that many such traditions are apocryphal and owe more to the mass psychological reaction of a population stunned by a sudden disaster than to a physical cause. However, since scientific instruments began to be scattered throughout the globe at the beginning of the twentieth century, a new list of precursory phenomena has been accumulated, returning the topic back to the realm of credibility.

The subject has remained nonetheless controversial, and its popularity as a pursuit that the ambitious scientist chooses to advance his or her career has fluctuated much more than others. As a subject that both is genuinely scientifically controversial and unfailingly attracts the attention of the public, it frequently prompts two contrasting responses among scientists: overzealousness and blind skepticism. The serious scientist, in considering this subject, therefore needs to pay careful attention to balance and to avoid such visceral reactions. The scientific data remain so scattered and ambiguous that they require particularly cool evaluation.

7.1.2 *Strategy for a difficult subject*

We need to be careful at the outset about what we mean by earthquake prediction. Usually, it is described as meaning the accurate forecasting of the place, size, and time of an impending earthquake. "Accurate" is the key word in this definition. The prediction has to be specific enough that when and if the predicted earthquake occurs there is no doubt that it was the one specified.

Prediction is usually broken into three categories: long-term, intermediate-term, and short-term. The first involves a prediction made years to decades in advance, the second with a warning of a few weeks to a few years in advance, and the third, hours to a few weeks. Accuracy of the time predicted must be a small fraction of the time scale for each type of prediction, but the size and location of the predicted earthquake should be well known for each time scale (i.e., we must know which earthquake we are talking about in each case). From the point of view of social response, these types of prediction are very different, and that is one reason for making this type of classification (Wallace, Davis, and McNally, 1984). The long-term time scale allows for major engineering works to be undertaken to reduce damage and loss of life. On the intermediate scale, emergency plans and services can be organized and mobilized and preparations can be made, whereas the short-term prediction scale is appropriate for issuing warnings to the general population. However, there also seem to be physical reasons for making this distinction between the three types of prediction. That is, they may involve different physical processes and hence different scientific tasks.

Long-range prediction amounts to determining the recurrence time of earthquakes on a certain fault segment and predicting the approximate time of the next earthquake from the known time of the previous one. The first part of this problem has been discussed in detail in Chapter 5. Because it is based on a primary feature of the earthquake process, the repetition of the seismic cycle, this problem is well posed, and considerable progress has been made. When implemented on a regional scale it constitutes seismic hazard analysis in which the time origin is specified within the recurrence intervals of all faults in the region. A methodology for proceeding toward that goal is outlined in Section 7.4.3.

Intermediate- and short-term prediction, by contrast, both depend on secondary processes, the identification of precursory phenomena of various kinds that indicate that the loading cycle has reached some advanced, perhaps imminent, stage. The types of precursors that have been observed appear, in some cases, to define categories of intermediate- and short-term precursors, independently of the societal needs mentioned above. In other cases the distinction between the two types is not so clear. Such precursory phenomena are simply anomalous signals that are observed in the vicinity of and some time prior to the occurrence of an earthquake. Quite apart from the question of whether such signals are truly anomalous, such precursory phenomena are not, ipso facto, premonitory to the earthquake. While some precursors may indicate the initiation of processes directly involved with the breakdown that irreversibly culminates in the earthquake, others simply may reflect an overall advanced state of the loading cycle in the broad tectonic process of which the earthquake is only a part. Ishibashi (1988) calls these two cases physical and tectonic precursors. Thus the doughnut seismicity pattern (Sec. 5.3.4) is a tectonic precursor, whereas foreshocks, in the way defined in Section 4.4.2, are physical precursors. The first warns that the conditions for earthquake occurrence are present, the other that the instability process is underway. Still other "precursors" may be unrelated to the occurrence of the earthquake, being merely coincidental.

This situation underlines the need to understand both the types of physical processes that can precede earthquakes and the types of geophysically observable phenomena that may be manifested by them. With such a model in mind, even if it is only conceptual, one may better be able to evaluate the meaning of various precursory phenomena. For this reason we concentrate here on models of precursory phenomena. Descriptions of observed precursors are limited to a brief summary and illustrative examples. Thorough compendia of precursory phenomena may be found elsewhere (Rikitake, 1976, 1982; Mogi, 1985).

Long-term earthquake prediction is amenable to direct scientific investigation and so has been progressing steadily over the past twenty years. This has not been the case with intermediate- and short-term prediction. Owing to the paucity of instrument deployment, the observation of precursors has, in the past, been a matter of chance, and their identification has been made ex post facto. Such observations as there have been are generally of a fragmentary nature and subject to many interpretations. So, too, models that have been developed to explain such phenomena have risen and fallen in their acceptability. The data have not been sufficient to make definitive cases, new precursors are slow in coming, and old precursors are continually subject to new doubts. The bulk of the evidence supports the existence of premonitory phenomena, but it is difficult to point to a single precursor that can be accepted universally as such.

A fundamental problem is that a proper scientific search for precursors cannot be done without a long-term prediction being made first, so that the area of investigation may be narrowed. It is only recently that this step has been taken, and earthquake prediction experiments have been initiated in areas where earthquakes are expected, based on a long-term prediction. These are in the region of the Tokai gap, near Shizuoka, Japan, and at Parkfield, California. Extensive arrays of instruments have been deployed in both places, but the expected earthquakes have not yet occurred.

7.2   PRECURSORY PHENOMENA

In this section various precursory phenomena are briefly reviewed. In this context it is not usually specified whether the observed phenomenon represents a physical or tectonic precursor, because that can be done only by interpreting it with a model. I am not even able to rule out the possibility that some of the precursors described are spurious. All I can do in that regard is to restrict the discussion to precursors that either have been repeatedly reported by many workers or for which the data appear particularly reliable.

7.2.1   *Preinstrumental observations*

Because the instrumental period is short and instrumental deployment sparse, we cannot afford to ignore historical descriptions of precursory phenomena. Such accounts are anecdotal and often, even when they can be confirmed from several sources, mysterious to the physical scientist trying to understand the physical processes responsible. In other cases, though, the observations clearly indicate a tectonic phenomenon that warrants serious study.

Striking illustrations of this latter type are given by four cases of crustal uplift that preceded earthquakes in Japan, first discussed in detail by Imamura (1937). These cases are:

1. the Adigasawa earthquake of 1793, in which the land rose 1 m 4 hr before the earthquake;
2. the Sado earthquake of 1802, in which there was also a 1-m uplift 4 hr prior to the earthquake;
3. the earthquake at Hamada in 1872, where a 2-m uplift preceded the earthquake by 15–20 min;
4. the Tango earthquake of 1927, preceded by a 1.5-m uplift $2\frac{1}{2}$ hr before the earthquake.

All of these cases were on the Japan Sea coast, where they were observed as sudden withdrawals of the sea from the land. On this coast the daily tidal range scarcely exceeds 30 cm, so these changes are large with respect to normal tides. Subsequent analysis also has shown that these cases each involved large earthquakes on faults that either crossed the coast (Hamada, Tango) or were within a few kilometers of it (Sado, Adigasawa). Each involved reverse faulting, except Tango, which was strike slip.

In general, when working with historical documents such as letters or diaries, the older they are, the more difficult they are to confirm. For example, in the Adigasawa case, Usami (1987) was unable to find mention of the precursory uplift in the contemporary diary of the local lord, and considers this case doubtful.

The Sado case is not disputed, having been established from several sources. As a visit to the site attracted the curiosity of the author, it is singled out for description here. The earthquake occurred off the Ogi peninsula on the southwest tip of Sado Island at about 2 p.m. on December 9, 1802 (see Fig. 7.5, Sec. 7.2.2, for location). At 10 o'clock in the morning of the same day, a strong shock occurred that dislodged plaster in some of the houses in the town of Ogi. At this time the water in the harbor receded 300 m, leaving some parts of the harbor dry. Four hours later the mainshock occurred, destroying most of the houses in the town and producing further uplift in which the sea retreated 500 m. Imamura estimated the preseismic uplift to have been about 1 m and the coseismic uplift the same amount, for a total of 2 m.

The Ogi peninsula has been subjected to repeated earthquakes of this type, as recorded in a flight of seven uplifted marine terraces (Ota, Matsuda, and Naganuma, 1976). They showed that the terraces were uplifted by earthquakes on a reverse fault off the coast. This mechanism is similar to more recent earthquakes in the Sea of Japan, such as that in 1964 off Niigata. The 1802 terrace can be observed for about 25 km along the coast. From measurement of the tilt of this terrace and the magnitude of the earthquake (6.6, estimated from damage reports) they concluded that the earthquake could not have been more than a few kilometers from the coast.

It may be argued in this case that the preseismic uplift was the result of coseismic motion during the foreshock of the morning, in which case the precursor was nothing more than a foreshock – in itself not at all remarkable. However, from the damage account the foreshock was much smaller than the mainshock, so it hardly is credible that the two would have produced the same amount of uplift. It seems more likely that the foreshock accompanied the precursory uplift, which was itself largely aseismic.

In the Hamada and Tango cases there were no strong foreshocks that could render the interpretation ambiguous. At Hamada there were 2–3 days of small foreshocks. About 15–20 min before the mainshock, water withdrew, so that at one point on the coast a small island, formerly separated from the mainland by water at a depth of about 2 m, became connected by dry land. Just before the earthquake, the water returned. In this case, both coseismic uplift and subsidence were observed along the coast, the fault evidently crossing the coast nearly tangentially. In the Tango case the uplift was observed in a localized region just to the south of where the left-lateral Gomura fault crosses the coast, suggesting that precursory motion on that fault, which slipped coseismically, may have been responsible.

These cases are of interest because they suggest that a large amount of preseismic slip may have occurred a few hours prior to these earthquakes. In reviewing these cases, Kanamori (1973) suggested that perhaps Sea of Japan earthquakes are special in having this characteristic. However, it seems just as likely that these observations arose because of the optimal viewing opportunities there – a large earthquake occurring very close to a densely populated coastline of low tidal range. Unfortunately, more recent earthquakes in the Sea of Japan, which will be discussed later, were much further offshore, so we have had no chance to check these old observations with modern methods.

By contrast to these types of historical observations, it is of interest to cite a few examples of the more "mysterious" type. Examples of these also are found related to the Sado earthquake and are recounted by Musha (1943) from the diary of a traveler who was visiting Ogi at the time. According to this diary, early on the morning of the earthquake a sea captain could not decide about embarking from the port of Ogi because of the peculiar weather that prevailed. A ground fog had lifted to halfway up the mountains, the tops of which were still clear. The captain had never seen such weather before and so could not decide on the safety of sailing. Perhaps more astonishing, the same traveler reported that upon visiting the local gold mines a few days later and inquiring about casualties during the earthquake, he was told that there were none. The reason given was that three days prior to the earthquake the temperature underground increased, and a fog appeared in the workings. According to Japanese folklore, this indicated an impending earthquake. Consequently, work was suspended so that no one was underground when the earthquake occurred. In Japanese tradition, both the ground fog at Ogi and the hot air in the mine are caused by the same thing, the *chiki* (literally, air from the earth).

For a fascinating account of such traditional earthquake precursory effects, including earthquake fogs, lights, and anomalous behavior of animals, the reader is directed to a book of Tributsch (1983).

7.2.2   *Intermediate-term precursors*

Anomalous effects that persist for periods of a few weeks to a few years prior to an earthquake are called intermediate-term precursors. Spatially they may extend over a region encompassing or somewhat larger than the entire rupture zone of the impending earthquake. Here we provide a brief review of these, with selected examples.

Seismicity patterns   The most frequently reported precursory phenomena involve patterns of seismicity. This probably reflects the fact that seismicity is the only earthquake-related process that is routinely monitored on a worldwide basis and constitutes the largest data set that may be examined for evidence of precursors. As noted in Section 5.3.4, the seismic cycle is accompanied by characteristic patterns in seismicity, first pointed out by Fedotov (1965) and enlarged upon by Mogi (1977, 1985). This is illustrated in Figure 7.1, which is modified from Mogi (1985) with a few additions and deletions. It should be emphasized that this is a summary of the various kinds of seismicity patterns that have been described; few, if any, earthquake cycles have been observed that contain all of these features.

The "principal rupture" is followed by an aftershock sequence A, which commonly is concentrated near the ends of the rupture and decays hyperbolically in time into a postseismic period of quiescence Q_1. Q_1 occupies an appreciable portion, typically 50–70% of the recurrence period T, and usually extends over the entire region surrounding the rupture zone. This is followed by a general increase in "background" seismicity B over the whole region (Mogi, 1981, refers to this as the "active period" and considers these a type of foreshock, but we avoid this terminology). This is sometimes followed by an intermediate-term quiescence Q_2, which typically extends over the entire zone surrounding the rupture and lasts for several years. If Q_2 does not extend to the

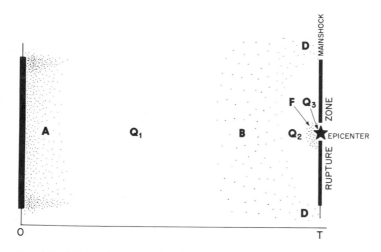

Fig. 7.1 Schematic space–time diagram illustrating various patterns of seismicity that may be recognized during the seismic cycle. (From Scholz, 1988.)

surrounding regions, where seismicity may be augmented, a doughnut pattern **D** appears.

The succeeding principal rupture is preceded by the immediate fore-shocks **F**, which typically begin a few weeks to days prior to the mainshock and which usually are concentrated close to the hypocentral region. A pronounced lull in foreshock activity often is observed just prior to the final rupture, defining a short-term quiescence Q_3. F and Q_3 are short-term precursors and the patterns **B**, **D**, and Q_2 are intermediate-term precursors. One example of the pattern Q_1, **B**, and **D**, for the San Francisco Bay area has been described (Sec. 5.3.4, Fig. 5.31). According to the interpretation given in that section, these are all tectonic precursors. However the quiescence Q_2 may be, as we shall see, a physical precursor and so deserves additional attention here.

This type of quiescence has been reported many times. We discuss here the case of the 1978 Oaxaca, Mexico, earthquake, both because it is well documented and because it served as the basis for a successful prediction (Ohtake, Matumoto, and Latham, 1977, 1981). The time history of this quiescence is shown in Figure 7.2 and its spatial development in Figure 7.3. In mid-1973 the quiescence began abruptly and simultaneously over a region 720 km in linear extent with an area of approximately 7×10^4 km². This area is much larger than the aftershock zone of the $M = 7.8$ earthquake that followed in 1978, hence the quiescence is a regional phenomenon that is not restricted to the eventual rupture plane. The quiescence was terminated by a period of renewed activity that

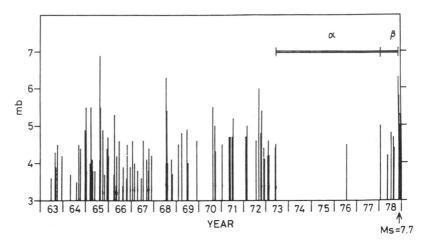

Fig. 7.2 Seismicity of the Oaxaca, Mexico, region, showing the development of the quiescence (α), followed by a return to normal activity (β), just before the earthquake of 1978. (From Ohtake et al., 1981.)

began 10 months before the mainshock. Ohtake remarks that this period of renewed activity prior to the mainshock is typical for earthquakes on the Pacific coast of Mexico, and Mogi (1985) describes the same behavior prior to some cases in Japan. The final foreshocks occurred as a cluster, beginning 1.8 days before the mainshock, followed by a short-term quiescence in the last 12 hr (McNally, 1981).

It can be noticed in Figure 7.2 that there was a sharp reduction of events of $M < 4.5$ beginning in 1967. This corresponds to a global reduction of earthquake perceptibility due to the closing of several large seismic arrays in that year (Habermann, 1981, 1988). Such change in catalogue completeness may introduce spurious rates of seismicity change into the record, but in the Oaxaca case, Habermann concluded that correcting for this effect did not change the pattern noted by Ohtake. One can verify this simply by ignoring the smaller events in Figure 7.2. In general, precautions need to be taken regarding heterogeneities in seismicity catalogues (e.g., Perez and Scholz, 1984). In many other cases the validation of a quiescence requires rigorous statistical testing and is a controversial subject. A collection of papers debating this topic may be found in the volume edited by Stuart and Aki (1988).

There have been a number of attempts to relate the duration of these and other intermediate-term precursors to the size of the earthquake that follows. The duration of intermediate-term quiescences for a large number of cases is plotted against the magnitude of the following earthquake in Figure 7.4. Although there does appear to be a tendency for duration to increase with magnitude, the great scatter precludes any definite

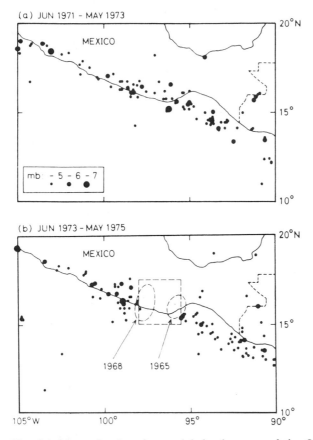

Fig. 7.3 Maps showing the spatial development of the Oaxaca quiescence. (From Ohtake et al., 1981.)

conclusion being reached as to the nature of the relationship. The line drawn through the data is an empirical relationship obtained from other types of precursors (Fig. 7.27).

Another intermediate-term seismicity pattern that has been noticed is a tendency for earthquake swarms to occur in and around the focal region several years before the earthquake (Evison, 1977). A particularly well-documented example was observed for the Coalinga, California, earthquake of 1983 (Eaton, Cockerham, and Lester, 1983).

Crustal deformation Cases of anomalous crustal deformation prior to earthquakes have also been reported frequently. Many of these were recorded on individual tiltmeters or strainmeters, and in these cases it is difficult to determine if they resulted from local site effects or reflected broad-scale phenomena. The two examples we shall describe,

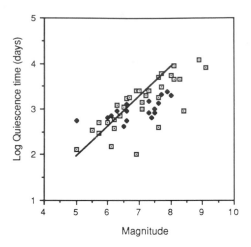

Fig. 7.4 The duration of intermediate-term quiescences plotted as a function of the magnitude of the shock that followed. Open symbols are data from Kanamori (1981) with additions by Mogi (1985), and solid symbols are data from Wyss and Habermann (1988). The line is the empirical relation obtained from other precursors, shown in Fig. 7.27.

however, were obtained from conventional geodetic surveys of regional extent.

The two cases to be described were for large reverse fault events that occurred in the Sea of Japan off the west coast of Honshu. These are the Niigata earthquake of 1964 ($M = 7.5$) and the Sea of Japan earthquake of 1983 ($M = 7.7$). The locations are shown in Figure 7.5.

The Niigata earthquake ruptured a west-dipping reverse fault just off the coast and produced uplift on the offshore island of Awashima. Repeated leveling along the Honshu coast prior to the earthquake produced the pattern shown in Figure 7.6. A steady rate of uplift and subsidence was observed up to 1955, but between 1955 and 1959 there was a rapid uplift of several centimeters all along the coast opposite the earthquake. This activity stabilized from 1959 to the time of the coseismic motions of 1964. These data therefore indicate that a broad region surrounding the rupture zone experienced a rapid uplift some five years prior to the earthquake.

Mogi (1985) has pointed out that there possibly may be problems with these data, however. He notes that there is a correlation between the 1955 survey results and topography and that this may indicate that topography-related surveying errors have contaminated this survey. If one ignores the 1955 survey on this basis, the evidence for an anomalous uplift from 1955–59 no longer exists. It is difficult to prove this one way

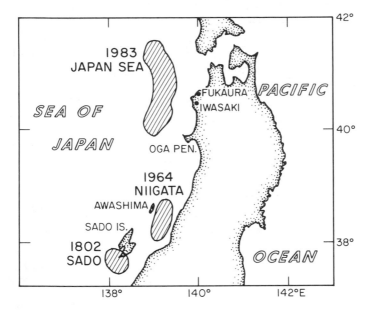

Fig. 7.5 Location map, Sea of Japan.

or the other, so this case illustrates the problem of confidently distin-
guishing precursory phenomena from spurious noise, because there is
often a low ratio of signal to noise. A similar problem arose regarding a
case of anomalous uplift in southern California that was reported on the
basis of leveling data and that was vigorously debated (e.g., Castle et al.,
1984).

The anomalous crustal movement preceding the 1983 earthquake, also
discussed in Mogi (1985), was very similar to the Niigata case, but has
not been challenged. Leveling around the Oga Peninsula and the cape of
Iwasaki showed a rapid increase in the rate of uplift beginning in the late
1970s. Tide gauges in these places showed that this began in 1978 and
continued steadily up to the earthquake, amounting to about 5 cm (Fig.
7.7). A tiltmeter at Oga also showed anomalous behavior beginning in
1978, and a volume strainmeter at the same locality showed anomalous
strain episodes from 1981 to 1984 (Linde et al., 1988). The 1978–83
period corresponded to a quiescence in the zone of the earthquake
(called a seismic gap in Fig. 7.7, in Mogi's terminology). These observations indicate, as in the Niigata case, a broad uplift of several centimeters some five years before the earthquake. Because the 1983 earthquake
occurred on an east-dipping reverse fault, the position of the uplifted
region with respect to the faults was different in the two cases, and this
can allow for a different interpretation of the mechanism (Sec. 7.3.4).

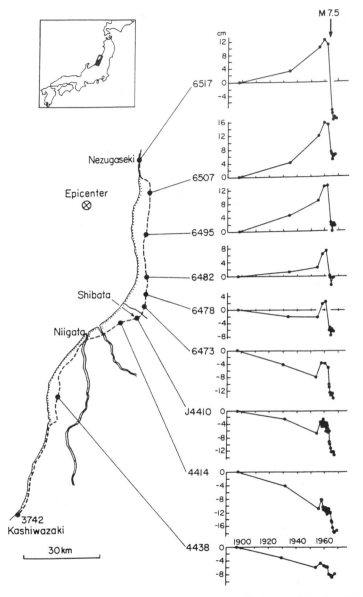

Fig. 7.6 Uplift pattern preceding the 1964 Niigata earthquake. (From Mogi, 1985.)

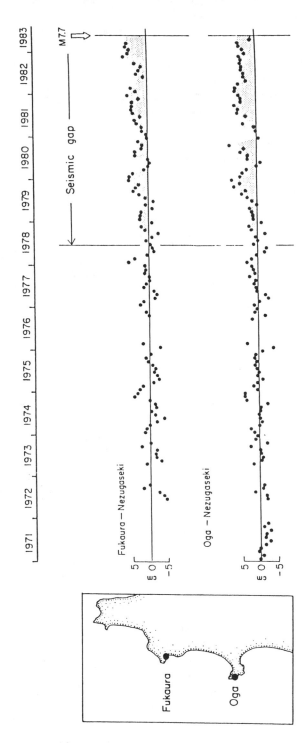

Fig. 7.7 Data from two tide gauges, differenced to Nezugaseki (Fig. 7.6), showing slow uplift prior to the 1983 Sea of Japan earthquake. The term "seismic gap" refers to a seismic quiescence. (From Mogi, 1985.)

347

Seismic wave propagation One way to sense any change in physical properties within the region about to rupture is to study the propagation of seismic waves through that region. The first study of that type to report anomalous precursory changes was by Semenov (1969). He found a reduction in the ratio V_P/V_S of compressional and shear velocities prior to a number of earthquakes near Garm, in Tadjikistan, USSR. This result was repeated for earthquakes at Blue Mountain Lake, New York (Fig. 7.8; Aggarwal et al., 1973). The observation was that the V_P/V_S ratio decreased by 10–15% within a zone surrounding the rupture and then recovered its normal value just prior to the earthquake.

These observations were a key to the development of the dilatancy–diffusion theory of earthquake precursors (Sec. 7.3.2). Subsequent to the announcement of this theory, a rash of papers were published that reported finding these premonitory velocity changes. More recently, however, a number of studies have reported negative results (McEvilly and Johnson, 1974; Mogi, 1985), and some of the earlier observations have been criticized. This led this technique and the theory based on it to fall into disfavor.

If there is a reduction of seismic wave velocity within the rupture preparation zone, as indicated by the V_P/V_S observations, then the study of body waves is not well suited for investigating it, because such waves would be refracted away from such a region. A potentially better method is to study the shear wave coda, a scattered wave-train that samples a volume (Aki, 1985; see Sato, 1988, for a recent review). Observations of this type indicate an increase in scattering and attenuation prior to, and sometimes for a period just following, a large earthquake (Fig. 7.9). These changes evidently occur within a substantial volume containing the rupture zone of the earthquake.

Crampin (1987) reviews a number of studies that indicate widespread seismic anisotropy in the crust. The anisotropy is of a type that produces S-wave birefringence, which he argues is the result of the presence of cracks aligned with the stress field. He reports cases in which this anisotropy changes prior to earthquakes, which he interprets as due to the growth or opening of the cracks (dilatancy).

Hydrological and geochemical Changes in pressure, flow rate, color, taste, smell, and chemical composition of surface or subsurface water, oil, or gas, repeatedly have been reported to have occurred prior to earthquakes. In terms of their temporal occurrence, these fall into both the short- and intermediate-term categories as defined here. Whereas many of these phenomena were observed within a few tens of kilometers of the earthquake, others have been reported from great distances, sometimes several hundred kilometers away, even for earthquakes whose rupture dimensions could not have exceeded ten kilometers.

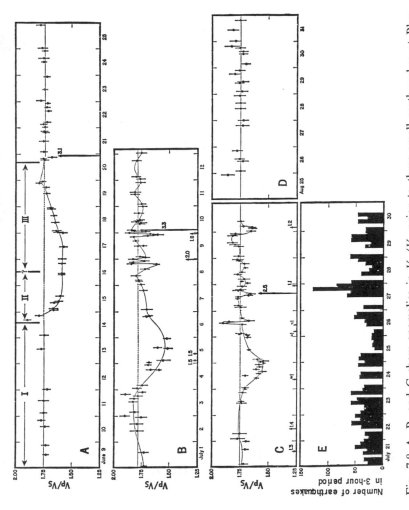

Fig. 7.8 A, B, and C show anomalies in V_P/V_S prior to three small earthquakes at Blue Mountain Lake, New York. E shows a slight quiescence on the same time scale as C. D shows variations in V_P/V_S during normal times. (From Aggarwal et al., 1973.)

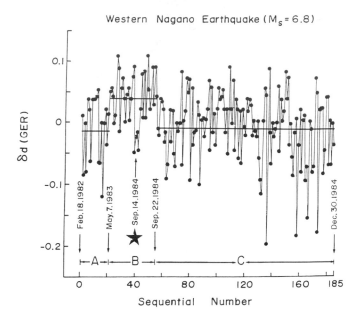

Fig. 7.9 An anomaly in coda Q for a period just before and just after an earthquake in Western Nagano Prefecture, Japan. The parameter plotted is the residual of the logarithm of coda duration from a linear regression before and after the earthquake. Bold lines indicate the mean during each period. (From Sato, 1988.)

The purely hydrologic precursors, in which water level changes were observed, are reviewed by Roeloffs (1988). She considers a variety of models, all based on the assumption that these changes result from the coupled poroelastic effect (Sec. 6.5.2) in response to preseismic strain changes. She adopts the hypothesis that preseismic strain changes are not likely to be larger than the coseismic volume strain change, which may be estimated. In Figure 7.10 are shown observed precursory water level changes as a function of distance from the epicenter. The curves are the calculated effect due to the coseismic strain change, using a semiempirical water-well response function. The anomalies at distances shorter than 150 km are consistent with the hypothesis assumed, but the more distant ones are not.

The first geochemical precursor reported was a change in radon content of a well in the immediate vicinity of the Tashkent ($M = 5.5$) earthquake of 1966 (Ulomov and Mavashev, 1971). A threefold increase in the radon content of the well water was observed for a period of at least a year prior to the earthquake. This observation spawned a number of subsequent studies. Wakita (1988) reviews a number of such observations in Japan. A particularly well-defined case is shown in Figure 7.11. There, changes in radon content, temperature, water level, and strain are

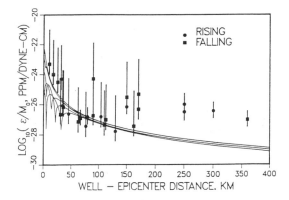

Fig. 7.10 The magnitude of hydrological precursors as a function of epicentral distance. The curves are changes predicted from the coupled poroelastic effect of the coseismic strain changes. (From Roeloffs, 1988.)

shown from different locations in the vicinity of the Izu–Oshima-kinkai earthquake of 1978 ($M = 7.0$). A similar temperature change prior to the nearby Izu–Hanto-toho-oki earthquake ($M = 6.7$, 1980) is recounted by Mogi (1985). Wakita notes that the temporal signature of these precursors fall into two categories, those lasting two or three months, and short spikes just before the earthquake.

Thomas (1988) reviews geochemical precursors and discusses a number of mechanisms that have been proposed to account for them. He concludes that only two possible mechanisms are consistent with the observations: an increase in reactive surface area due to cracking, or a change in the mixing of fluids from several sources, either because of aquifer breaching or strain-induced fluid pressure changes.

Electrical resistivity Changes in electrical resistivity and, in some cases, of magnetic field have been reported to have occurred prior to a number of earthquakes in the USSR, China, Japan, and the United States. Mogi (1985) gives a brief review. They typically involve a decrease in resistivity, often for several months prior to the earthquake. One of the most credible cases of this type is that of the Tangshan, China, earthquake of 1976 ($M = 7.8$). In that case the anomaly was clear and was observed at a number of sites.

7.2.3 *Short-term precursors*

Many of the precursory phenomena described in the previous section continue up to the time of the earthquake. In other cases the nature of the anomaly changes shortly before the earthquake, either by a rapid

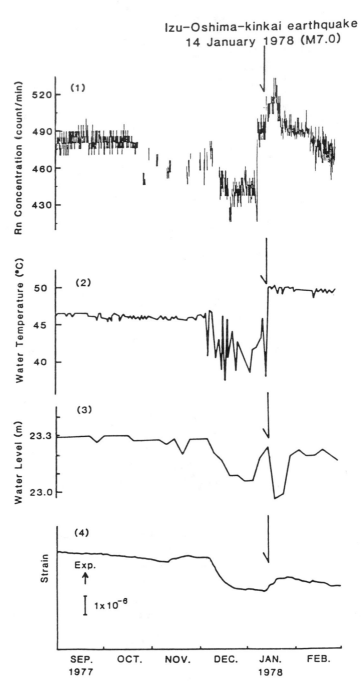

Fig. 7.11 Records of four intermediate- to short-term precursors to the Izu–Oshima-kinkai earthquake of 1978, recorded at several sites in the Izu Peninsula. (From Wakita, 1988.)

acceleration of the anomalous behavior, a return to normal, or, in some cases, a change in sign. These differences suggest that a change in physical processes may be occurring at this late stage. Other phenomena only appear during this late stage, so we unambiguously may call them short-term precursors. Here we give a few examples of the less ambiguous type.

Seismicity Foreshocks are the most obvious premonitory phenomenon preceding earthquakes. Their occurrence has, in a few cases, led to successful earthquake predictions, such as the Haicheng, China, earthquake of 1975 ($M = 7.5$; Raleigh et al., 1977). There are many cases in the historical record in which a local populace spontaneously abandoned their homes because of alarm over foreshocks thereby avoiding harm in the mainshock. Aside from these considerations, foreshocks are the strongest clue that an accelerating phase of deformation immediately precedes the earthquake instability.

A global survey of foreshock activity was made by Jones and Molnar (1979). They found that since 1950, when the perceptibility of worldwide seismic networks became sufficient, that 60–70% of all earthquakes of $M \geq 7$ were preceded by foreshocks (defined as events within 100 km at a rate greater than background). Foreshock activity in individual cases varies greatly, ranging from single events to swarms. However, when considered collectively they define a clear temporal signature, as shown in Figure 7.12a. Foreshock activity typically becomes evident 5–10 days before the mainshock and rapidly accelerates up to its occurrence. There seems to be no dependence of this on the size of the mainshock. Jones and Molnar also found no relation between the size of the largest foreshock and the magnitude of the mainshock (other than that the former is smaller). In their data treatment they did not separate out compound earthquakes (Sec. 4.4.3) or use our stricter definition of foreshocks (Sec. 4.4.2). It is possible that such a relation may appear if these distinctions are made.

They found that the time sequence of foreshocks could be fit with an empirical relation

$$n = at^{-\varpi} \tag{7.1}$$

where t is the time before the origin time of the mainshock, n is the frequency of foreshocks, and a and ϖ are constants. The exponent ϖ is close to one, as was found also by Papazachos (1975) and Kagan and Knopoff (1978). Jones and Molnar explained this relation with a model in which a population of asperities fail by static fatigue. Using Equation 1.50 for the fatigue law and the assumption that failed asperities transfer stress to the remaining ones, they obtained good agreement with the data shown in Figure 7.12.

Voight (1989) showed that the behavior of materials in the terminal stages of failure under conditions of constant stress and temperature can

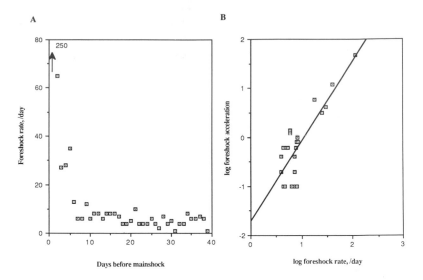

Fig. 7.12 (a) The frequency of foreshocks collected for a large number of sequences, as a function of time before the mainshock. (Data from Jones and Molnar, 1979.) (b) The data from (a) plotted versus its derivative. The line is the fit to Equation 7.2.

be described by an empirical relation,

$$\dot{\Omega}^{-\alpha}\ddot{\Omega} - A = 0 \tag{7.2}$$

where Ω is a measurable quantity such as strain, the dots indicate time derivatives, and α and A are constants. He showed that this relation provides a good description of tertiary creep and deformation preceding landslides, and that it may be related to some general constitutive laws used in damage mechanics (Rabotnov, 1969). For the cases studied by Voight, α is close to 2. If n in Equation 7.1 is identified with the rate term $\dot{\Omega}$ in Equation 7.2, then for the case of $\varpi = 1$, Equation 7.1 is equivalent to Equation 7.2 with $\alpha = 2$. In Figure 7.12b, the acceleration of foreshock activity is plotted versus foreshock rate, using the data from Figure 7.12a. The line through the data is for Equation 7.2, where it is found that $\alpha = 1.63$ and $A = 0.182$.

Jones and Molnar also found evidence for a lull in foreshock activity 4–8 hr prior to the mainshock, not accounted for in Equation 7.1. In individual cases, such short-term quiescences, labeled Q_3 in Figure 7.1, often have been seen and are particularly evident when foreshock swarms occur. A notable case was Haicheng, 1975. Several cases of this phenomenon for earthquakes in the Izu Peninsula, Japan, were collected by Mogi (1985) and are shown in Figure 7.13. The magnitude range for these three mainshocks is considerable: 7.0 for the Izu–Oshima-kinkai

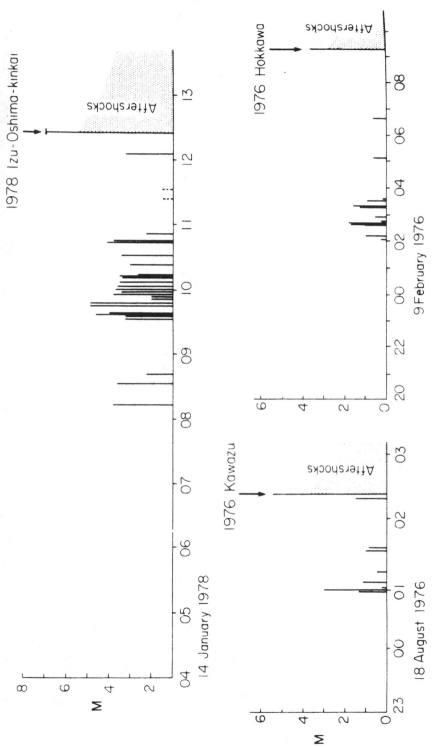

Fig. 7.13 Foreshock swarms followed by short-term quiescences before three earthquakes in the Izu Peninsula, Japan. (From Mogi, 1985.)

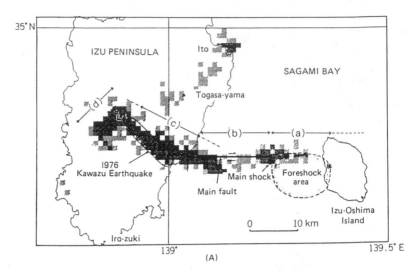

Fig. 7.14 Foreshock, mainshock, aftershock pattern of the 1978 Izu–Oshima-kinkai earthquake. The mainshock occurred at the west edge of the foreshock swarm (Fig. 7.13). Immediate aftershocks occurred on the mainshock rupture zone (b), followed by a secondary extension of the fault (c). Activity on the right-lateral conjugate section (d) did not begin until 15 hr after the mainshock. See Figure 3.2 for location. (After Tsumura et al., 1978.)

earthquake, 5.4 for the Kawazu, and 3.6 for the Hokkawa shocks. In each case, the premonitory swarm consisted of earthquakes about two magnitude units smaller than the mainshock, but the time scale for neither the swarm nor the quiescence depended on the size of the mainshock. The swarm preceding the Izu–Oshima-kinkai earthquake was tightly restricted to the hypocentral region of the mainshock (Tsumura et al., 1978), which is a characteristic of earthquake swarms before Izu earthquakes (Mogi, 1985). The development of the foreshock–mainshock–aftershock sequence for this earthquake is shown in Figure 7.14.

In terms of their utility in earthquake prediction, the difficulty with foreshocks lies in their lack of recognizable distinction from other earthquakes. It has been noticed that foreshock sequences are characterized by a smaller b-value (or B-value, Eq. 4.31) than aftershocks or other earthquakes (Suyehiro, Asada, and Ohtake, 1964). Experimental studies have shown that the b-value of acoustic emissions decreases prior to rock fracture, whether that is achieved by an increase of stress or, in creep tests, of time (Scholz, 1968; Mogi, 1981). This indicates that the mean fracture size increases as gross failure is approached. However, there is yet to be discovered some diagnostic character of individual foreshocks that will allow them to be recognized.

Crustal deformation   Several cases of intermediate-term precursors detected by conventional geodetic measurements have been described in the last section. Short-term crustal deformation precursors have also been recorded by more continuous monitoring systems such as tilt- and strainmeters and tide gauges. A notable case, from a leveling survey in progress at the time of an earthquake, has been recounted by Mogi (1982, 1985). The survey was being carried out near the cape of Omaezaki, which is on the south coast of Honshu just adjacent to the northeast end of the Tonankai earthquake of December 7, 1944 ($M = 8.1$). Very unusual discrepancies in elevation were observed in repeated occupations during the two days prior to the earthquake, and the instrument could not be leveled in the last few minutes. These events, as reconstructed by Mogi, are shown in Figure 7.15.

If we compare this precursory deformation with that described before the Sea of Japan earthquakes (Sec. 7.2.1), this example has the same sense but is much smaller. It is more similar to a precursory uplift prior to the Nankaido earthquake observed from a tide gauge record at Tosashimizu (see Fig. 7.26). Omaezaki and Tosashimizu are located similarly relative to the rupture zones of their respective earthquakes: they are just off the end of the rupture at the opposite end from the

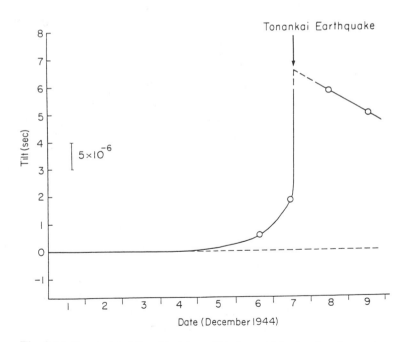

Fig. 7.15 Synopsis of the short-term tilt observed by leveling just prior to the 1944 Tonankai earthquake on the Nankai trough, Japan. (From Mogi, 1985.)

hypocenter (see Kanamori, 1973). This observation is relevant to any discussion of the mechanism of these precursors. If they are related to nucleation, then this cannot be restricted to the immediate hypocentral region. We reconsider this topic in Section 7.3.4.

7.3   MECHANISMS OF PRECURSORY PHENOMENA

In order to evaluate precursory phenomena properly and to be able to use them confidently for predictive purposes, we must understand the physical processes that give rise to them. It is of the nature of rock fracture and friction that the breakdown instability does not occur without some preceding phase of accelerated deformation. We may study these processes in the laboratory in order to understand the physics involved and then, by means of models, predict the geophysical phenomena that are likely to result. Such models, based on known constitutive properties of rock, may be termed physical models. This distinguishes them from kinematic models, in which fault slip behavior, say, is prescribed, usually in order to fit observational data. The latter are not predictive, and hence will not be useful in predicting future behavior. The deductive approach explored here differs philosophically from inductive methods, such as the pattern recognition technique of Keilis-Borok et al. (1988).

In this section we emphasize physical models of precursory phenomena. These fall into two broad categories: those based on fault constitutive relations, which predict fault slip behavior but no change in properties in the material surrounding the fault, and those based on bulk rock constitutive relations, which predict physical property changes in a volume surrounding the fault. The most prominent of the first type are nucleation and lithospheric loading models and of the second type, the dilatancy models.

7.3.1   *Nucleation models*

In Section 4.2.2 it was pointed out that fault rupture may be described in two ways: in terms of a crack model in which the energy dissipation at the crack edge is paramount, and in terms of a friction-slider model in which the effects of the edges are not explicitly considered. The energy balance problem discussed in Section 4.2.1 prevents these two approaches from being joined fully, but we have presented a number of applications in which these two approaches yield analogous results. They are often complementary, because crack models can describe explicitly crack propagation, whereas the friction models explicitly allow for the calculation of the evolution of stresses in terms of material properties of the fault.

Crack and friction models both predict that instability will not occur until slip has occurred over a fault patch of a critical radius, which is a function of the fault strength, state of stress, and elastic constants of the surrounding rock. Expressions for this critical radius have been given in Equations 4.13, 2.31, and 4.15 for crack, friction-slider, and hybrid crack models, respectively. Implicit in these results is that stable slip must occur as the patch grows to this critical radius. Experimental observations of this precursory stable slip have been shown in Figure 2.26 and 2.27. The growth of the slipping patch up to the point of instability is called nucleation. This process is therefore a potential mechanism for generating earthquake precursory phenomena.

Nucleation has been modeled using both crack and friction-slider models. A crack nucleation model was explored by Das and Scholz (1981). They assumed that, under slow loading conditions, a crack first begins to propagate subcritically because of a stress-corrosion type process. They used Equation 1.47 to describe this, and combined it with the expression for the stress intensity factor (Eq. 1.25). This leads to an instability because K increases with crack length, and the crack-tip velocity increases exponentially with K. Thus the crack tip accelerates until it reaches an unbounded velocity, which defines the instability. An example of one of their model results is shown in Figure 7.16, which shows the crack growth as a function of time. The corresponding slip on the crack may be calculated using Equation 4.24 with an assumed value of stress drop.

A nucleation model based on the friction constitutive law of Equation 2.27 is described by Dieterich (1986). In this case the sliding patch radius was held fixed, reducing the problem to one dimension, and slip was calculated until instability occurred, defined by an unbounded slip rate. Examples of the slip history, during late times, is shown in Figure 7.17. One may note the general similarity of this to Figure 7.16 (recalling that slip will be proportional to crack length, for a given stress drop). Both indicate a very strong acceleration of fault slip just prior to instability, so that the possibility of detecting precursory phenomena resulting from nucleation will be significant only for a short period before the earthquake. The time scales are such that nucleation is therefore a potential mechanism for short-term precursors, but not for those of intermediate term.

In Dieterich's model a stress jump, above steady state, is required to initiate nucleation. This condition is artificial. In two-dimensional models it is not required, as in the example shown in Figure 5.12. It is interesting, however, that the time to nucleation as a function of this

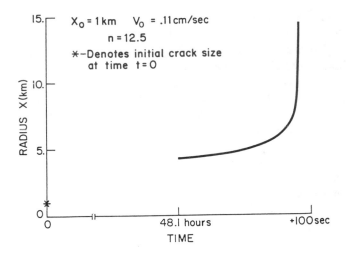

Fig. 7.16 Crack radius as a function of time, from a crack model of nucleation. This behavior results from a stress-corrosion mechanism. Notice that only 100 s of calculated growth is shown, after 48 hr of growth since initiation. (From Das and Scholz, 1981.)

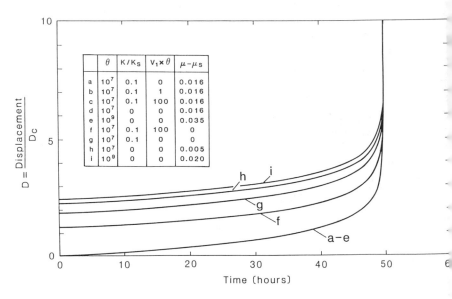

Fig. 7.17 Slip as a function of time, from a friction slider model of nucleation. (From Dieterich, 1986.) Slip is normalized to $D_c = \mathscr{L}$.

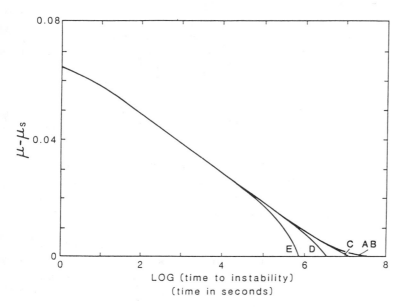

Fig. 7.18 The dependence of time to instability on the stress (friction) jump for various choices of parameters in Dieterich's nucleation model. (From Dieterich, 1986.)

stress jump, shown in Figure 7.18, is the same form as the static fatigue law (Eq. 1.50), and the model of Das and Scholz shows the same type of dependence on stress drop, the equivalent parameter in their model. Thus the results are quite similar, although the physics on which they are based seem very different. However, as noted in Section 2.2.2, the underlying cause of the time dependence of friction and subcritical crack growth are probably the same, so this agreement need not be as surprizing as it first looks.

Because nucleation is an essential part of the instability, if it can be detected by geophysical means, then short-term earthquake prediction may become a reliable possibility. The crucial questions are, therefore, how large is the nucleation zone and what is the slip in it. Dieterich's model indicates that both the nucleation slip and the critical patch radius depend linearly on $\mathscr{L}$, the critical slip distance, and so the nucleation moment depends on its cube. Values of these parameters, for some selected values of $\mathscr{L}$ are given in Table 7.1.

Stability studies in which fault behavior is simulated using this friction law show that $\mathscr{L}$ cannot exceed a value of about 100 mm or the fault behavior will become stable (Tse and Rice, 1986; Stuart, 1988). This sets an upper limit on the nucleation parameters. Our calculation of $\mathscr{L}$ with a contact model, based on fault topography (Sec. 2.4) indicates the $\mathscr{L}$ will decrease with depth, and will be in the range 1–10 mm at seismo-

Table 7.1. *Nucleation parameters*

$\mathcal{L}$ mm	Slip mm	Patch size km	Moment N-m
50	250	5.3	5.5×10^{17}
5	25	0.53	5.5×10^{14}
0.5	2.5	0.053	5.5×10^{11}

genic depths. Because large earthquakes usually nucleate near the base of the seismogenic layer, where $\mathcal{L}$ is at the small end of this range, Table 7.1 tells us nucleation is not likely to produce a signal that can be detected with surface measurements.

All of this discussion is, of course, very simplified. It assumes, for example, that faults are uniform in their frictional properties, which cannot be the case. If a fault has a uniform value of $\mathcal{L}$, then it could not produce earthquakes smaller than the nucleation moment corresponding to that value of $\mathcal{L}$. Such a lower cutoff in earthquake size is not observed, so $\mathcal{L}$ must be spatially variable (but see Sec. 4.3.2 for a debate on this point). If $\mathcal{L}$ is variable, then nucleation sizes will also be variable, which makes their detection, and predictions based on them, less reliable. On top of all this is the problem that even if nucleation were detected, it does not indicate how large the ensuing rupture will grow. This prediction depends on other considerations, such as the distribution of friction on the fault and its prior slip history.

7.3.2   *Dilatancy models*

Dilatancy models typify another class of models that is based on the constitutive properties of the bulk material, either within or without the fault zone. There are two classes of dilatancy models: volume dilatancy models, in which it is assumed that dilatancy occurs in a volume of rock surrounding the fault zone, and fault zone dilatancy models, in which it is assumed that dilatancy occurs only within the fault zone itself.

By far the best known of these is the dilatancy–diffusion model, a volume dilatancy model developed by Nur (1972) and expanded upon by Whitcomb, Garmony, and Anderson (1973) and Scholz, Sykes, and Aggarwal (1973). It assumes that dilatancy occurs within the stressed volume surrounding an impending rupture zone and develops at an accelerating rate, just as observed in laboratory fracture experiments (Sec. 1.2.1). As stress (or time – recall that failure may be approached along a time path, as in a creep test) increases, the rate of dilatancy increases (stage I). Eventually, the dilatancy rate becomes high enough that pore fluid diffusion cannot maintain the pore pressure (stage II).

This results in dilatancy hardening, which both strengthens the fault temporarily, delaying the earthquake, and inhibits further dilatancy. In an extreme case, the cracks may become unsaturated in this process. The next stage (III) involves reestablishing the pore pressure by fluid diffusion, followed by the rupture (stage IV). Following the earthquake the dilatancy recovers, at a time constant determined by the hydraulic diffusivity of the system (stage V).

This sequence of events is based on phenomena that are observed in the laboratory, but its application to earthquakes was motivated by certain types of precursors that indicate that before some earthquakes there were changes in physical properties in a volume surrounding the rupture. Most prominent among these are the velocity anomalies, such as shown in Figure 7.8. If such a change in elastic wave velocity occurs, the only plausible mechanism is if there is a change in the void space in the solid, and that is only likely to occur through dilatancy. In order to proceed, we must digress briefly to discuss the effect of cracks on the properties of solids.

The P wave velocity is a measure of bulk modulus $\mathcal{K}$, and the S wave velocity, of shear modulus $\mathcal{G}$. Now $\mathcal{K}_{\text{rock}} \approx 3\mathcal{K}_{\text{water}} \gg \mathcal{K}_{\text{air}}$ whereas $\mathcal{G}_{\text{rock}} \gg \mathcal{G}_{\text{water}} \approx \mathcal{G}_{\text{air}}$, so that the introduction of saturated cracks will reduce V_S but will have little effect on V_P, whereas if the cracks are dry both will be reduced. The effect of cracks on elastic wave velocity is summarized in Figure 7.19, which shows V_P/V_S versus V_S, both normalized to their intrinsic (crack free) values, as functions of crack density and percent saturation. Notice that V_P/V_S increases with crack density if the cracks are saturated, and decreases if they are dry. A path is indicated that corresponds to the velocity anomaly of the type shown in Figure 7.8. The results of laboratory experiments, in which these kind of velocity anomalies are observed, are shown in Figure 7.20.

Once the dilatancy path has been established from the velocity anomalies, many other types of precursory phenomena are predicted by this model, as shown in Figure 7.21. This model also predicts a scaling relationship between the duration of the precursors and the volume of the dilatant region, because the duration depends on the hydraulic diffusivity. If the dilatant volume is proportional to the size of the impending earthquake, this leads to a precursor time–magnitude relation, which will be discussed in Section 7.3.4.

Another volume dilatancy model is the dry-dilatancy, or IND model (from the acronym, in Russian, for crack instability avalanche). It is described by Mjachkin et al. (1975). It assumes that stages II and III result from dilatancy localization and stress reduction rather than from a pore pressure interaction as in the dilatancy–diffusion model.

One objection to dilatancy models is that in laboratory experiments dilatancy is not observed at stresses less than about half the fracture

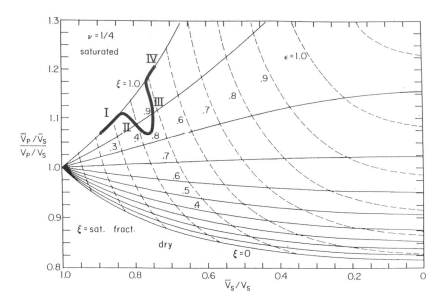

Fig. 7.19 The elastic wave velocities of cracked solids, normalized to their intrinsic values. ε is crack density and ξ is saturated fraction. Heavy curve indicates a path that would produce a velocity anomaly of the type shown in Figure 7.8. The stages of the dilatancy–diffusion model keyed to Figure 7.21, are indicated by Roman numerals. (After O'Connell and Budiansky, 1974.)

strength, which is greater than the frictional strength for normal stresses less than about 100 MPa (Hadley, 1973). However, as discussed in Sections 1.3.2 and 2.4, on a geological scale fracture strength should be much less than measured in the laboratory, whereas friction should be about the same, so dilatancy will be expected at lower stresses relative to the frictional strength.

Fault zone dilatancy models have been reviewed by Rice (1983) and Rudnicki (1988). In these models it is assumed that dilatancy occurs within the fault zone as slip occurs during nucleation. The dilatancy mechanism may be either joint dilatancy, in which the fault walls must move apart to accommodate slip, or dilatancy due to shear of granular materials, such as gouge or breccia, within the fault zone. Fault zone dilatancy is envisioned in the context of a slip-weakening model as shown in Figure 7.22. Dilatancy is assumed to be proportional to slip, so as slip accelerates in the postyield part of the stress–displacement curve dilatancy also accelerates. Two effects then occur. As dilatancy reduces the pore pressure within the fault zone, the stiffness of the fault zone materials increases as their modulus changes from the drained to the undrained modulus (Fig. 7.22a). There is also an increase in frictional resistance because of dilatancy hardening (Fig. 7.22b). These two effects

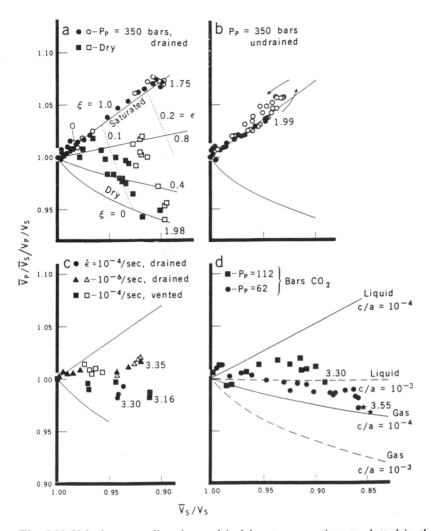

Fig. 7.20 Velocity anomalies observed in laboratory experiments plotted in the same format as Figure 7.19. (a) Westerly granite, deformed in triaxial compression at a constant strain rate under drained conditions, initially dry or saturated. In this case the saturated, drained experiments do not show any evidence of undersaturation during dilatancy. (b) The same, but undrained conditions. In this case, dilatancy hardening was observed, but undersaturation did not occur. (c) San Marcos gabbro, a rock with a much lower hydraulic diffusivity than Westerly granite, shows velocity anomalies characteristic of undersaturation under all conditions. (d) Westerly granite with CO_2 as the pore fluid. In this case, a velocity anomaly was observed due to a liquid–gas phase transition of the CO_2 induced by the dilatancy. (From Scholz, 1978.)

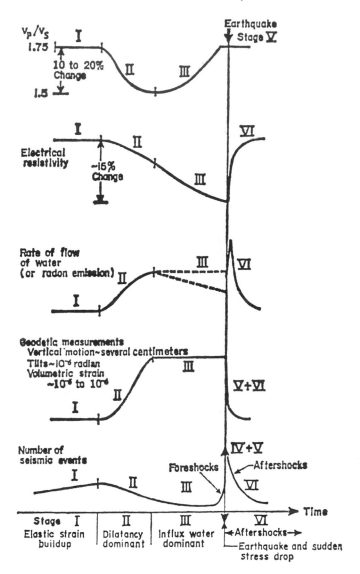

Fig. 7.21 Various phenomena predicted by the dilatancy–diffusion model. (From Scholz et al., 1973.)

will stabilize the fault temporarily and will inhibit the nucleation process. Just as in the volume dilatancy models, the fault will continue to instability after some time delay controlled by the hydraulic diffusivity of the fault zone.

Rice and Rudnicki (1979) considered two cases, one in which dilatancy occurred in a spherical inclusion and one in which it occurred in a

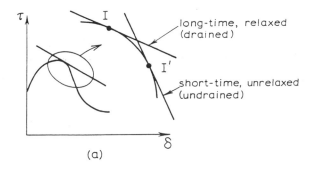

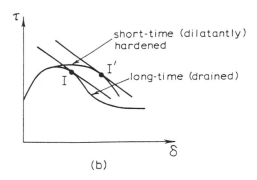

Fig. 7.22 The two effects expected from fault zone dilatancy models shown in schematic plots of shear stress (τ) versus slip (δ): (a) increase of the stiffness of the medium from drained to undrained modulus; (b) dilatancy hardening of the fault. In both cases the instability is delayed from I to I'. (From Rudnicki, 1988.)

narrow ellipsoid. For a range of constitutive parameters and an inclusion radius of 1 km, they obtained precursor times characterized by a period of accelerated slip of 15–240 days for the spherical inclusion case and times shorter by about a factor of 10 for the narrow ellipsoid case. The former is more like a volume dilatancy model, the latter a fault zone dilatancy model. The fault zone dilatancy case thus produces short term precursors, with a similar time period and shape as in the nucleation models.

7.3.3 *Lithospheric loading models*

In the above models, the effects of different constitutive relations have been examined in homogeneous loading situations. Other models have treated the problem more completely by considering the loading of a lithosphere that contains a fault with these properties. Such models can

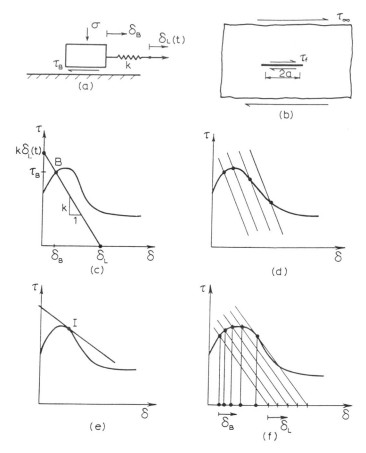

Fig. 7.23 Schematic representation of a slip-weakening model: (a) and (b) geometry of the model; (c) illustration of a graphical method of analysis whereupon a stiffness line, intersecting the constitutive law at B, is marched along at the rate of imposed slip at the load point δ_L; (d) a case where motion is always stable; (e) an instability point I is reached; (f) illustration of why slip of the block, δ_B, accelerates as δ_L increases steadily. (From Rudnicki, 1988.)

then include the effects of a depth-dependent strength and of coupling to a viscous asthenosphere. Simple slip-weakening constitutive laws often have been used to describe the fault behavior in these models. Slip-weakening laws contain a falling part in their stress–displacement function (Fig. 7.23) so they can produce an instability, but unlike the rate–state-variable friction laws they do not contain a mechanism for the rehealing of the fault. However they are consistent with these more elaborate laws under conditions of uniformly accelerated motion (Gu,

1984), and they are conceptually and computationally simpler to use. An essential point of these models is illustrated in Figure 7.23f. The ratio of increments of block motion, δ_B to imposed slip δ_L increases as the instability point is approached. This implies that some precursory stage of accelerated fault slip will occur.

Two-dimensional fault models with slip weakening have been studied by Stuart (1979), Stuart and Mavko (1979), and Li and Rice (1983a, b). The Li and Rice model is of the type shown in Figure 5.10 in which a lithosphere is coupled to an asthenosphere. The fault strength is assumed to increase with depth to a maximum at 7–10 km and then decrease. The result is that rupture is found to propagate gradually up from depth as a Mode III crack (for a strike-slip fault). This progressively increases the depth-averaged stress on the shallower parts on the fault, which has the effect of enhancing and stretching out the time of precursory surface deformation from that which would be expected from a one-dimensional model. The precursor time, measured from the time the near-surface strain rate is double the long-term rate, decreases with tectonic displacement rate and increases with rupture length (since that scales the recurrence time). Precursor times of a few months to five years are predicted, depending on the choice of these and other parameters. This model therefore provides a possible mechanism for intermediate-term precursors.

Finally, models based on rate- and state-variable friction exhibit both intermediate- and short-term precursory behavior. These are illustrated by Stuart (1988) with a model of the Nankaido earthquake sequence. Plots, from his model, of crustal uplift at various distances from the outcrop position of the thrust are shown in Figures 7.24 and 7.25. Crustal uplift becomes nonlinear in the last 5–10 yr prior to instability as slip penetrates to shallower depth. This effect is barely discernible except near the fault, where, at Nankaido, there are no observations. A much more pronounced acceleration of surface uplift accompanies nucleation in the last few days prior to instability. For points on land, which are 30 km or more from the fault outcrop, this short-term uplift is of the order of a few tens of millimeters or less, for the values of $\mathbf{a} - \mathbf{b}$ and $\mathscr{L}$ he chose. Because nucleation is localized, the shape and sign of the precursor are strongly dependent on position relative to the nucleation zone.

Although these various kinds of models have been discussed separately, there is nothing to prevent several of these types of behavior occurring in concert. Li and Rice (1983b), for example, remarked that the rapid preearthquake stress increase predicted by their slip-weakening model could induce dilatancy in near-surface rock and so produce velocity and other anomalies associated with that phenomenon. Fault zone dilatancy and pore-pressure stabilization could also accompany

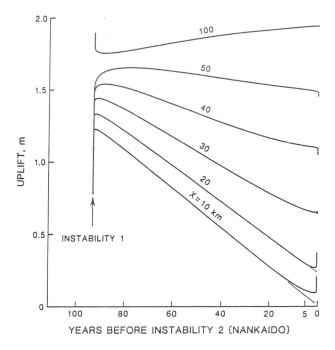

Fig. 7.24 Crustal deformation predicted through a seismic cycle at various sites adjacent to a thrust boundary modeled with a lithospheric loading model employing rate–state-variable frictions. X is distance from the surface outcrop of the thrust. Instability 1 represents the prior earthquake. (From Stuart, 1988.)

nucleation and therefore modify the behavior predicted by a nucleation model which does not contain that effect.

7.3.4 *Comparison of models and observations*

The host of models described previously all indicate that near-surface strain accumulation should become nonlinear near the end of the loading cycle as stress concentrates on the locked patch of fault, and that final nucleation of the instability should produce greatly accelerated effects of the same kind in the last 1–10 days. Dilatancy, volume or fault zone, may or may not accompany these stages. Thus intermediate- and short-term precursory phenomena are predicted as separate stages of the seismic cycle. In all the cases and choices of parameters studied, however, such precursory strain changes are small when compared to coseismic changes, and so will not be easy to detect with surface measurements.

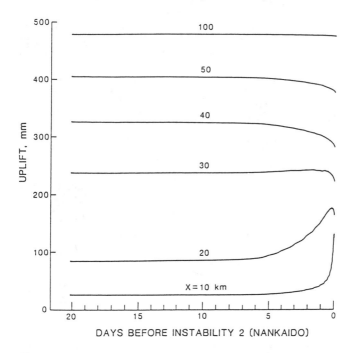

Fig. 7.25 Detail showing short-term predictions of Stuart's model. (From Stuart, 1988.)

Given this situation, we need to return to the observations of precursory effects to see if these correspond to any of the theoretical predictions. In doing so we are faced with several problems. First of all, as discussed before, many of these observations are themselves controversial. In some cases, individual measurements have been questioned. In other cases, such as that of the velocity anomalies, whole classes of phenomena have been cast into doubt (rightly or wrongly) because failures have been reported or individual cases criticized. This means that the observations usually cannot be used in the conventional hypothesis-testing mode in which a definitive test can be designed. Furthermore, the observations are typically fragmentary, which, even if they could be trusted absolutely, would not allow them to serve as adequate tests of the theories. Finally, except for dilatancy models, the models predict only crustal deformation precursors. Precursors of other types must be related to the predicted deformation patterns by inference.

Nonetheless, it is useful to reexamine the precursory phenomena in the light of the theories. In doing so, we can expect to relate the two in general ways only: quantitative comparisons are not warranted, except in terms of order of magnitude relative to coseismic changes. Also, we must be cognizant of the fact that the models are simplistic and so cannot be used to invalidate, as a precursor, an observation that disagrees with them.

Short-term precursors   The characteristics of foreshocks show that they are most likely to be a manifestation of nucleation. They typically begin about 10 days before the mainshock, rapidly increase in frequency as it is approached, and are confined to the immediate vicinity of the hypocenter (Fig. 7.14). If only cases of single nucleation are considered (i.e., compound earthquakes are excluded) they are typically several orders of magnitude smaller in moment than the mainshock (Fig. 7.13).

The spatial occurrence of foreshocks, their temporal development, and their size relative to the mainshock are consistent with the enhanced period of slip predicted by nucleation models. The models predict only stable slip, and do not predict foreshocks explicitly. However, as Das and Scholz (1981) point out, in a real, heterogeneous case, such nucleation is unlikely without accompanying local instabilities. In this interpretation, foreshocks are not an intrinsic part of nucleation, but a symptom of it. Because of this, foreshock activity is highly varied in individual cases. It is only when their collective behavior is analyzed that one can identify the pattern expressed in Equation 7.1 and recognize its similarity to the predictions of nucleation theory.

This means that foreshocks may serve as an important tool for studying nucleation, because, if they can be located accurately, they may allow the determination of the nucleation zone size and its temporal development. For example, in the case shown in Figure 7.14, nucleation seems to have occurred in a patch a few kilometers in diameter near the hypocentral region of the Izu–Oshima-kinkai earthquake. Any modeling of distant precursors, as recorded, say, on the Izu Peninsula, must take this into account.

The short-term quiescences observed in foreshock sequences (Fig. 7.13) probably indicate that fault zone dilatancy (with dilatancy hardening) is occurring during nucleation (Scholz, 1988). In this case the quiescence is restricted to earthquakes within the nucleation zone, and the fact that its duration does not increase with the size of the mainshock means that there is no corresponding increase in the minimum fluid diffusion path. This is consistent with the thin ellipsoidal inclusion model of Rice and Rudnicki (1979). If the dilatant zone is a thin wafer within the fault zone, the thickness of the wafer need not bear any relation to the size of the eventual earthquake, and neither does the quiescence duration.

The crustal deformation precursors may be compared more directly with the theories. Of these, the historic cases of uplift preceding earthquakes in the Sea of Japan (Sec. 7.2.1) are much too large, compared with the coseismic movements, to be consistent with any of the models. No permissible choice of parameters, in a homogeneous friction model, will allow such large relative movements to occur during nucleation. Recall, however, from our discussion of seismic coupling (Sec. 6.4.3), that the friction parameters may be highly variable over the fault plane, and there may be stable regions interspersed with unstable regions. Some combinations of heterogeneous friction parameters and stability conditions may allow for much larger preseismic motions. While maintaining this as a possibility, we are at the same time led to view these observations suspiciously. However, the models that have been suggested most certainly do not encompass all natural rupture phenomena. The great Chile earthquake of 1960, for example, clearly was preceded immediately by an aseismic slip event of a size nearly equal to the coseismic moment release (Kanamori and Cipar, 1974; Cifuentes and Silver, 1989).

The crustal deformation precursor of the Tonankai earthquake, shown in Figure 7.15, on the other hand, has both the form and an amplitude level consistent with nucleation models. Stuart (1988) compared the prediction of his model with precursory uplift, observed with a tide gauge, prior to the 1946 Nankaido earthquake (Fig. 7.26). The temporal shape and magnitude is about the same, but the sign is wrong. Stuart points out that this discrepancy could be removed with some modification of the geometric or frictional parameters used in his model (this sensitivity is apparent in Fig. 7.25). A more serious criticism, mentioned earlier, is that the Tosashimizu tide gauge is at the opposite end of the Nankaido rupture from the hypocenter, and so was probably several hundred kilometers from the most plausible zone of nucleation. Stuart's two-dimensional model does not take this factor into account. The same difficulty exists with the precursor of Figure 7.15.

Intermediate-term precursors   The crustal deformation precursor to the 1983 Sea of Japan earthquake (Fig. 7.7) can be compared directly with the model results of Figure 7.24. The two tide gauges that produced the data in Figure 7.7 are about 70 and 90 km from the fault outcrop. The observed anomaly is similar in form to that predicted in Figure 7.24, in that the signal gradually emerges from a very low long-term rate, but its sign is the opposite of that predicted. On the other hand, Mogi (1985) used a kinematic model of deep stable slip (1 m) to explain this precursor. Although this model does not agree with Stuart's physical model, the parameters in the latter are not sacrosanct either. The flexibility of possible models indicates that this is probably a legitimate intermediate-term precursor, but cannot be used definitively to identify the mechanism. Linde et al. (1988) also explained their observed strain

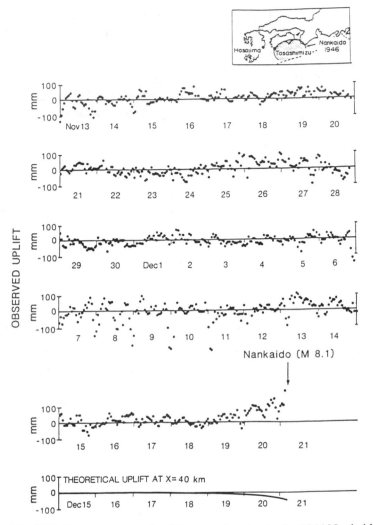

Fig. 7.26 Short-term crustal uplift anomaly prior to the 1946 Nankaido earthquake observed at the tide gauge at Tosashimizu. At the bottom is Stuart's prediction. The epicenter of the earthquake was near the arrow in the map, the tide gauge is at the other end of the rupture. (From Stuart, 1988.)

anomalies for the same earthquake with episodic slip of a downdip fault extension similar to that employed by Mogi.

If we consider the Niigata uplift data shown in Figure 7.6, we find it disagrees with both of the models suggested to explain the 1983 case. First, the uplift occurred abruptly five years before the earthquake, rather than gradually, and second it occurred on the footwall block, and

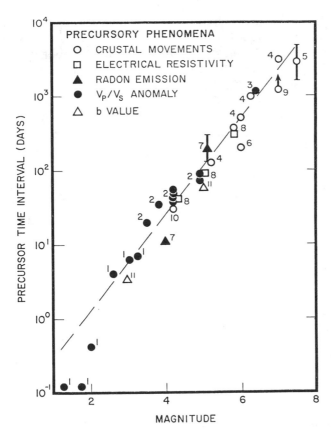

Fig. 7.27 An empirical precursor time–magnitude relation. The line through the data is consistent with the precursor time being diffusion controlled with a diffusivity of 10^4 cm^2/s. (From Scholz et al., 1973.)

so could not result from a deep slip mechanism. Originally, it was interpreted as evidence for volume dilatancy (Scholz et al., 1973), because that model predicts uplift on both sides of a dip-slip fault.

The several types of seismic wave propagation precursors, V_P/V_S, coda Q, and S wave anisotropy all indicate that a volume is affected and are diagnostic of volume dilatancy. Electrical resistivity anomalies are also diagnostic of dilatancy, and crustal uplift, if accompanied by precise gravity measurements, can also yield diagnostic results. [Dilatancy in the case of the Matsushiro, Japan, earthquake swarm was established by combined uplift and gravity data – see Nur (1974) and Kisslinger (1975).] Other phenomena consistent with the dilatancy–diffusion model, as shown in Figure 7.21, were identified by Scholz et al. (1973). Those data define a precursor duration–magnitude relation (Fig. 7.27). This relation was interpreted by Scholz et al. to indicate that the precursor duration was controlled by fluid diffusion. From the duration relation they obtained a value for hydraulic diffusivity of the crust of about $10^4 \, \text{cm}^2/\text{s}$. A similar value has been obtained independently from studies of reservoir-induced seismicity (Sec. 6.5.2).

However, several studies have failed to find evidence for precursory velocity anomalies. The cases were for a series of $4.5 < M < 5.1$ earthquakes on the creeping section of the San Andreas fault (McEvilly and Johnson, 1974), an $M = 5.2$ earthquake in southern California (Kanamori and Fuis, 1976), and the Izu–Oshima-kinkai earthquake (reviewed by Mogi, 1985). These cases were all for strike-slip earthquakes, whereas earlier reported cases were for thrust events. Because dilatant cracks will be oriented with the stress field, only ray paths striking at a high angle to σ_1 will be expected to show any effect for strike-slip situations, whereas all upward-directed rays will show some effect in thrust cases. This may be part of the reason for the discrepancy.

There is other evidence for dilatancy in the crust, such as from observations of S wave birefringence, reported by Crampin (1987 and earlier papers). He argues that this phenomenon is due to the presence of oriented and fluid-filled dilatant crack arrays, but has yet to demonstrate convincingly that the dilatancy changes temporally before earthquakes. Geological evidence also exists. Ramsay (1983) has shown that oriented microcracks commonly are found in rock, and that cementation within them indicates that the cracks have undergone many cycles of opening and sealing. He considers this *crack–seal* mechanism to be an important deformation process. In order for this process to operate, the cracks must be subject to cyclic loading. The only situation for this is if the cracks are within the fluctuating stress field of the seismic cycle. In that case, Ramsay's microcracks may be fossil evidence of many cycles of dilatancy and stress relief, followed by cementation during interseismic periods. Sibson (1987) has proposed similar dilatancy cycles to explain vein–fault relationships found in hydrothermal mineral deposits.

Various mechanisms invoked to explain intermediate-term seismicity quiescences are reviewed in Scholz (1988). The conclusion is that they must indicate either a temporal stress relaxation or hardening. Wyss, Klein, and Johnson (1981) and Cao and Aki (1985) have suggested that they may be due to stress relaxation as a result of slip weakening. However, the lithospheric loading models employing either slip weakening or rate–state-variable friction indicate an acceleration of near-surface loading during the last few years of the cycle, rather than a relaxation. The other possibility is dilatancy hardening. In the Oaxaca case (Figs. 7.2–7.3), quiescence extended over a volume surrounding the source. Volume dilatancy must be involved, if this is the mechanism. Quiescence is a much more sensitive indicator of dilatancy hardening than seismic velocity or resistivity anomalies, because it will occur at the onset of hardening, whereas the others require hardening to be sufficient to partially undersaturate cracks. If, on the other hand, quiescence is restricted to the fault zone, it may be the result of fault zone dilatancy hardening. Future studies of this phenomenon should strive to differentiate between these two cases.

Hydrological and geochemical precursors also may be caused by dilatancy, but they are much more difficult to interpret. As noted by Thomas (1988), geochemical precursors may be highly dependent on the local hydrological conditions. This point should be emphasized for all single-point measurements. The crust of the earth is not a uniform elastic continuum, as idealized in models, but is broken up into blocks of various sizes. The block boundaries are fractures, which have different constitutive properties from the blocks, and which may produce local tilts and strains that are not anticipated by continuum models (Bilham and Beavan, 1978). An example of such block motion is shown in Figure 5.9. This behavior implies that point measurements of precursory phenomena may be widely variable and not readily interpretable with simple models. Some places may show no anomalous signals, whereas at other places precursory phenomena may be greatly amplified.

The state of affairs at present is that observed precursory phenomena indicate that several types of precursory phenomena, of both intermediate- and short-term character, seem to precede some earthquakes, but no universal mechanism for these has been identified. In future earthquake prediction studies, more attention needs to be paid to designing experiments to test specific hypotheses, such as are presented in the models described before. Experiments can be designed to test, for example, whether precursory changes take place in a volume or are restricted to the fault zone alone. If dilatancy occurs, it must exhibit postearthquake recovery (Scholz, 1974), which is testable with experiments conducted in the weeks or months following a large earthquake.

7.4 EARTHQUAKE HAZARD ANALYSIS

A more straightforward goal of earthquake prediction research is to estimate the hazard presented by earthquakes. This is related to long-term prediction, which, when carried out thoroughly for a region, provides a basis upon which seismic hazard can be assessed and expressed in probabilistic terms.

7.4.1 *Traditional methods*

Seismic hazard maps have been produced for many years. In their simplest form they are representations of the past historic and/or instrumentally recorded seismicity of a region, which may be indicated by maps of intensity distributions or contours of elastic energy release. Such maps assume that future seismicity will be the same as past activity. If the data set upon which the map is based is complete for a period long compared to the recurrence time of the slowest-moving fault in the region, such a map will represent accurately the long-term hazard. In practice, this condition rarely is met. Usually a complete record of damaging earthquakes is available for only the last one or two centuries, which is shorter than the complete seismic cycle for most plate boundaries, including the secondary faults associated with them. Even regions with exceptionally long written histories, such as China, Japan, and Italy, have insufficient records because they contain faults with recurrence times of thousands to tens of thousands of years.

Maps constructed in this way may give an erroneous picture of present-day hazard. A quiet zone on such a map, representing low hazard, may delineate a seismic gap and actually be a place of high present hazard. On the other hand, a region that recently has experienced a damaging earthquake, and hence is represented as high hazard on the map, actually may be a region of low hazard in the near future because it is now at an early stage in a new seismic cycle. There are therefore two problems with such maps: incompleteness of geographical and temporal coverage in the record and a lack of identification of a time datum upon which to base the hazard estimation.

The frequency–magnitude relation has been used often in attempts to mitigate the effects of incomplete data coverage. This relation may be determined by recording small earthquakes in a region, and then extrapolated to calculate the recurrence time of potentially damaging earthquakes of larger magnitude. This method often has been used to estimate hazard at sites for critical facilities, which otherwise have no record of destructive earthquakes. However, as pointed out in Section 4.3.2, large, potentially damaging earthquakes belong to a different fractal set than small earthquakes and cannot be predicted with this extrapolation. The large earthquake that ruptures a given fault segment will typically be one to two orders of magnitude larger than that predicted by an extrapolation of the size distribution of small earthquakes on the same segment. It is only for a very large region that contains many active faults that this extrapolation will give accurate results.

These difficulties have been overcome in recent years. The incompleteness of the historic record may be remedied by the incorporation of geological data on fault slip rates bolstered with dates of paleoseismic events determined by the excavation of faults. The true nature of the earthquake size distribution has been recognized and empirical earthquake scaling laws allow recurrence times to be calculated even in the absence of data on prehistoric events (Sec. 4.3.2). These developments form a basis for the calculation of *long-term seismic hazard*. This type of hazard analysis provides an estimate of the geographical distribution of seismic hazard that is independent of a time datum. With additional data on the occurrence time of the last large earthquake on each fault segment and with proper consideration of the seismic cycle, an analysis of *instantaneous seismic hazard* can be made, in which the hazard estimation is specific to a particular time datum, namely the present.

7.4.2 *Long-term hazard analysis*

This method will be illustrated by outlining the steps leading to the development of a map of long-term seismic hazard of Japan, which is the first place that this method was employed systematically on a regional basis. The details may be found in Wesnousky, Scholz, and Shimazaki (1982) and Wesnousky et al. (1983, 1984).

Japan's seismic hazard arises from two main types of sources. Great interplate earthquakes occur on the subduction zone interfaces of the Sagami and Nankai troughs, which constitute the Eurasian–Philippine Sea plate boundary, and of the Japan trench on the Eurasian–Pacific plate boundary. These earthquakes, typically $M \geq 8.0$, have recurrence times $T \sim 60$–200 yr. In addition, intraplate earthquakes occur frequently onshore and just offshore in the Sea of Japan. These earthquakes, with few exceptions, are typically $M \leq 7.5$, and occur on faults with recurrence times of thousands to tens of thousands of years. However, there are many such faults, and although the intraplate earthquakes are smaller, they are often more hazardous because of their proximity to human habitation.

The catalog of destructive earthquakes in Japan is complete for the last 400 yr. This is long enough to estimate an average recurrence time for most parts of the plate boundaries adjacent to Japan, but is insufficient to characterize the seismicity of the intraplate faults. However, these faults have been mapped and their mean slip rate estimated for the late Quaternary (Research Group for Active Faults in Japan, 1980). Because the geological data provide both fault lengths and slip rates, a moment release rate can be calculated for each fault, which is an average over the late Quaternary. If fault creep is negligible and seismic coupling total, these data are nearly the equivalent of having a 100,000-yr earthquake history.

To check this assertion, moment release rate was calculated from the geological data and from the 400-yr seismic record, and strain rates were determined for both by using Equation 6.2. This comparison, for all of Honshu and various subregions, showed agreement within a factor of 2 (considered to be the uncertainty in the measurements). This agrees with the lack of any observed aseismic fault slip in Japan and allows one to assume the faults are fully coupled.

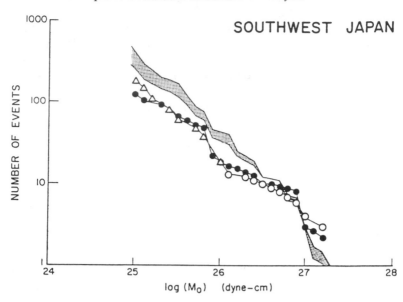

Fig. 7.28 Seismicity of southwest Japan for the last 400 yr: open symbols, observed; closed symbols, predicted from active fault data using the M_0^{max} model; stippled, predicted by the b-value model. (From Wesnousky et al., 1983.)

The next step is to determine the appropriate way to partition the moment release rate of a fault into individual earthquakes. Two starting assumptions are made: Either the size distribution of earthquakes on a fault obeys Equation 4.31 from the smallest to the largest (the b-value model), or the fault is ruptured in one large earthquake and smaller earthquakes on the fault follow Equation 4.31 beginning 1.5 orders of magnitude below the largest event (the M_0^{max} model). Large intraplate earthquakes in Japan commonly rupture over the complete length of preexisting mapped faults (Matsuda, 1977). Therefore the maximum moment earthquake in either model can be estimated from an empirical relation between M_0 and L, established for large Japanese intraplate earthquakes ($\log M_0 = 23.5 + 1.94 \log L$). We can then calculate, from the geological data, the number of earthquakes of different sizes expected in 400 yr, and determine whether it agrees with the historical catalogue. As shown in Figure 7.28, the M_0^{max} model gives good results, whereas the b-value model does not.

This shows that almost all the moment release is contained in the M_0^{max} earthquakes. We may then estimate T_{ave} for each fault by dividing its M_0^{max} by the fault's geological moment release rate. Using an empirical relation between M_0 and area of different intensity levels, we may then calculate the frequency of shaking at that level, from any source, for any point in Japan. The resulting map is shown in Figure 7.29. To

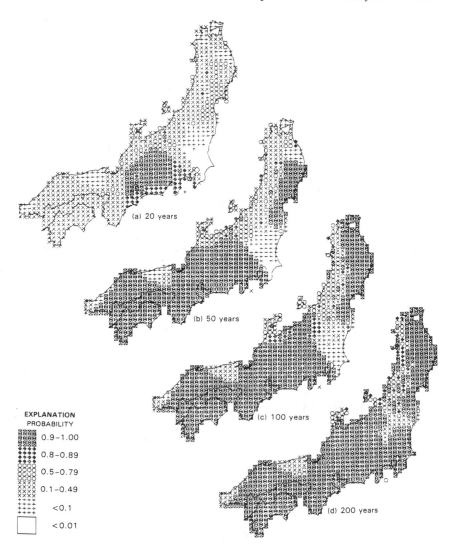

Fig. 7.29 A representation of the long-term seismic hazard of Japan. The maps give the probability of shaking at JMA intensity $\geq V$, from all sources, during the next 20, 50, 100, and 200 yr. (From Wesnousky et al., 1984.)

determine shaking from interplate earthquakes, their size and recurrence time is taken from the historical record, and a different scaling relation to shaking is used. If we assume that each fault acts independently, the earthquake activity may be described by a Poisson process, and a map of probabilities of shaking over different time intervals may be constructed.

Once such a map has been produced, it can be updated easily with new data, or modified to see the effects of different assumptions concerning, say, fault activity. A similar map has been produced of long-term seismic hazard of California (Wesnousky, 1986). Because much of the hazard there is due to long faults like the San Andreas that do not rupture in their entirety, a segmentation model must be used to divide the fault into different rupture segments. This is based largely on historical and paleoseismic data.

7.4.3   Analysis of instantaneous hazard

Long-term hazard analysis gives the average probability of shaking for some arbitrary time interval ΔT. If additional information is available giving the last date of rupture of each fault segment, then a time-dependent assessment of hazard can be made that takes into account the cyclicity of rupture. This determination of instantaneous seismic hazard is clearly more valuable in attempts to mitigate the damage from future earthquakes.

This type of analysis requires one to assume a probability density function for recurrence time, $f(T)$. The probability that an earthquake will occur at some time t in some interval $(T, T + \Delta T)$ is then

$$P(T \le t \le T + \Delta T) = \int_T^{T+\Delta T} f(t)\, dt \qquad (7.3)$$

If the date of the previous earthquake is known, we can determine the conditional probability that the earthquake will occur in the next interval $(T, T + \Delta T)$, provided that it has not occurred in the time T since the last earthquake

$$P(T \le t \le T + \Delta T | t > T) = \frac{\int_T^{T+\Delta T} f(t)\, dt}{\int_T^{\infty} f(t)\, dt} \qquad (7.4)$$

Various investigators have assumed Gaussian, Weibull, and lognormal probability density functions (Hagiwara, 1974; Lindh, 1983; Wesnousky et al., 1984; Sykes and Nishenko, 1984; Nishenko, 1985; Nishenko and Buland, 1987). The Nishenko and Buland results were described in Section 5.3.5.

This exercise has been carried out for the San Andreas fault system of California (Working Group on California Earthquake Probabilities, 1988). Following Nishenko and Buland, they assumed a lognormal probability density function,

$$f(t) = \frac{1}{T\sigma\sqrt{2\pi}} \exp\left\{ \frac{-\left[\ln(T) - (\ln \overline{T} + \varpi)\right]^2}{2\sigma^2} \right\} \qquad (7.5)$$

where σ is the standard deviation of $\ln(T/\bar{T})$ and $\bar{T}$ is related to the observed average recurrence time T_{ave} by $\ln \bar{T} = \ln(T_{\text{ave}}) + \varpi$. $\bar{T}$ is the recurrence time calculated from the time-predictable model (see Sec. 5.3.5). The value of ϖ, from the global survey of Nishenko and Buland, is -0.0099. The standard deviation is

$$\sigma = \sqrt{\sigma_m^2 + \sigma_D^2} \qquad (7.6)$$

where $\sigma_D = 0.21$, as determined by Nishenko and Buland, is the intrinsic standard deviation, reflecting variation from perfect periodicity of the process. The standard deviation of $\ln \bar{T}$, σ_m, depends on the quality and quantity of input data and varies from segment to segment.

The expected recurrence time from Equation 7.5 is

$$T_{\text{exp}} = \bar{T} e^{\varpi + \sigma^2/2} \qquad (7.7)$$

Note that, except where σ_m is exceptionally large, $T_{\text{ave}} \approx \bar{T} \approx T_{\text{exp}}$.

The results of this study are shown in Figure 7.30. In order to execute this analysis, the faults first were divided into segments that are expected to slip in individual large earthquakes. For each segment the time of the previous earthquake was determined, and T_{exp} and σ_m were calculated. Because different types and quality of data were available for different segments, different methods were used to estimate these parameters in each case. Each segment therefore required an analysis specific to it.

These instantaneous hazard estimates may be compared with estimates based on long-term hazard analysis (L. R. Sykes, unpublished work, 1988). If we assume a Poisson process, the probability, for any 30-yr period, for an $M \geq 7$ earthquake in the greater San Francisco Bay Area is 0.55, and 0.29 for an event of $M \geq 7.5$ anywhere on the main faults in Southern California. The conditional probabilities for these events, from the results of Figure 7.30, are 0.5 and 0.9, respectively. This comparison indicates that the San Francisco Bay region is at about an average state of the seismic cycle but that Southern California is, on the whole, at an advanced state, because the fault segments in Southern California were last ruptured a longer time ago.

The results shown in Figure 7.30 are of course only as good as the data that produced them. As new data are gathered, these results can be updated and improved. These results do not summarize all the seismic hazard of California; there are many secondary faults that can produce potentially damaging earthquakes. In principle, sufficient data could be obtained for those faults as well, and the analysis extended to them.

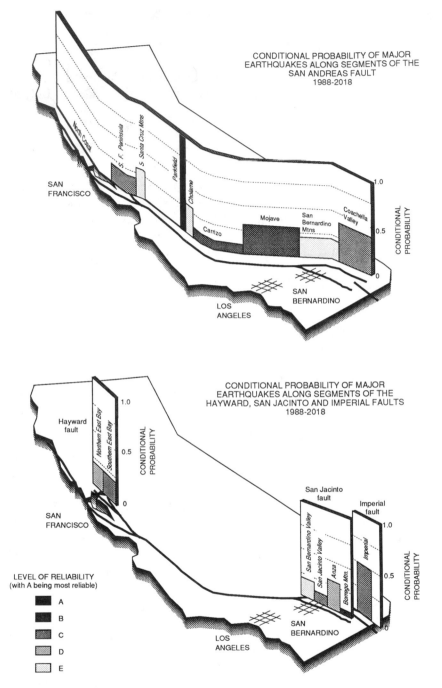

Fig. 7.30 An illustration of instantaneous seismic hazard analysis. The conditional probability of rupture of various segments of the San Andreas and other major faults in California. (From Working Group on California Earthquake Probabilities, 1988.)

7.5 FUTURE PROSPECTS AND PROBLEMS

The methodology for instantaneous seismic hazard analysis (ISHA) is now well established and there is no reason that it cannot be extended to other, secondary, faults in California, and applied to other tectonic regions. As data sets improve, σ_m will decrease and the estimated time windows will narrow.

The intrinsic standard deviation σ_D, which reflects the basic aperiodicity of earthquake recurrence, may be a fundamental barrier to precise long-term prediction. Nothing is known at present about what gives rise to this parameter. Much more work needs to be done on the nonlinear dynamics of the system, as introduced in Section 5.4. It seems likely that the value of σ_D found by Nishenko and Buland can be improved upon if more precise information on the initial conditions at the start of each cycle is known. It is only by investigation of the dynamic system, with full treatment of inhomogeneity and with the proper friction functions and modeling of dynamic crack propagation, that the types of information needed can be identified.

This remark also applies to fault segmentation, the lack of understanding of which is another weak point in the analysis. As outlined in Section 5.3.2, fault segmentation models are presently rules of thumb, and the models of fault segmentation that go into such ISHA analyses as shown in Figure 7.30 are little more than enlightened scenarios. More work, both theoretical and observational, needs to be done on the factors that stop earthquake rupture. At present only qualitative ideas are available.

The application of ISHA degenerates when applied to intraplate regions. Because the standard deviation of recurrence time scales with recurrence time, probabilities will be reduced and the prediction windows lengthened for the slower-moving faults in those regions. This is true even if the same quality of data is available for intraplate faults; it is simply a consequence of the expanded time scale, which is proportional to the recurrence time. This is not to say that the analysis has no value. In the case of the Borah Peak earthquake, for example, the Lost River fault is clearly one of the faster-moving faults in Idaho. If it had been known, before the earthquake of 1983, that the previous rupture had been circa 6,000 yr b.p., it certainly would have been recognized as a site of enhanced seismic risk. It is just that the probability numbers would have been comparatively smaller than in the San Andreas cases shown in Figure 7.30. This problem can be remedied only if the value for the intrinsic standard deviation in T can be reduced. The problem is more serious in midplate regions, like the eastern United States, where the active faults cannot be identified readily (Sec. 6.3.5). The recognition of active faults in midplate regions requires the development of new methodology and a level of effort higher than that used in more tectonically active regions. Such an effort, in the United States, is justified because of the greater urban development of the eastern part of the country.

Another area that will improve hazard analysis greatly is the improvement of models for estimating strong ground motion. Within the framework of specific segmentation models, such as those in Figure 7.30, not only is the approximate rupture time predicted, but also the focal mechanism and moment of each segment-breaking earthquake. It is a reasonable goal to predict the strong ground motions for each expected event. The high-frequency details will depend on the direction of rupture propagation and its heterogeneity, as illustrated by case of the Imperial Valley, 1979, earthquake (Sec. 4.4.1). We cannot expect to know the heterogeneity beforehand, so this probably will have to be treated in a statistical manner by considering it a random process (e.g., Hanks and McGuire, 1981). On the other hand, the rupture propagation direction sometimes may be guessed (as in the case at Parkfield, where the last two earthquakes ruptured from north to south).

Progress in understanding both site and building response can allow such predicted ground motion models to be converted into estimates of damage. The ground motion model can be convolved through specific site responses, derived, say, from microzonation maps of soil and bedrock conditions for the particular locale. The response of critical structures to the predicted event then can be evaluated. Thus the ISHA process provides a basis for all the activities that can lead to the mitigation of seismic hazard to society.

In contrast, the fate of intermediate- and short-term earthquake prediction remains far from certain. The progress in long-term prediction means, for the first time, that meaningful experiments can be set up to study these problems. The Parkfield prediction experiment, described by Bakun and Lindh (1985), and the Tokai gap experiment (see Mogi, 1985) are current examples. From theory and observation it seems that to detect immediate-term precursors one needs to identify a soon-to-rupture gap and to make close-in measurements. If short-term prediction depends entirely on detecting nucleation, then the probable nucleation zone within the gap also needs to be identified and instrumented, a far more exacting requirement. This is possible at Parkfield, because the last two earthquakes there are known to have nucleated within a small region at the north end of the gap. At Tokai, on the other hand, there is no basis for guessing the nucleation zone, so that instruments have to be spread evenly over the entire gap. Progress in this area will await developments in these and similar experiments. The era of relying on serendipity is over.

References

Chapter 1

Allegre, C. J., Le Mouel, J. L., and Provost, A. 1982. Scaling rules in rock fracture and possible implications for earthquake prediction. *Nature* 297: 47–9.

Atkinson, B. K. 1984. Subcritical crack growth in geological materials. *J. Geophys. Res.* 89: 4077–114.

Barenblatt, G. I. 1962. The mathematical theory of equilibrium cracks in brittle fracture. *Adv. Appl. Mech.* 7: 55–80.

Boland, J. N. and Tullis, T. E. 1986. Deformation behavior of wet and dry clinopyroxenite in the brittle to ductile transition region. In *Mineral and Rock Deformation: Laboratory Studies. AGU Geophys. Mono. 36*, ed. B. E. Hobbs and H. C. Heard. Washington, D.C.: American Geophysical Union, pp. 35–50.

Brace, W. F. 1960. An extension of the Griffith theory of fracture to rocks. *J. Geophys. Res.* 65: 3477–80.

Brace, W. F. 1961. Dependence of the fracture strength of rocks on grain size. *Penn. State Univ. Min. Ind. Bull.* 76: 99–103.

Brace, W. F. and Bombalakis, E. G. 1963. A note on brittle crack growth in compression. *J. Geophys. Res.* 68: 3709–13.

Brace, W. F. and Martin, R. J. 1968. A test of the law of effective stress for crystalline rocks of low porosity. *Int. J. Rock Mech. Min. Sci.* 5: 415–26.

Brace, W. F., Paulding, B. W., and Scholz, C. H. 1966. Dilatancy in the fracture of crystalline rocks. *J. Geophys. Res.* 71: 3939–53.

Brace, W. F. and Walsh, J. B. 1962. Some direct measurements of the surface energy of quartz and orthoclase. *Am. Mineral.* 47: 1111–22.

Brady, B. T. 1969. A statistical theory of brittle fracture of rock materials, I. *Int. J. Rock Mech. Min. Sci.* 6: 21–42.

Byerlee, J. D. 1967. Frictional characteristics of granite under high confining pressure. *J. Geophys. Res.* 72: 3639–48.

Byerlee, J. D. 1978. Friction of rock. *Pageoph* 116: 615–26.

Carter, N. L. and Kirby, S. 1978. Transient creep and semi-brittle behavior of crystalline rocks. *Pure Appl. Geophys.* 116: 807–39.

Cook, R. F. 1986. Crack propagation thresholds: A measure of surface energy. *J. Mater. Res.* 1: 852–60.

Cox, S. J. D. and Scholz, C. H. 1988a. An experimental study of shear fracture in rocks: Mechanical observations. *J. Geophys. Res.* 93: 3307–20.

Cox, S. J. D. and Scholz, C. H. 1988b. On the formation and growth of faults: An experimental study. *J. Struct. Geol.* 10: 413–30.

Cruden, D. M. 1970. A theory of brittle creep in rock under uniaxial compression. *J. Geophys. Res.* 75: 3431–42.

Donath, F. A. 1961. Experimental study of shear failure in anisotropic rocks. *Bull. Geol. Soc. Am.* 72: 985–90.

Dugdale, D. S. 1960. Yielding of steel sheets containing slits. *J. Mech. Phys. Solids* 8: 100–15.

Dunning, J. D., Petrovski, D., Schuyler, J., and Owens, A. 1984. The effects of aqueous chemical environments on crack growth in quartz. *J. Geophys. Res.* 89: 4115–24.

Edmond, J. M. and Paterson, M. S. 1972. Volume changes during the deformation of rocks at high pressure. *Int. J. Rock Mech. Min. Sci.* 9: 161–82.

Etheridge, M. A., Wall, V. J., Cox, S. F., and Vernon, R. H. 1984. High fluid pressures during regional metamorphism: Implications for mass transport and deformation mechanisms. *J. Geophys. Res.* 89: 4344–58.

Evans, A. G., Heuer, A. H., and Porter, D. L. 1977. The fracture toughness of ceramics. *Proc. 4th Int. Conf. on Fracture*, Waterloo, Canada, 529–56.

Frank, F. C. 1965. On dilatancy in relation to seismic sources. *Rev. Geophys.* 3: 484–503.

Freiman, S. W. 1984. Effects of chemical environments on slow crack growth in glasses and ceramics. *J. Geophys. Res.* 89: 4072–76.

Friedman, M., Handin, J., and Alani, G. 1972. Fracture-surface energy of rocks. *Int. J. Rock Mech. Min. Sci.* 9: 757–66.

Griffith, A. A. 1920. The phenomena of rupture and flow in solids. *Phil. Trans. Roy. Soc., Ser. A* 221: 163–98.

Griffith, A. A. 1924. The theory of rupture. In *Proc. 1st. Int. Congr. Appl. Mech.*, ed. C. B. Biezeno and J. M. Burgers. Delft: Tech. Boekhandel en Drukkerij J. Walter Jr., pp. 54–63.

Griggs, D. T. and Blacic, J. D. 1965. Quartz-anomalous weakness of synthetic crystals. *Science* 147: 292–95.

Hadley, K. 1975. Azimuthal variation of dilatancy. *J. Geophys. Res.* 80: 4845–50.

Irwin, G. R. 1958. Fracture. In *Handbuch der Physik*, vol. VI, ed. S. Flugge. Berlin: Springer-Verlag, pp. 551–90.

Jaeger, J. C. and Cook, N. G. W. 1976. *Fundamentals of Rock Mechanics*, 2nd ed. London: Chapman and Hall.

Kerrich, R., Beckinsdale, R. D., and Durham, J. J. 1977. The transition between deformation regimes dominated by intercrystalline diffusion and intracrystalline creep evaluated by oxygen isotope thermometry. *Tectonophysics* 38: 241–57.

Kirby, S. 1983. Rheology of the lithosphere. *Rev. Geophys.* 21: 1458–87.

Kranz, R. L. 1979. Crack growth and development during creep of Barre granite. *Int. J. Rock Mech. Min. Sci.* 16: 23–35.

Kranz, R. L. 1980. The effects of confining pressure and stress difference on static fatigue of granite. *J. Geophys. Res.* 85: 1854–66.

Kranz, R. L. and Scholz, C. H. 1977. Critical dilatant volume of rocks at the onset of tertiary creep. *J. Geophys. Res.* 82: 4893–98.

Lawn, B. R. and Wilshaw, T. R. 1975. *Fracture of Brittle Solids*. Cambridge: Cambridge University Press.

Li, V. C. 1987. Mechanics of shear rupture applied to earthquake zones. In *Fracture Mechanics of Rock*, ed. B. Atkinson. London: Academic Press, pp. 351–428.

Lockner, D. and Byerlee, J. D. 1977. Acoustic emission and fault formation in rocks. In *Proc. 1st. Conf. Microseismic Activity in Geological Structures and Materials*, ed. H. R. Hardy and F. W. Leighton. Claustal: Trans-tech Publ., pp. 99–107.

Mainprice, D. H. and Paterson, M. S. 1984. Experimental studies of the role of water in the plasticity of quartzites. *J. Geophys. Res.* 89: 4257–70.

McClintock, F. A. and Walsh, J. B. 1962. Friction of Griffith cracks in rock under pressure. In *Proc. 4th U.S. Natl. Congr. Appl. Mech.*, vol. II. New York: American Society of Mechanical Engineering, pp. 1015–21.

Mogi, K. 1962. Study of elastic shocks caused by the fracture of heterogeneous material and its relation to earthquake phenomena. *Bull. Earthquake Res. Inst., Univ. Tokyo* 40: 125–73.

Mogi, K. 1966. Some precise measurements of fracture strength of rocks under uniform compressive stress. *Felsmech. und Ingenieurgeol.* 4: 41–55.

Mogi, K. 1977. Dilatancy of rocks under general triaxial stress states with special reference to earthquake precursors. *J. Phys. Earth* 25, suppl.: S203–17.

Nemat-Nasser, S. and Horii, H. 1982. Compression-induced nonplanar crack extension with application to splitting, exfoliation, and rockburst. *J. Geophys. Res.* 87: 6805–21.

Nicolas, A. and Poirier, J.-P. 1976. *Crystalline Plasticity and Solid State Flow in Metamorphic Rocks*. New York: John Wiley.

Nur, A. and Byerlee, J. D. 1971. An exact effective stress law for elastic deformation of rock with fluids. *J. Geophys. Res.* 76: 6414–19.

Obriemoff, J. W. 1930. The splitting strength of mica. *Proc. Roy. Soc. London, Ser. A* 127: 290–302.

Orowan, E. 1944. The fatigue of glass under stress. *Nature* 154: 341–43.

Orowan, E. 1949. Fracture and strength of solids. *Rep. Prog. Phys.* 12: 48–74.

Paterson, M. S. 1978. *Experimental Rock Deformation – Brittle Field*. Berlin: Springer-Verlag.

Paterson, M. S. 1987. Problems in the extrapolation of laboratory rheological data. *Tectonophysics* 133: 33–43.

Paterson, M. S. and Weaver, C. W. 1970. Deformation of polycrystalline MgO under pressure. *J. Am. Ceram. Soc.* 53: 463–71.

Peck, L., Nolen Hoeksema, R. C., Barton, C. C., and Gordon, R. B. 1985. Measurement of the resistance of imperfectly elastic rock to the propagation of tensile cracks. *J. Geophys. Res.* 90: 7827–36.

Poirier, J.-P. 1985. *Creep of Crystals*. Cambridge: Cambridge University Press.

Pratt, H. R., Black, A. D., Brown, W. S., and Brace, W. F. 1972. The effect of specimen size on the strength of unjointed diorite. *Int. J. Rock Mech. Min. Sci.* 9: 513–29.

Procter, B. A., Whitney, I., and Johnson, J. W. 1967. The strength of fused silica. *Proc. Roy. Soc. London, Ser. A* 297: 534–47.

Rice, J. R. 1968. A path-independent integral and the approximate analysis of strain concentration by notches and cracks. *J. Appl. Mech.* 35: 379–86.

Rice, J. R. 1980. The mechanics of earthquake rupture. In *Physics of the Earth's Interior*, ed. A. Dziewonski and E. Boschi. Amsterdam: North Holland, pp. 555–649.

Robin, P.-Y. 1973. Note on effective stress. *J. Geophys. Res.* 78: 2434–37.

Rutter, E. H. 1986. On the nomenclature of mode of failure transitions in rocks. *Tectonophysics* 122: 381–87.

Scholz, C. H. 1968a. Microfracturing and the inelastic deformation of rock in compression. *J. Geophys. Res.* 73: 1417–32.

Scholz, C. H. 1968b. Experimental study of the fracturing process in brittle rock. *J. Geophys. Res.* 73: 1447–54.

Scholz, C. H. 1968c. Mechanism of creep in brittle rock. *J. Geophys. Res.* 73: 3295–302.

Scholz, C. H. 1972. Static fatigue of quartz. *J. Geophys. Res.* 77: 2104–14.

Scholz, C. H., Boitnott, G. A., and Nemat-Nasser, S. 1986. The Bridgman ring paradox revisited. *Pageoph* 124: 587–600.

Scholz, C. H. and Koczynski, T. A. 1979. Dilatancy anisotropy and the response of rock to large cyclic loads. *J. Geophys. Res.* 84: 5525–34.

Soga, N., Mizutani, H., Spetzler, H., and Martin, R. J. 1978. The effect of dilatancy on velocity anisotropy in Westerly granite. *J. Geophys. Res.* 83: 4451–8.

Sondergeld, C. H. and Estey, L. H. 1982. Source mechanisms and microfracturing during axial cycling of rock. *Pure Appl. Geophys.* 120: 151–66.

Swanson, P. L. 1984. Subcritical crack growth and other time- and environment-dependent behavior in crustal rocks. *J. Geophys. Res.* 89: 4137–52.

Swanson, P. L. 1987. Tensile fracture resistance mechanisms in brittle polycrystals: An ultrasonic and microscopic investigation. *J. Geophys. Res.* 92: 8015–36.

Tada, H., Paris, P., and Irwin, G. 1973. *The Stress Analysis of Cracks Handbook*. Hellertown Pa.: Del Research Corp.

Tapponnier, P. and Brace, W. F. 1976. Development of stress-induced microcracks in Westerly granite. *Int. J. Rock Mech. Min. Sci.* 13: 103–12.

Tullis, J. and Yund, R. A. 1977. Experimental deformation of dry Westerly granite. *J. Geophys. Res.* 82: 5705–18.

Tullis, J. and Yund, R. A. 1980. Hydrolytic weakening of experimentally deformed Westerly granite and Hale albite rock. *J. Struct. Geol.* 2: 439–51.

Tullis, J. and Yund, R. A. 1987. Transition from cataclastic flow to dislocation creep of feldspar: Mechanisms and microstructure. *Geology* 15: 606–9.

Voll, G. 1976. Recrystallization of quartz, biotite, and feldspars from Erstfeld to the Levantina nappe, Swiss Alps, and its geological implications. *Schweiz. Miner. Petrogr. Mitt.* 56: 641–7.

Wawersik, W. and Brace, W. F. 1971. Post-failure behavior of a granite and a diabase. *Rock Mech.* 3: 61–85.

Weiderhorn, S. M. and Bolz, L. H. 1970. Stress corrosion and static fatigue of glass. *J. Am. Ceram. Soc.* 53: 543–51.

White, S. 1975. Tectonic deformation and recrystallization of oligoclase. *Contr. Mineral. Petrol.* 50: 287–304.

Wong, T.-f. 1982. Shear fracture energy of Westerly granite from postfailure behavior. *J. Geophys. Res.* 87: 990–1000.

Wong, T.-f. 1986. On the normal stress dependence of the shear fracture energy. In *Earthquake Source Mechanics. AGU Geophys. Mono. 37*, ed. S. Das, J. Boatwright, and C. Scholz. Washington, D.C.: American Geophysical Union, pp. 1–12.

Yanigadani, T., Ehara, S., Nishizawa, O., Kusenose, K., and Terada, M. 1985. Localization of dilatancy in Oshima granite under constant uniaxial stress. *J. Geophys. Res.* 90: 6840–58.

Chapter 2

Andrews, J. 1976. Rupture velocity of plane strain shear cracks. *J. Geophys. Res.* 81: 5679–87.

Archard, J. F. 1953. Contact and rubbing of flat surfaces. *J. Applied Physics* 24: 981–8.

Archard, J. F. 1957. Elastic deformation and the laws of friction. *Proc. R. Soc. London, Ser. A* 243: 190–205.

Barton, N. and Choubey, V. 1977. The shear strength of rocks in theory and practice. *Rock Mech.* 10: 1–54.

Bowden, F. P. and Tabor, D. 1950. *The Friction and Lubrication of Solids. Part I.* Oxford: Clarenden Press.

Bowden, F. P. and Tabor, D. 1964. *The Friction and Lubrication of Solids. Part II.* Oxford: Clarenden Press.

Brace, W. F. 1972. Laboratory studies of stick-slip and their application to earthquakes. *Tectonophysics* 14: 189–200.

Brace, W. F. and Byerlee, J. D. 1966. Stick slip as a mechanism for earthquakes. *Science* 153: 990–2.

Brace, W. F. and Byerlee, J. D. 1970. California earthquakes – why only shallow focus? *Science* 168: 1573–5.

Brown, S. R. and Scholz, C. H. 1985a. Closure of random elastic surfaces in contact. *J. Geophys. Res.* 90: 5531–45.

Brown, S. R. and Scholz, C. H. 1985b. Broad bandwidth study of the topography of natural rock surfaces. *J. Geophys. Res.* 90: 12575–82.

Brown, S. R. and Scholz, C. H. 1986. Closure of rock joints. *J. Geophys. Res.* 91: 4939–48.

Byerlee, J. D. 1967a. Frictional characteristics of granite under high confining pressure. *J. Geophys. Res.* 72: 3639–48.

Byerlee, J. D. 1967b. Theory of friction based on brittle fracture. *J. Appl. Phys.* 38: 2928–34.

Byerlee, J. D. 1970. The mechanics of stick-slip. *Tectonophysics* 9: 475–86.

Byerlee, J. D. 1978. Friction of rocks. *Pure Appl. Geophys.* 116: 615–26.

Byerlee, J. D. and Brace, W. F. 1968. Stick-slip, stable sliding, and earthquakes – effect of rock type, pressure, strain rate, and stiffness. *J. Geophys. Res.* 73: 6031–7.

Cao, T. and Aki, K. 1985. Seismicity simulation with a mass-spring model and a displacement hardening–softening friction law. *Pageoph* 122: 10–24.

Challen, J. M. and Oxley, P. L. B. 1979. An explanation of the different regimes of friction and wear using asperity deformation models. *Wear* 53: 229–43.

Dieterich, J. 1972. Time-dependent friction in rocks. *J. Geophys. Res.* 77: 3690–7.

Dieterich, J. 1978. Time dependent friction and the mechanics of stick slip. *Pure Appl. Geophys.* 116: 790–806.

Dieterich, J. 1979a. Modelling of rock friction: 1. Experimental results and constitutive equations. *J. Geophys. Res.* 84: 2161–8.

Dieterich, J. 1979b. Modelling of rock friction: 2. Simulation of preseismic slip. *J. Geophys. Res.* 84: 2169–75.

Dieterich, J. 1981. Constitutive properties of faults with simulated gouge. In *Mechanical Behavior of Crustal Rocks. AGU Geophys. Mono. 24.* Washington, D.C.: American Geophysical Union, pp. 103–20.

Dieterich, J. 1986. A model for the nucleation of earthquake slip. In *Earthquake Source Mechanics. AGU Geophys. Mono. 37,* ed. S. Das, J. Boatwright, and C. Scholz. Washington, D.C.: American Geophysical Union, pp. 37–49.

Dieterich, J. and Conrad, G. 1984. Effect of humidity on time- and velocity-dependent friction in rocks. *J. Geophys. Res.* 89: 4196–202.

Engelder, T. 1974. Microscopic wear grooves on slickensides: Indicators of paleoseismicity. *J. Geophys. Res.* 79: 4387.

Engelder, T., Logan, J., and Handin, J. 1975. The sliding characteristics of sandstone on quartz fault gouge. *Pure Appl. Geophys.* 113: 69–86.

Engelder, T. and Scholz, C. H. 1976. The role of asperity indentation and ploughing in rock friction – II. Influence of relative hardness and normal load. *Int. J. Rock Mech. Min. Sci. & Geomech. Abstr.* 13: 155–63.

Frank, 1965. On dilatancy in relation to seismic sources. *Rev. Geophys. Space Phys.* 3: 485–503.

Friedman, M., Logan, J., and Rigert, J. 1974. Glass-indurated quartz gouge in sliding-friction experiments on sandstone. *Bull. Geol. Soc. Am.* 85: 937–42.

Goetze, C. 1971. High temperature rheology of Westerly granite. *J. Geophys. Res.* 76: 1223–30.

Greenwood, J. A. and Williamson, J. 1966. Contact of nominally flat surfaces. *Proc. Roy. Soc. London, Ser. A* 295: 300–19.

Gu, J. C., Rice, J. R., Ruina, A. L., and Tse, S. T. 1984. Slip motion and stability of a single degree of freedom elastic system with rate and state dependent friction. *J. Mech. Phys. Sol.* 32: 167–96.

Hobbs, B. Chaotic behavior of frictional shear instabilities. In *2nd Int. Symp. Rockbursts and Seismicity in Mines.* Minneapolis, Minn., in press.

Horowitz, F. and Ruina, A. 1985. Frictional slip patterns generated in a spatially homogeneous elastic fault model. *Eos* 66: 1069.

Hundley-Goff, E. and Moody, J. 1980. Microscopic characteristics of ortho-quartzite from sliding friction experiments, I. Sliding surfaces. *Tectonophysics* 62: 279–99.

Ishlinski, A. Y. and Kraghelsky, I. V. 1944. On stick-slip in friction. *Zhur. Tekhn. Fiz.* 14: 276–82.

Jaeger, J. C. and Cook, N. G. W. 1976. *Fundamentals of Rock Mechanics*. London: Chapman and Hall.

Johnson, T. L. 1981. Time dependent friction of granite: Implications for precursory slip on faults. *J. Geophys. Res.* 86: 6017–28.

Johnson, T. L. and Scholz, C. H. 1976. Dynamic properties of stick-slip friction in rock. *J. Geophys. Res.* 81: 881–8.

Johnson, T. L., Wu, F. T., and Scholz, C. H. 1973. Source parameters for stick-slip and for earthquakes. *Science* 179: 278–80.

Kessler, D. W. 1933. Wear resistance of natural stone flooring. *U.S. Bureau of Standards, Res. Pap.* RP612: 635–648.

Kranz, R. L., Frankel, A., Engelder, T., and Scholz, C. 1979. The permeability of whole and jointed Barre granite. *Int. J. Rock Mech. Min. Soc.* 16: 225–34.

Lawn, B. R. 1967. Partial cone crack formation in a brittle material loaded with a sliding spherical indenter. *Proc. Roy. Soc. London, Ser. A* 299: 307–16.

Logan, J. M., Higgs, N., and Friedman, M. 1981. Laboratory studies on natural gouge from the U.S.G.S. Dry Lake Valley No. 1 Well, San Andreas fault zone. In *Mechanical Behavior of Crustal Rocks. AGU Geophys. Mono. 24*, ed. N. Carter, M. Friedman, J. Logan, and D. Sterns. Washington, D.C.: American Geophysical Union, pp. 121–34.

Logan, J. M. and Teufel, L. W. 1986. The effect of normal stress on the real area of contact during frictional sliding in rocks. *Pure Appl. Geophys.* 124: 471–86.

Macelwane, J. 1936. Problems and progress on the geologico-seismological frontier. *Science* 83: 193–8.

Mandelbrot, B. B. 1983. *The Fractal Geometry of Nature*. San Francisco: Freeman.

Marone, C., Raleigh, C.B., and Scholz, C.1990. Frictional behavior and constitutive modelling of simulated fault gouge. *J. Geophys. Res.*, 95: 7007–26.

Marone, C. and Scholz, C. 1988. The depth of seismic faulting and the upper transition from stable to unstable slip regimes. *Geophys. Res. Lett.* 15: 621–4.

Nur, A. 1978. Nonuniform friction as a basis for earthquake mechanics. *Pageoph* 116: 964–89.

Ohnaka, M. 1975. Frictional characteristics of typical rocks. *J. Phys. Earth* 23: 87–112.

Ohnaka, M., Kuwahana, Y., Yamamoto, K., and Hirasawa, T. 1986. Dynamic breakdown processes and the generating mechanism for high frequency elastic radiation during stick-slip instabilities. In *Earthquake Source Mechanics. AGU Geophys. Mono. 37*, ed. S. Das, J. Boatwright, and C. Scholz. Washington, D.C.: American Geophysical Union, pp. 13–24.

Okubo, P. and Dieterich, J. H. 1984. Effects of physical fault properties on frictional instabilities produced on simulated faults. *J. Geophys. Res.* 89: 5815–27.

Okubo, P. and Dieterich, J. H. 1986. State variable fault constitutive relations for dynamic slip. In *Earthquake Source Mechanics. AGU Geophys. Mono. 37*, ed. S. Das, J. Boatwright, and C. Scholz. Washington, D.C.: American Geophysical Union, pp. 25–36.

Parks, G. 1984. Surface and interfacial free energies of quartz. *J. Geophys. Res.* 89: 3997–4008.

Paterson, M. S. 1978. *Experimental Rock Deformation – the Brittle Field*. Berlin: Springer-Verlag.

Power, W. L., Tullis, T. E., and Weeks, J. D. 1988. Roughness and wear during brittle faulting. *J. Geophys. Res.* 93: 15268–78.

Queener, C. A., Smith, T. C., and Mitchell, W. L. 1965. Transient wear of machine parts. *Wear* 8: 391–400.

Rabinowicz, E. 1951. The nature of the static and kinetic coefficients of friction. *J. Appl. Phys.* 22: 1373–9.

Rabinowicz, E. 1956. Autocorrelation analysis of the sliding process. *J. Appl. Phys.* 27: 131–5.

Rabinowicz, E. 1958. The intrinsic variables affecting the stick-slip process. *Proc. Phys. Soc. (London)* 71: 668–75.

Rabinowicz, E. 1965. *Friction and Wear of Materials*. New York: John Wiley.

Raleigh, C. B., Healy, J., and Bredehoeft, J. 1972. Faulting and crustal stress at Rangely, Colorado. In *Flow and Fracture of Rocks. AGU Geophys. Mono. 16*, ed. H. Heard, I. Borg, N. Carter, and C. Raleigh. Washington, D.C.: American Geophysical Union, pp. 275–84.

Raleigh, C. B., Healy, J., and Bredehoeft, J. 1976. An experiment in earthquake control at Rangely, Colorado. *Science* 191: 1230–7.

Raleigh, C. B. and Marone, C. 1986. Dilatancy of quartz gouge in pure shear. In *Mineral and Rock Deformation: Laboratory Studies. AGU Geophys. Mono. 36*, ed. B. Hobbs and H. Heard. Washington, D.C.: American Geophysical Union, pp. 1–10.

Rice, J. 1980. The mechanics of earthquake rupture. In *Physics of the Earth's Interior, Proc. Int. Sch. Phys. Enrico Fermi*, ed. A. Dziewonski and E. Boschi. Amsterdam: North-Holland, pp. 555–649.

Rice, J. R. and Gu, J. C. 1983. Earthquake aftereffects and triggered seismic phenomena. *Pure Appl. Geophys.* 121: 187–219.

Rice, J. R. and Ruina, A. L. 1983. Stability of steady frictional slipping. *J. Appl. Mech.* 105: 343–9.

Ruina, A. L. 1983. Slip instability and state variable friction laws. *J. Geophys. Res.* 88: 10359–70.

Sammis, C. G., Osborne, R., Anderson, J., Banerdt, M., and White, P. 1986. Self-similar cataclasis in the formation of fault gouge. *Pure Appl. Geophys.* 124: 53–78.

Scholz, C. H. 1987. Wear and gouge formation in brittle faulting. *Geology* 15: 493–5.

Scholz, C. H. 1988. The critical slip distance for seismic faulting. *Nature* 336: 761–3.

Scholz, C. H. and Aviles, C. A. 1986. The fractal geometry of faults and faulting. In *Earthquake Source Mechanics. AGU Geophys. Mono. 37*, ed. S. Das, J. Boatwright, and C. Scholz. Washington, D.C.: American Geophysical Union, pp. 147–56.

Scholz, C. H. and Engelder, T. 1976. Role of asperity indentation and ploughing in rock friction, 1: Asperity creep and stick-slip. *Int. J. Rock Mech. Min. Sci.* 13: 149–54.

Scholz, C. H. and Martin, R. J. 1971. Crack growth and static fatigue in quartz. *J. Am. Ceram. Soc.* 54: 474.

Scholz, C. H., Molnar, P., and Johnson, T. 1972. Detailed studies of frictional sliding of granite and implications for earthquake mechanism. *J. Geophys. Res.* 77: 6392–404.

Shimamoto, T. 1986. A transition between frictional slip and ductile flow undergoing large shearing deformation at room temperature. *Science* 231: 711–4.

Shimamoto, T. and Logan, J. 1986. Velocity-dependent behavior of simulated halite shear zones: An analog for silicates. In *Earthquake Source Mechanics. AGU Geophys. Mono. 37,* ed. S. Das, J. Boatwright, and C. Scholz. Washington, D.C.: American Geophysical Union, pp. 49–64.

Sibson, R. 1982. Fault zone models, heat flow, and the depth distribution of seismicity in the continental crust of the United States. *Bull. Seismol. Soc. Am.* 72: 151–63.

Stesky, R. 1975. The mechanical behavior of faulted rock at high temperature and pressure. PhD thesis, Massachusetts Institute of Technology, Cambridge, Mass.

Stesky, R. 1978. Mechanisms of high temperature frictional sliding in Westerly granite. *Can. J. Earth Sci.* 15: 361–75.

Stesky, R., Brace, W., Riley, D., and Robin, P-Y. 1974. Friction in faulted rock at high temperature and pressure. *Tectonophysics* 23: 177–203.

Stesky, R. and Hannan, G. 1987. Growth of contact areas between rough surfaces under normal stress. *Geophys. Res. Lett.* 14: 550–3.

Suh, N. P. and Sin, H.-C. 1981. The genesis of friction. *Wear* 69: 91–114.

Summers, R. and Byerlee, J. 1977. Summary of results of frictional sliding studies, at confining pressures up to 6.98 kb, in selected rock materials. *U.S. Geol. Surv. Open-file Rept. 77-142.*

Sundaram, P., Goodman, R., and Wang, C-Y. 1976. Precursory and coseismic water pressure variations in stick-slip experiments. *Geology* 4: 108–10.

Swan, G. and Zongqi, S. 1985. Prediction of shear behavior of joints using profiles. *Rock Mech. Rock Eng.* 18: 183–212.

Teufel, L. W. and Logan, J. M. 1978. Effect of displacement rate on the real area of contact and temperatures generated during frictional sliding of Tennessee sandstone. *Pure Appl. Geophys.* 116: 840–72.

Tse, S. and Rice, J. 1986. Crustal earthquake instability in relation to the depth variation of frictional slip properties. *J. Geophys. Res.* 91: 9452–72.

Tullis, T. E. and Weeks, J. D. 1986. Constitutive behavior and stability of frictional sliding of granite. *Pageoph* 124: 383–414.

Villaggio, P. 1979. An elastic theory of Coulomb friction. *Arch. Rational Mech. Anal.* 70: 135–43.

Walsh, J. B. 1965. The effect of cracks on the compressibility of rocks. *J. Geophys. Res.* 70: 381–9.

Wang, C., Mau, N., and Wu, F. 1980. Mechanical properties of clays at high pressure. *J. Geophys. Res.* 85: 1462–8.

Westbrook, J. and Jorgensen, P. 1968. Effects of water desorption on indentation hardness. *Am. Mineral.* 53: 1899–1904.

Yamada, T., Takida, N., Kagami, J., and Naoi, T. 1978. Mechanisms of elastic contact and friction between rough surfaces. *Wear* 48: 15–34.

Yoshioka, N. 1986. Fracture energy and the variation of gouge and surface roughness during frictional sliding of rocks. *J. Phys. Earth* 34: 335–55.

Yoshioka, N. and Scholz, C. 1989. Elastic properties of contacting surfaces under normal and shear loads. *J. Geophys. Res.*, 97: 17, 681–700.

Chapter 3

Anderson, E. M. 1905. The dynamics of faulting. *Trans. Edinburgh Geol. Soc.* 8: 387–402.

Anderson, E. M. 1936. The dynamics of the formation of cone-sheets, ring-dykes, and cauldron-subsidences. *Proc. Roy. Soc. Edinburgh* 56: 128–56.

Anderson, E. M. 1942. *The Dynamics of Faulting*, 1st ed. Edinburgh: Oliver and Boyd.

Anderson, E. M. 1951. *The Dynamics of Faulting*, 2nd ed. Edinburgh: Oliver and Boyd.

Anderson, J. L., Osborne, R., and Palmer, D. 1983. Cataclastic rocks of the San Gabriel fault zone – and expression of deformation at deeper crustal levels in the San Andreas fault zone. *Tectonophysics* 98: 209–51.

Angellier, J. 1984. Tectonic analysis of fault slip data sets. *J. Geophys. Res.* 89: 5835–48.

Barka, A. A. and Kadinsky-Cade, C. 1988. Strike-slip fault geometry in Turkey and its influence on earthquake activity. *Tectonics* 7: 663–84.

Barnett, J. A., Mortimer, J., Rippon, J. H., Walsh, J. J., and Watterson, J. 1987. Displacement geometry in the volume containing a single normal fault. *Am. Assoc. Pet. Geol. Bull.* 71: 925–37.

Barr, T. D. and Dahlen, F. A. 1989. Brittle frictional mountain building 2. Thermal structure and heat budget. *J. Geophys. Res.* 94: 3923–47.

Bell, T. H. and Etheridge, M. A. 1973. Microstructures of mylonites and their descriptive terminology. *Lithos* 6: 337–48.

Berthe, D., Choukroune, P., and Jegouzo, P. 1979. Orthogneiss, mylonite and non-coaxial deformation of granites: The example of the South Armorican shear zone. *J. Struct. Geol.* 1: 31–42.

Boatwright, J. 1985. Characteristics of the aftershock sequence of the Borah Peak, Idaho, earthquake determined from digital recordings of the events. *Bull. Seismol. Soc. Am.* 75: 1265–84.

Bowden, F. and Tabor, D. 1964. *The Friction and Lubrication of Solids. Part II.* Oxford: Clarendon Press.

Brace, W. F. and Kohlstedt, D. 1980. Limits on lithospheric stress imposed by laboratory experiments. *J. Geophys. Res.* 85: 6248–52.

Brace, W. F. and Walsh, J. B. 1962. Some direct measurements of the surface energy of quartz and orthoclase. *Am. Mineral.* 47: 1111–22.

Brun, J. P. and Cobbold, P. R. 1980. Strain heating and thermal softening in continental shear zones: A review. *J. Struct. Geol.* 2: 149–58.

Brune, J., Henyey, T., and Roy, R. 1969. Heat flow, stress, and rate of slip along the San Andreas fault, California. *J. Geophys. Res.* 74: 3821–7.

Cardwell, R. K., Chinn, D. S., Moore, G. F., and Turcotte, D. L. 1978. Frictional heating on a fault zone with finite thickness. *Geophys. J. Roy. Astron. Soc.* 52: 525–30.

Chen, W. P. and Molnar, P. 1983. Focal depths of intracontinental and intraplate earthquakes and their implications for the thermal and mechanical properties of the lithosphere. *J. Geophys. Res.* 88: 4183–214.

Chester, F. M., Friedman, M., and Logan, J. M. 1985. Foliated cataclasites. *Tectonophysics* 111: 134–46.

Chester, F. M. and Logan, J. M. 1986. Implications for mechanical properties of brittle faults from observations of the Punchbowl fault zone, California. *Pageoph* 124: 79–106.

Chinnery, M. A. 1964. The strength of the earth's crust under horizontal shear stress. *J. Geophys. Res.* 69: 2085–9.

Christie, J. M. 1960. Mylonitic rocks of the Moine thrust zone in the Assynt district, northwest Scotland. *Trans. Geol. Soc. Edinburgh* 18: 79–93.

Christie-Blick, N. and Biddle, K. T. 1985. Deformation and basin formation along strike-slip faults. In *Strike-slip Deformation, Basin Formation, and Sedimentation. Soc. Econ. Pal. Miner. Spec. Publ. 37*, ed. K. Biddle and N. Christie-Blick, pp. 1–34.

Cox, S. J. D. and Scholz, C. H. 1988a. Rupture initiation in shear fracture of rocks: An experimental study. *J. Geophys. Res.* 93: 3307–20.

Cox, S. J. D. and Scholz, C. H. 1988b. On the formation and growth of faults: An experimental study. *J. Struct. Geol.* 10: 413–30.

Crone, A. J. and Machette, M. N. 1984. Surface faulting accompanying the Borah Peak earthquake, central Idaho. *Geology* 12: 664–7.

Crowell, J. C. 1984. Origin of late Cenozoic basins in southern California. In *Tectonics and Sedimentation. Soc. Econ. Pal. Miner. Spec. Publ. 22*, ed. W. Dickinson, pp. 190–204.

Dahlen, F. A. and Barr, T. D. 1989. Frictional mountain building 1. Deformation and mechanical energy budget. *J. Geophys. Res.* 94: 3906–22.

Das, S. 1982. Appropriate boundary conditions for modeling very long earthquakes and physical consequences. *Bull. Seismol. Soc. Am.* 72: 1911–26.

Das, S. and Scholz, C. H. 1983. Why large earthquakes do not nucleate at shallow depths. *Nature* 305: 621–3.

Deng, Q., Wu, D., Zhang, P., and Chen, S. 1986. Structure and deformational character of strike-slip fault zones. *Pageoph* 124: 203–24.

Doser, D. and Kanamori, H. 1986. Depth of seismicity in the Imperial Valley region (1977–83) and its relationship to heat flow, crustal structure and the October 15, 1979 earthquake. *J. Geophys. Res.* 91: 675–88.

Durney, D. W. and Ramsay, J. G. 1973. Incremental strains measured by syntectonic crystal growths. In *Gravity and Tectonics*, ed. K. A. de Jong and R. Scholten. New York: John Wiley, pp. 67–96.

Eaton, J. P., O'Neill, M. E., and Murdock, J. N. 1970. Aftershocks of the 1966 Parkfield–Cholame, California, earthquake: A detailed study. *Bull. Seismol. Soc. Am.* 60: 1151–97.

Einarsson, P. and Eiriksson, J. 1982. Earthquake fractures in the districts Land and Rangarvellir in the South Iceland seismic zone. *Jokull* 32: 113–20.

Elliott, D. 1976. The energy balance and deformation mechanisms of thrust sheets. *Phil. Trans. Roy. Soc. London, Ser. A* 283: 289–312.

Engelder, J. T. 1974. Cataclasis and the generation of fault gouge. *Geol. Soc. Am. Bull.* 85: 1515–22.

Erismann, T., Heuberger, H., and Preuss, E. 1977. Der Bimstein von Kofels (Tirol), ein Bergsturz-"Frictionit". *Tschermaks Mineral. Petrogr. Mitt.* 24: 67–119.

Etchecopar, A., Granier, T., and Larroque, J.-M. 1986. Origine des fentes en echelon: Propagation des failles. *C.R. Acad. Sci. Paris* 302(II): 479–84.

Etheridge, M. A., Wall, V. J., Cox, S. F., and Vernon, R. H. 1984. High fluid pressures during regional metamorphism and deformation: Implications for mass transport and deformation mechanisms. *J. Geophys. Res.* 89: 4344–58.

Feng, R. and McEvilly, T. V. 1983. Interpretation of seismic reflection profiling data for the structure of the San Andreas fault zone. *Bull. Seismol. Soc. Am.* 73: 1701–20.

Fleitout, L. and Froidevaux, J. C. 1980. Thermal and mechanical evolution of shear zones. *J. Struct. Geol.* 2: 159–64.

Fleuty, M. J. 1975. Slickensides and slickenlines. *Geol. Mag.* 112: 319–22.

Fyfe, W. S., Price, N. J., and Thompson, A. B. 1978. *Fluids in the Earth's Crust.* Amsterdam: Elsevier.

Gamond, J.-F. and Giraud, A. 1982. Identification des zones de faille a l'aide des associations de fractures de second ordre. *Bull. Soc. Geol. France* 24: 755–62.

Gephardt, J. W. and Forsyth, D. W. 1984. An improved method for determining the regional stress tensor using earthquake focal mechanism data. *J. Geophys. Res.* 89: 9305–20.

Granier, T. 1985. Origin, damping and pattern of development of faults in granite. *Tectonics* 4: 721–37.

Gretener, P. E. 1977. On the character of thrust sheets with particular reference to the basal tongues. *Bull. Can. Pet. Geol.* 25: 110–22.

Grocott, J. 1981. Fracture geometry of pseudotachylyte generation zones: A study of shear fractures formed during seismic events. *J. Struct. Geol.* 3: 169–78.

Hafner, W. 1951. Stress distributions and faulting. *Bull. Geol. Soc. Am.* 62: 373–98.

Hanks, T. C. 1977. Earthquake stress drops, ambient tectonic stress, and the stresses that drive plate motion. *Pageoph* 115: 441–58.

Hanks, T. C. and Raleigh, C. B. 1980. Stress in the lithosphere. *J. Geophys. Res.* 85: 6083–435.

Hobbs, B. E., Ord, A., and Teyssier, C. 1986. Earthquakes in the ductile regime? *Pageoph* 124: 309–36.

Hubbert, M. K. and Rubey, W. W. 1959. Role of fluid pressure in the mechanics of overthrust faulting. *Bull. Geol. Soc. Am.* 70: 115–66.

Hull, J. 1988. Thickness–displacement relationships for deformation zones. *J. Struct. Geol.* 10: 431–5.

Irwin, W. P. and Barnes, I. 1975. Effects of geological structure and metamorphic fluids on seismic behavior of the San Andreas fault system in central and northern California. *Geology* 3: 713–16.

Jaeger, J. C. and Cook, N. G. W. 1969. *Fundamentals of Rock Mechanics.* London: Chapman and Hall.

Jaoul, O., Tullis, J. A., and Kronenberg, A. K. 1984. The effect of varying water content on the creep behavior of Heavitree quartzite. *J. Geophys. Res.* 89: 4289–312.

Jones, L. M. 1988. Focal mechanisms and the state of stress on the San Andreas fault in Southern California. *J. Geophys. Res.* 93: 8869–92.

Jordan, P. 1988. The rheology of polymineralic rocks – an approach. *Geol. Runds.* 77: 285–94.

Kerrich, R. 1986. Fluid infiltration into fault zones: Chemical, isotopic, and mechanical effects. *Pageoph* 124: 225–68.

King, G. C. P. 1986. Speculations on the geometry of the initiation and termination processes of earthquake rupture and its relation to morphology and geological structure. *Pageoph* 124: 567–86.

King, G. C. P. and Nabelek, J. 1985. Role of fault bends in the initiation and termination of earthquake rupture. *Science* 228: 984–7.

Kirby, S. 1980. Tectonic stress in the lithosphere: Constraints provided by the experimental deformation of rock. *J. Geophys. Res.* 85: 6353–63.

Kirby, S. 1983. Rheology of the lithosphere. *Rev. Geophys. Space Phys.* 21: 1458–87.

Knipe, R. J. and White, S. H. 1979. Deformation in low grade shear zones in the Old Red Sandstone, SW Wales, *J. Struct. Geol.* 1: 53–66.

Kohlstedt, D. L. and Weathers, M. S. 1980. Deformation-induced microstructures, paleopiezometers, and differential stresses in deeply eroded fault zones. *J. Geophys. Res.* 85: 6269–85.

Lachenbruch, A. 1980. Frictional heating, fluid pressure, and the resistance to fault motion. *J. Geophys. Res.* 85: 6097–112.

Lachenbruch, A. and Sass, J. 1973. Thermo-mechanical aspects of the San Andreas. In *Proc. Conf. on the Tectonic Problems of the San Andreas Fault System. Stanford Univ. Publ. Geol. Sci. 13*, ed. R. Kovach and A. Nur. Palo Alto, Calif.: Stanford University Press, pp. 192–205.

Lachenbruch, A. and Sass, J. 1980. Heat flow and energetics of the San Andreas fault zone. *J. Geophys. Res.* 85: 6185–222.

Lachenbruch, A. and Sass, J. 1988. The stress heat-flow paradox and thermal results from Cajon Pass. *Geophys. Res. Lett.* 15: 981–4.

Lapworth, C. 1885. The highland controversy in British geology; its causes, course and consequence. *Nature* 32: 558–9.

Lensen, G. J. 1981. Tectonic strain and drift. *Tectonophysics* 71: 173–88.

Lister, G. and Snoke, A. 1984. S-C mylonites. *J. Struct. Geol.* 6: 617–38.

Lister, G. S. and Williams, P. F. 1979. Fabric development in shear zones: Theoretical controls and observed phenomena. *J. Struct. Geol.* 1: 283–97.

Logan, J. M., Friedman, M., Higgs, N., Dengo, C., and Shimamoto, T. 1979. Experimental studies of simulated fault gouge and their application to studies of natural fault zones. *Proc. Conf. VIII – Analysis of Actual*

Fault-zones in Bedrock. U.S. Geol. Surv. Open-file Rept. 79-1239, pp. 305–43.

Macelwane, J. 1936. Problems and progress on the Geologico-seismological frontier. *Science* 83: 193–8.

Maddock, R. H. 1983. Melt origin of fault-generated pseudotachylytes demonstrated by textures. *Geology* 11: 105–8.

Marone, C. and Scholz, C. H. 1988. The depth of seismic faulting and the upper transition from stable to unstable slip regimes. *Geophys. Res. Lett.* 15: 621–4.

McGarr, A. and Gay, N. C. 1978. State of stress in the earth's crust. *Ann. Rev. Earth Planet. Sci.* 6: 405–36.

McGarr, A., Pollard, D. D., Gay, N. C., and Ortlepp, W. D. 1979. Observations and analysis of structures in exhumed mine-induced faults. *Proc. Conf. VIII – Analysis of Actual Fault-zones in Bedrock. U.S. Geol. Surv. Open-file Rept. 79-1239*, pp. 101–20.

McGarr, A., Zoback, M. D., and Hanks, T. C. 1982. Implications of an elastic analysis of in situ stress measurements near the San Andreas fault. *J. Geophys. Res.* 87: 7797–806.

McKenzie, D. P. 1969. The relation between fault plane solutions for earthquakes and the directions of the principal stresses. *Bull. Seismol. Soc. Am.* 59: 591–601.

McKenzie, D. P. and Brune, J. N. 1972. Melting on fault planes during large earthquakes. *Geophys. J. Roy. Astron. Soc.* 29: 65–78.

Means, W. D. 1984. Shear zones of type I and II and their significance for reconstructing rock history. *Geol. Soc. Am. Prog. Abstr.* 16: 50.

Meissner, R. and Strelau, J. 1982. Limits of stress in continental crust and their relation to the depth–frequency relation of shallow earthquakes. *Tectonics* 1: 73–89.

Mitra, G. 1984. Brittle to ductile transition due to large strains along the White Rock thrust, Wind River Mountains, Wyoming. *J. Struct. Geol.* 6: 51–61.

Molnar, P., Chen, W.-P., and Padovani, E. 1983. Calculated temperatures in overthrust terrains and possible combinations of heat sources responsible for the Tertiary granites in the Greater Himalaya. *J. Geophys. Res.* 88: 6415–29.

Mooney, W. D. and Ginsberg, A. 1986. Seismic measurements of the internal properties of fault zones. *Pageoph* 124: 141–58.

Mount, V. and Suppe, J. 1987. State of stress near the San Andreas fault: Implications for wrench tectonics. *Geology* 15: 1143–6.

Muraoka, H. and Kamata, H. 1983. Displacement distribution along minor fault traces. *J. Struct. Geol.* 4: 483–95.

Nabelek, J. 1988. Planar vs. listric faulting: The rupture process and fault geometry of the 1983 Borah Peak, Idaho, earthquake from inversion of teleseismic body waves. *J. Geophys. Res.* in press.

Nakamura, K. 1969. Arrangement of parasitic cones as a possible key to a regional stress field. *Bull. Volcanol. Soc. Japan* 14: 8–20.

Nakamura, K., Jacob, K., and Davies, J. 1977. Volcanoes as possible indicators of tectonic stress indicators – Aleutians and Alaska. *Pageoph* 115: 86–112.

Nakamura, K., Shimazaki, K., and Yonekura, N. 1984. Subduction, bending, and eduction. Present and Quaternary tectonics of the northern border of the Philippine Sea plate. *Bull. Soc. Geol. France* 26: 221–43.

Naylor, M. A., Mandl, G., and Sijpesteijn, C. H. 1986. Fault geometries in basement-induced wrench faulting under different initial stress states. *J. Struct. Geol.* 8: 737–52.

O'Neil, J. R. and Hanks, T. C. 1980. Geochemical evidence for water–rock interaction along the San Andreas and Garlock faults of California. *J. Geophys. Res.* 85: 6286–92.

Oppenheimer, D. H., Reasenberg, P. A., and Simpson, R. W. 1988. Fault plane solutions for the 1984 Morgan Hill, California, earthquake sequence: Evidence for the state of stress on the Calaveras fault. *J. Geophys. Res.* 93: 9007–26.

Ord, A. and Christie, J. M. 1984. Flow structures from microstructures of mylonitic quartzites of the Moine thrust zone, Assynt area, Scotland. *J. Struct. Geol.* 6: 639–54.

Passchier, C. 1984. The generation of ductile and brittle deformation bands in a low-angle mylonite zone. *J. Struct. Geol.* 6: 273–81.

Pollard, D. D., Segall, P., and Delaney, P. T. 1982. Formation and interpretation of dilatant echelon cracks. *Geol. Soc. Am. Bull.* 93: 1291–303.

Power, W. L., Tullis, T. E., Brown, S., Boitnott, G. N., and Scholz, C. H. 1987. Roughness of natural fault surfaces. *Geophys. Res. Lett.* 14: 29–32.

Power, W. L., Tullis, T. E., and Weeks, J. D. 1988. Roughness and wear during brittle faulting. *J. Geophys. Res.* 93: 15268–78.

Rabinowicz, E. 1965. *Friction and Wear of Materials*. New York: John Wiley.

Ramsay, J. G. and Graham, R. H. 1970. Strain variation in shear belts. *Can. J. Earth Sci.* 7: 786–813.

Reynolds, S. J. and Lister, G. S. 1987. Structural aspects of fluid–rock interactions in detachment zones. *Geology* 15: 362–6.

Riedel, W. 1929. Zur Mechanik geologischer Brucherscheinungen. *Centralbl. f. Mineral. Geol. u. Pal.* 1929B: 354–68.

Rippon, J. H. 1985. Contoured patterns of the throw and hade of normal faults in the Coal Measures (Westphalian) of northwest Derbyshire. *Proc. Yorkshire Geol. Soc.* 45: 147–61.

Robertson, E. C. 1982. Continuous formation of gouge and breccia during fault displacement. In *Issues in Rock Mechanics, Proc. Symp. Rock Mech. 23rd*, ed. R. E. Goodman and F. Hulse. New York: Am. Inst. Min. Eng. pp. 397–404.

Rutter, E. H., Maddock, R. H., Hall, S. H., and White, S. H. 1986. Comparative microstructures of natural and experimentally produced clay-bearing fault gouges. *Pageoph* 124: 3–30.

Sanford, A. R. 1959. Analytical and experimental study of simple geologic structures. *Bull. Geol. Soc. Am.* 70: 19–51.

Scholz, C. H. 1980. Shear heating and the state of stress on faults. *J. Geophys. Res.* 85: 6174–84.

Scholz, C. H. 1985. The Black Mountain asperity: Seismic hazard on the San Francisco peninsula, California. *Geophys. Res. Lett.* 12: 717–19.

Scholz, C. H. 1987. Wear and gouge formation in brittle faulting. *Geology* 15: 493–5.

Scholz, C. H. 1988. The brittle–plastic transition and the depth of seismic faulting. *Geol. Runds.* 77: 319–28.

Scholz, C. H., Beavan, J., and Hanks, T. C. 1979. Frictional metamorphism, argon depletion, and tectonic stress on the Alpine fault, New Zealand. *J. Geophys. Res.* 84: 6770–82.

Scholz, C. H., Wyss, M., and Smith, S. W. 1969. Seismic and aseismic slip on the San Andreas fault. *J. Geophys. Res.* 74: 2049–69.

Schmid, S. M., Boland, J. N., Paterson, M. S. 1977. Superplastic flow in finegrained limestone. *Tectonophysics* 43: 257–91.

Schwartz, D. P. and Coopersmith, K. J. 1984. Fault behavior and characteristic earthquakes: Examples from the Wasatch and San Andreas fault zones. *J. Geophys. Res.* 89: 5681–98.

Segall, P. and Pollard, D. D. 1980. Mechanics of discontinuous faults. *J. Geophys. Res.* 85: 4337–50.

Segall, P. and Pollard, D. D. 1983. Nucleation and growth of strike-slip faults in granite. *J. Geophys. Res.* 88: 555–68.

Segall, P. and Simpson, C. 1986. Nucleation of ductile shear zones on dilatant fractures. *Geology* 14: 56–9.

Sharp, R. V. 1979. Implications of surficial strike-slip fracture patterns for simplification and widening with depth. *Proc. Conf. VIII – Analysis of Actual Fault-zones in Bedrock. U.S. Geol. Surv. Open-file Rept. 79-1239*, pp. 66–78.

Sharp, R. V. and Clark, M. M. 1972. Geologic evidence of previous faulting near the 1968 rupture on the Coyote Creek fault. *The Borrego Mountain, California Earthquake of April 9, 1968. U.S. Geol. Surv. Prof. Pap. 787*, pp. 131–40.

Sharp, R., Lienkaemper, J., Bonilla, M., Burke, D., Fox, B., Herd, D., Miller, D., Morton, D., Ponti, D., Rymer, M., Tionsley, J., Yount, J., Kahle, J., Hart, E., and Sieh, K. 1982. Surface faulting in the central Imperial Valley. In *The Imperial Valley, California, Earthquake of October 15, 1979. U.S. Geol. Surv. Prof. Paper 1254*, pp. 119–44.

Sibson, R. H. 1973. Interactions between temperature and pore fluid pressure during an earthquake faulting and a mechanism for partial or total stress relief. *Nature* 243: 66–8.

Sibson, R. H. 1975. Generation of pseudotachylyte by ancient seismic faulting. *Geophys. J. Roy. Astron. Soc.* 43: 775–94.

Sibson, R. H. 1977. Fault rocks and fault mechanisms. *J. Geol. Soc. London* 133: 191–213.

Sibson, R. H. 1980a. Transient discontinuities in ductile shear zones. *J. Struct. Geol.* 2: 165–71.

Sibson, R. H. 1980b. Power dissipation and stress levels on faults in the upper crust. *J. Geophys. Res.* 85: 6239–47.

Sibson, R. H. 1981. Fluid flow accompanying faulting: Field evidence and models. In *Earthquake Prediction, an International Review. M. Ewing*

Ser. 4, ed. D. Simpson and P. Richards. Washington, D.C.: American Geophysical Union, pp. 593–604.

Sibson, R. H. 1982. Fault zone models, heat flow, and the depth distribution of earthquakes in the continental crust of the United States. *Bull. Seismol. Soc. Am.* 72: 151–63.

Sibson, R. H. 1984. Roughness at the base of the seismogenic zone: Contributing factors. *J. Geophys. Res.* 89: 5791–9.

Sibson, R. H. 1985. Stopping of earthquake ruptures at dilatational jogs. *Nature* 316: 248–51.

Sibson, R. H. 1986a. Brecciation processes in fault zones: Inferences from earthquake rupturing. *Pageoph* 124: 159–76.

Sibson, R. H. 1986b. Earthquakes and rock deformation in crustal fault zones. *Ann. Rev. Earth Planet. Sci.* 14: 149–75.

Sibson, R. H. 1986c. Rupture interaction with fault jogs. In *Earthquake Source Mechanics. AGU Geophys. Mono. 37*, ed. S. Das, J. Boatwright, and C. Scholz. Washington, D.C.: American Geophysical Union, pp. 157–68.

Sibson, R. H. 1987. Earthquake rupturing as a mineralizing agent in hydrothermal systems. *Geology* 15: 701–4.

Simpson, C. 1984. Borrego Springs–Santa Rosa mylonite zone: A late Cretaceous west-directed thrust in southern California. *Geology* 12: 8–11.

Simpson, C. 1985. Deformation of granitic rocks across the brittle–ductile transition. *J. Struct. Geol.* 5: 503–12.

Simpson, C. and Schmid, S. M. 1983. An evaluation of criteria to deduce the sense of movement in sheared rock. *Geol. Soc. Am. Bull.* 94: 1281–8.

Smith, R. B. and Bruhn, R. L. 1984. Intraplate extensional tectonics of the eastern Basin–Range: Inferences on structural style from seismic reflection data, regional tectonics, and thermo-mechanical models of brittle–ductile transition. *J. Geophys. Res.* 89: 5733–62.

Somerville, P. 1978. The accommodation of plate collision by deformation in the Izu block, Japan. *Bull. Earthquake Res. Inst., Univ. Tokyo* 53: 629–48.

Spray, J. G. 1987. Artificial generation of pseudotachylyte using friction welding apparatus: Simulation of melting on a fault plane. *J. Struct. Geol.* 9: 49–60.

Stel, H. 1981. Crystal growth in cataclasites: Diagnostic microstructures and implications. *Tectonophysics* 78: 585–600.

Stel, H. 1986. The effect of cyclic operation of brittle and ductile deformation on the metamorphic assemblage in cataclasites and mylonites. *Pageoph* 124: 289–307.

Strelau, J. 1986. A discussion of the depth extent of rupture in large continental earthquakes. In *Earthquake Source Mechanics. AGU Geophys. Mono. 37*, ed. S. Das, J. Boatwright, and C. Scholz. Washington, D.C.: American Geophysical Union, pp. 131–46.

Suppe, J. 1985. *Principles of Structural Geology*. Englewood Cliffs, New Jersey: Prentice-Hall.

Thatcher, W. 1975. Strain accumulation and release mechanism of the 1906 San Francisco earthquake. *J. Geophys. Res.* 80: 4862–72.

Thatcher, W. and Lisowski, M. 1987. Long-term seismic potential of the San Andreas fault southeast of San Francisco, California. *J. Geophys. Res.* 92: 4771–84.

Tchalenko, J. S. 1970. Similarities between shear zones of different magnitudes. *Bull. Geol. Soc. Am.* 81: 1625–40.

Tchalenko, J. S. and Berberian, M. 1975. Dasht-e-Bayez fault, Iran: Earthquake and earlier related structures in bed rock. *Geol. Soc. Am. Bull.* 86: 703–9.

Tse, S. T. and Rice, J. R. 1986. Crustal earthquake instability in relation to the depth variation of frictional slip processes. *J. Geophys. Res.* 91: 9452–72.

Tsuboi, C. 1933. Investigation of deformation of the crust found by precise geodetic means. *Japan J. Astron. Geophys.* 10: 93–248.

Tsuneishi, Y., Ito, T., and Kano, K. 1978. Surface faulting associated with the 1978 Izu–Oshima-kinkai earthquake. *Bull. Earthquake Res. Inst., Univ. Tokyo* 53: 649–74.

Voight, B. 1976. *Mechanics of Thrust Faults and Decollements.* Stroudsburg, Penn.: Dowden, Huchinson, and Ross.

Wallace, R. E. 1973. Surface fracture patterns along the San Andreas fault. In *Proc. Conf. Tectonic Problems of the San Andreas Fault System. Spec. Publ. Geol. Sci.* 13, ed. R. Kovach and A. Nur. Palo Alto, Calif.: Stanford University Press, pp. 248–50.

Wallace, R. E. Segmentation in fault zones. 1989. In *Proc. USGS Conference of Segmentation of Faults, U.S. Geol. Surv. Open-file Rept.* 89–315.

Wallace, R. E. and Morris, H. T. 1986. Characteristics of faults and shear zones in deep mines. *Pageoph* 124: 107–25.

Walsh, J. J. and Watterson, J. 1988. Analysis of the relationship between displacements and dimensions of faults. *J. Struct. Geol.* 10: 238–347.

Wang, C.-Y. 1984. On the constitution of the San Andreas fault zone in central California. *J. Geophys. Res.* 89: 5858–66.

Wang, C.-Y., Mao, N., and Wu, F. T. 1980. Mechanical properties of clays at high pressure. *J. Geophys. Res.* 85: 1462–71.

Wang, C.-Y., Rui, F., Zhengshen, Y., and Xingjue, S. 1986. Gravity anomaly and density structure of the San Andreas fault zone. *Pageoph* 124: 127–40.

Watterson, J. 1986. Fault dimensions, displacements and growth. *Pageoph* 124: 365–73.

Weertman, J. and Weertman, J. R. 1964. *Elementary Dislocation Theory.* New York: MacMillan.

Wenk, H. R. 1978. Are pseudotachylites products of fracture or fusion? *Geology* 6: 507–11.

Wenk, H. and Weiss, L. 1982. Al-rich calcic pyroxene in pseudotachylyte: An indicator of high pressure and temperature? *Tectonophysics* 84: 329–41.

Wernicke, B. and Burchfiel, B. C. 1982. Modes of extensional tectonics. *J. Struct. Geol.* 4: 105–15.

Wesnousky, S. 1988. Seismological and structural evolution of strike-slip faults. *Nature* 335: 340–2.

White, S., Burrows, S., Carreras, J., Shaw, N., and Humphreys, F. 1980. On mylonites in ductile shear zones. *J. Struct. Geol.* 2: 175–87.

Williams, C. F. and Narasimhan, T. N. 1989. Hydrogeologic constraints on heat flow along the San Andreas fault: A testing of hypotheses. *Earth Planet. Sci. Lett.* 92, 131–43.

Wise, D. U., et al. 1984. Fault-related rocks: Suggestions for terminology. *Geology* 12: 391–4.

Woodcock, N. H. and Fischer, M. 1986. Strike-slip duplexes. *J. Struct. Geol.* 8: 725–35.

Yoshioka, N. 1986. Fracture energy and the variation of gouge and surface roughness during frictional sliding of rocks. *J. Phys. Earth* 34: 335–55.

Zoback, M. D. and Healy, J. 1984. Friction, faulting, and in situ stress. *Annales Geophysicae* 2: 689–98.

Zoback, M. D., Tsukahara, H., and Hickman, S. 1980. Stress measurements at depth in the vicinity of the San Andreas fault: Implications for the magnitude of shear stress at depth. *J. Geophys. Res.* 85: 6157–73.

Zoback, M. D., Zoback, M. L., Mount, V., Eaton, J., Healy, J., Oppenheimer, D., Reasonberg, P., Jones, L., Raleigh, B., Wong, I., Scotti, O., and Wentworth, C. 1987. New evidence on the state of stress of the San Andreas fault system. *Science* 238: 1105–11.

Chapter 4

Achenbach, J. D. 1973. Dynamic effects in brittle fracture. In *Mechanics Today*. New York: Pergamon, pp. 1–57.

Adams, F. D. 1938. *The Growth and Development of the Geological Sciences*. Baltimore: Williams & Wilkins.

Aki, K. 1967. Scaling law of seismic spectrum. *J. Geophys. Res.* 72: 1217–31.

Aki, K. 1979. Characterization of barriers on an earthquake fault. *J. Geophys. Res.* 84: 6140–8.

Aki, K. 1981. A probabilistic synthesis of precursory phenomena. In *Earthquake Prediction, an International Review. M. Ewing Ser. 4*, ed. D. Simpson and P. Richards. Washington, D.C.: American Geophysical Union, pp. 566–74.

Aki, K. 1984. Asperities, barriers and characteristics of earthquakes. *J. Geophys. Res.* 89: 5867–72.

Aki, K. 1987. Magnitude–frequency relation for small earthquakes: A clue for the origin of f_{max} of large earthquakes. *J. Geophys. Res.* 92: 1349–55.

Aki, K. and Richards, P. 1980. *Quantitative Seismology: Theory and Methods*. San Francisco: W. H. Freeman.

Allen, C. R. 1969. Active faulting in northern Turkey. Contr. No. 1577, Div. Geol. Sci. Calif. Inst. Tech. p. 32.

Allen, C., Wyss, M., Brune, J., Grantz, A., and Wallace, R. 1972. Displacements on the Imperial, Superstition Hills, and San Andreas faults triggered by the Borrego Mountain earthquake. In *The Borrego Mountain Earthquake of April 9, 1968. U.S. Geol. Surv. Prof. Paper 787*, pp. 87–104.

Anderson, J. and Hough, S. 1984. A model for the shape of the Fourier amplitude spectrum of acceleration at high frequencies. *Bull. Seismol. Soc. Am.* 74: 1969–94.

Ando, M. 1975. Source mechanisms and tectonic significance of historical earthquakes along the Nankai trough. *Tectonophysics* 27: 119–40.

Andrews, D. 1976a. Rupture propagation with finite stress in antiplane strain. *J. Geophys. Res.* 81: 3575–82.

Andrews, D. 1976b. Rupture velocity of plane strain shear cracks. *J. Geophys. Res.* 81: 5679–87.

Andrews, D. 1980. A stochastic fault model, static case. *J. Geophys. Res.* 85: 3867–87.

Andrews, D. 1985. Dynamic plane strain shear rupture with a slip-weakening friction law calculated by a boundary integral method. *Bull. Seismol. Soc. Am.* 75: 1–21.

Archuleta, R. 1984. A faulting model for the 1979 Imperial Valley earthquake. *J. Geophys. Res.* 89: 4559–85.

Bakun, W. H., King, G. C. P., and Cockerham, R. S. 1986. Seismic slip, aseismic slip, and the mechanics of repeating earthquakes on the Calaveras fault, California. In *Earthquake Source Mechanics. AGU Geophys. Mono. 37*, ed. S. Das, C. Scholz, and J. Boatwright. Washington, D.C.: American Geophysical Union, pp. 195–207.

Barka, A. and Kadinsky-Cade, K. 1988. Strike-slip fault geometry in Turkey and its influence on earthquake activity. *Tectonics* 7: 663–84.

Berberian, M. 1982. Aftershock tectonics of the 1978 Tabas-e-Golshan (Iran) earthquake sequence: A documented active 'thin- and thick-skinned tectonic' case. *Geophys. J. R.A.S.* 68: 499–530.

Boatwright, J. 1980. A spectral theory for circular seismic sources: Simple estimates of source dimension, dynamic stress-drop and radiated seismic energy. *Bull. Seismol. Soc. Am.* 70: 1–27.

Bolt, B. A. 1978. *Earthquakes: A Primer*. San Francisco: Freeman.

Burdick, L. and Mellman, G. 1976. Inversion of the body waves from the Borrego Mountain earthquake to the source mechanism. *Bull. Seismol. Soc. Am.* 66: 1485–99.

Burridge, R. 1973. Admissible speeds for plane strain self-similar shear cracks with friction but lacking cohesion. *Geophys. J. R.A.S.* 35: 439–55.

Brune, J. 1970. Tectonic stress and the spectra of seismic shear waves from earthquakes. *J. Geophys. Res.* 75: 4997–5009.

Cao, T. and Aki, K. 1986. Effect of slip rate on stress drop. *Pageoph* 124: 515–30.

Clark, M. 1972. Surface rupture along the Coyote Creek fault. In *The Borrego Mountain Earthquake of April 9, 1968. U.S. Geol. Surv. Prof. Paper 787*, pp. 55–86.

Cockerham, R. S. and Eaton, J. P. 1984. The April 24, 1984 Morgan Hill earthquake and its aftershocks. In *The 1984 Morgan Hill, California, Earthquake. Calif. Div. Mines and Geol. Spec. Publ. 68*, ed. J. Bennett and R. Sherburne. Sacramento, Calif.: Calif. Div. of Mines, pp. 209–13.

Cottrell, A. H. 1953. *Dislocations and Plastic Flow in Crystals*. Oxford: Clarendon Press.

Crone, A. and Machette, M. 1984. Surface faulting accompanying the Borah Peak earthquake, central Idaho. *Geology* 12: 664–7.

Crone, A., Machette, M., Bonilla, M., Lienkaemper, J., Pierce, K., Scott, W., and Bucknam, R. 1987. Surface faulting accompanying the Borah Peak earthquake and segmentation of the Lost River fault, central Idaho. *Bull. Seismol. Soc. Am.* 77: 739–70.

Das, S. 1981. Three-dimensional rupture propagation and implications for the earthquake source mechanism. *Geophys. J. R.A.S.* 67: 375–93.

Das, S. and Aki, K. 1977. Fault planes with barriers: A versatile earthquake model. *J. Geophys. Res.* 82: 5658–70.

Das, S. and Scholz, C. 1981. Theory of time-dependent rupture in the earth. *J. Geophys. Res.* 86: 6039–51.

Das, S. and Scholz, C. 1982. Off-fault aftershock clusters caused by shear stress increase? *Bull. Seismol. Soc. Am.* 71: 1669–75.

Das, S. and Kostrov, B. 1983. Breaking of a single asperity: Rupture process and seismic radiation. *J. Geophys. Res.* 88: 4277–88.

Davison, F. and Scholz, C. 1985. Frequency–moment distribution of earthquakes in the Aleutian Arc: A test of the characteristic earthquake model. *Bull. Seismol. Soc. Am.* 75: 1349–62.

Day, S. 1982. Three-dimensional simulation of spontaneous rupture: The effect of nonuniform prestress. *Bull. Seismol. Soc. Am.* 72: 1881–1902.

Doser, D. I. 1986. Earthquake processes in the Rainbow Mountain–Fairview Peak–Dixie Valley, Nevada, region 1954–1959. *J. Geophys. Res.* 91: 12572–86.

Doser, D. and Kanamori, H. 1986. Depth of seismicity in the Imperial Valley region (1977–1983) and its relation to heat flow, crustal structure, and the October 15, 1979 earthquake. *J. Geophys. Res.* 91: 675–88.

Eaton, J., O'Neill, M., and Murdock, J. 1970. Aftershocks of the 1966 Parkfield–Cholame, California, earthquake: A detailed study. *Bull. Seismol. Soc. Am.* 60: 1151–97.

Einarsson, P., Bjornsson, S., Foulger, G., Stefansson, R., and Skaftadottir, T. 1981. Seismicity pattern in the south Iceland seismic zone. In *Earthquake Prediction, an International Review*. *M. Ewing Ser. 4*, ed. D. Simpson and P. Richards. Washington, D.C.: American Geophysical Union, pp. 141–52.

Eshelby, J. 1957. The determination of the elastic field of an ellipsoidal inclusion and related problems. *Proc. Roy. Soc. London, Series A* 241: 376–96.

Fuis, G., Mooney, W., Healey, J., McMechan, G., and Lutter, W. 1982. Crustal structure of the Imperial Valley region. In *The Imperial Valley, California, Earthquake of October 15, 1979. U.S. Geol. Surv. Prof. Paper 1254*, pp. 25–50.

Gilbert, G. K. 1884. A theory of the earthquakes of the Great Basin, with a practical application. *Am. J. Sci.* XXVII: 49–54.

Hamilton, R. 1972. Aftershocks of the Borrego Mountain earthquake from April 12 to June 12, 1968. In *The Borrego Mountain Earthquake of April 9, 1968. U.S. Geol. Surv. Prof. Paper 787*, pp. 31–54.

Hanks, T. C. 1974. The faulting mechanism of the San Fernando earthquake. *J. Geophys. Res.* 79: 1215–29.

Hanks, T. C. 1977. Earthquake stress-drops, ambient tectonic stresses, and stresses that drive plates. *Pure Appl. Geophys.* 115: 441–58.

Hanks, T. C. 1979. *b* values and $\omega^{-\gamma}$ seismic source models: Implications for tectonic stress variations along active crustal fault zones and the estimation of high frequency strong ground motion. *J. Geophys. Res.* 84: 2235–42.

Hanks, T. C. 1982. f_{max}. *Bull. Seismol. Soc. Am.* 72: 1867–80.

Hanks, T. C. and Johnson, D. A. 1976. Geophysical assessment of peak accelerations. *Bull. Seismol. Soc. Am.* 66: 959–68.

Hanks, T. and Kanamori, H. 1979. A moment–magnitude scale. *J. Geophys. Res.* 84: 2348–52.

Hanks, T. and McGuire, R. 1981. The character of high-frequency strong ground motion. *Bull. Seismol. Soc. Am.* 71: 2071–95.

Hanks, T. and Schwartz, D. 1987. Morphological dating of the pre-1983 fault scarp on the Lost River fault at Doublesprings Pass road, Custer County, Idaho. *Bull. Seismol. Soc. Am.* 77: 837–46.

Haskell, N. 1964. Total energy and energy spectral density of elastic wave radiation from propagating faults. *Bull. Seismol. Soc. Am.* 54: 1811–42.

Hirata, T. 1989. Fractal dimension of fault systems in Japan: fractal structure in rock fracture geometry at various scales. *Pageoph*, 131: 157–70.

Hudnut, K., Seeber, L., and Pacheco, J. 1989. Cross-fault triggering in the November 1987 Superstition Hills earthquake sequence, southern California. *Geophys. Res. Lett.* 16: 199–202.

Husseini, M. 1977. Energy balance for motion along a fault. *Geophys. J. R.A.S.* 49: 699–714.

Ida, Y. 1972. Cohesive force across tip of a longitudinal shear crack and Griffith's specific energy balance. *J. Geophys. Res.* 77: 3796–805.

Ida, Y. 1973. Stress concentration and unsteady propagation of longitudinal shear cracks. *J. Geophys. Res.* 78: 3418–29.

Johnson, C. and Hutton, K. 1982. Aftershocks and preearthquake seismicity. In *The Imperial Valley, California, Earthquake of October 15, 1979. U.S. Geol. Surv. Prof. Paper 1254*, pp. 59–76.

Kanamori, H. 1977. The energy release in great earthquakes. *J. Geophys. Res.* 82: 2981–87.

Kanamori, H. and Allen, C. 1986. Earthquake repeat time and average stress drop. In *Earthquake Source Mechanics. AGU Geophys. Mono. 37*, ed. S. Das, J. Boatwright, and C. Scholz. Washington, D.C.: American Geophysical Union, pp. 227–36.

Kanamori, H. and Anderson, D. 1975. Theoretical basis of some empirical relations in seismology. *Bull. Seismol. Soc. Am.* 65: 1073–95.

Kasahara, K. 1981. *Earthquake Mechanics*. Cambridge: Cambridge University Press.

Kisslinger, C. 1975. Processes during the Matsushiro swarm as revealed by leveling, gravity, and spring-flow observations. *Geology* 3: 57–62.

Knopoff, L. 1958. Energy release in earthquakes. *Geophys. J. R.A.S.* 1: 44–52.

Kostrov, B. V. 1964. Self-similar problems of propagation of shear cracks. *J. Appl. Math. Mech.* 28: 1077–87.

Kostrov, B. 1966. Unsteady propagation of longitudinal shear cracks. *J. Appl. Math. Mech.* 30: 1241–8.

Kostrov, B. 1974. Seismic moment and energy of earthquakes, and seismic flow of rock. *Izv. Acad. Sci. USSR Phys. Solid Earth* 1: 23–40.

Kostrov, B. and Das, S. 1988. *Principles of Earthquake Source Mechanics.* Cambridge: Cambridge University Press.

Koto, B. 1893. On the cause of the Great Earthquake in central Japan, 1891. *J. Coll. Sci. Imp. Univ.* 5 (pt 4): 294–353.

Kristy, M., Burdick, L., and Simpson, D. 1980. The focal mechanisms of the Gazli, USSR, earthquakes. *Bull. Seismol. Soc. Am.* 70: 1737–50.

Lay, T. and Kanamori, H. 1980. Earthquake doublets in the Solomon Islands. *Phys. Earth Planet. Inter.* 21: 283–304.

Lay, T. and Kanamori, H. 1981. An asperity model of great earthquake sequences. In *Earthquake Prediction, an International Review. M. Ewing Ser. 4,* ed. D. Simpson and P. Richards. Washington, D.C.: American Geophysical Union, pp. 579–92.

Lay, T., Ruff, L., and Kanamori, H. 1980. The asperity model and the nature of large earthquakes in subduction zones. *Earthquake Pred. Res.* 1: 3–71.

Li, V. C. 1987. Mechanics of shear rupture applied to earthquake zones. In *Fracture Mechanics of Rock,* ed. B. Atkinson. London: Academic Press, pp. 351–428.

Li, V. C. and Rice, J. R. 1983. Preseismic rupture progression and great earthquake instabilities at plate boundaries. *J. Geophys. Res.* 88: 4231–46.

Li, V. C., Seale, S. H., and Cao, T. 1987. Postseismic stress and pore pressure readjustment and aftershock distributions. *Tectonophysics* 144: 37–54.

Lomnitz, C., Mooser, F., Allen, C., Brune, J., and Thatcher, W. 1970. Seismicity and tectonics of the northern Gulf of California region, Mexico: Preliminary results. *Geofisica Internacional* 10: 37–48.

Lyell, Sir Charles. 1868. *Principles of Geology,* vol. II. London: John Murray.

Madariaga, R. 1976. Dynamics of an expanding circular fault. *Bull. Seismol. Soc. Am.* 66: 636–66.

Mandelbrot, B. B. 1977. *Fractals: Form, Chance, and Dimension.* San Francisco: W. H. Freeman.

Mandelbrot, B. B. 1983. *The Fractal Geometry of Nature.* San Francisco: W. H. Freeman.

McGarr, A. 1984. Scaling of ground motion parameters, state of stress, and focal depth. *J. Geophys. Res.* 89: 6969–79.

McKay, A. 1890. On earthquakes of September, 1888, in the Amuri and Marlborough Districts of the South Island, New Zealand. *New Zealand Geol. Surv. Geol. Exploration 1888–1889 Rept.* 20: 1–16.

Mendoza, C. and Hartzell, S. H. 1988. Aftershock patterns and main shock faulting. *Bull. Seismol. Soc. Am.* 78: 1438–49.

Mogi, K. 1962. Study of elastic shocks caused by the fracture of heterogenous materials and their relation to earthquake phenomena. *Bull. Earthquake Res. Inst. Univ. Tokyo* 40: 125–73.

Mogi, K. 1963. Some discussions on aftershocks, foreshocks and earthquake swarms – the fracture of a semi-infinite body caused by inner stress origin and its relation to the earthquake phenomena (3). *Bull. Earthquake Res. Inst. Univ. Tokyo,* 41: 615–58.

Mogi, K. 1969. Some features of recent seismicity in and near Japan (2). Activity before and after great earthquakes. *Bull. Earthquake Res. Inst. Univ. Tokyo* 47: 395–417.

Molnar, P. and Deng, Q. 1984. Faulting associated with large earthquakes and the average rate of deformation in central and east Asia. *J. Geophys. Res.* 89: 6203–14.

Nabelek, J. Planar vs. listric faulting: The rupture process and fault geometry of the 1983 Borah Peak, Idaho, earthquake from inversion of teleseismic body waves. *J. Geophys. Res.*, in press.

Nabelek, J., Chen, W.-P., and Ye, H. 1987. The Tangshan earthquake sequence and its implications for the evolution of the North China basin. *J. Geophys. Res.* 92: 12615–28.

Nakano, H. 1923. Notes on the nature of the forces which give rise to the earthquake motions. *Seismol. Bull. Central Meteorol. Obs., Japan* 1: 92–120.

Nur, A. 1974. Matsushiro, Japan earthquake swarm: Confirmation of the dilatancy–fluid diffusion model. *Geology* 2: 217–21.

Nur, A. 1981. Rupture mechanics of plate boundaries. In *Earthquake Prediction, an International Review. M. Ewing Ser. 4*, ed. D. Simpson and P. Richards. Washington, D.C.: American Geophysical Union, pp. 629–34.

Nur, A. and Booker, J. R. 1972. Aftershocks caused by pore fluid flow? *Science* 175: 885–7.

Okubo, P. 1988. Rupture propagation in a nonuniform stress field following a state variable friction model. *Eos* 69: 400–1.

Pareto, V. 1896. *Cours d'Economie Politique*. Reprinted as a volume of *Oeuvres Complètes* (1965). Geneva: Droz.

Purcaru, G. and Berckhemer, H. 1978. A magnitude scale for very large earthquakes. *Tectonophysics* 49: 189–98.

Quin, H. 1990. Dynamic stress drop and rupture dynamics of the October 15, 1979 Imperial Valley California earthquake. *Tectonophysics.* 175: 93–118.

Reid, H. F. 1910. The mechanics of the earthquake. *Rept. State Earthquake Inv. Comm., The California Earthquake of April 18, 1906.* Washington, D.C.: Carnegie Inst.

Richards, P. 1976. Dynamic motions near an earthquake fault: A three-dimensional solution. *Bull. Seismol. Soc. Am.* 66: 1–32.

Richins, W., Pechmann, J., Smith, R., Langer, C., Goter, S., Zollweg, J., and King, J. 1987. The 1983 Borah Peak, Idaho, earthquake and its aftershocks. *Bull. Seismol. Soc. Am.* 77: 694–723.

Ryall, A. and Malone, S. D. 1971. Earthquake distribution and mechanism of faulting in the Rainbow Mountain–Dixie Valley–Fairview Peak area, central Nevada. *J. Geophys. Res.* 76: 7241–8.

Savage, J. and Wood, M. 1971. The relation between apparent stress and stress drop. *Bull. Seismol. Soc. Am.* 61: 1381–8.

Scholz, C. 1968a. The frequency–magnitude relation of microfracturing in rock and its relation to earthquakes. *Bull. Seismol. Soc. Am.* 58: 399–415.

Scholz, C. 1968b. Microfractures, aftershocks, and seismicity. *Bull. Seismol. Soc. Am.* 58: 1117–30.

Scholz, C. 1972. Crustal movements in tectonic areas. *Tectonophysics* 14: 201–17.

Scholz, C. H. 1977. A physical interpretation of the Haicheng earthquake prediction. *Nature* 267: 121–4.

Scholz, C. H. 1982a. Scaling laws for large earthquakes: Consequences for physical models. *Bull. Seismol. Soc. Am.* 72: 1–14.

Scholz, C. H. 1982b. Scaling relations for strong ground motions in large earthquakes. *Bull. Seismol. Soc. Am.* 72: 1903–9.

Scholz, C. H. and Aviles, C. 1986. The fractal geometry of faults and faulting. In *Earthquake Source Mechanics. AGU Geophys. Mono. 37*, ed. S. Das, J. Boatwright, and C. Scholz. Washington, D.C.: American Geophysical Union, pp. 147–55.

Scholz, C. H., Aviles, C., and Wesnousky, S. 1986. Scaling differences between large intraplate and interplate earthquakes. *Bull. Seismol. Soc. Am.* 76: 65–70.

Scholz, C., Molnar, P., and Johnson, T. 1972. Detailed studies of frictional sliding of granite and implications for the earthquake mechanism. *J. Geophys. Res.* 77: 6392–406.

Scholz, C., Wyss, M., and Smith, S. 1969. Seismic and aseismic slip on the San Andreas fault. *J. Geophys. Res.* 74: 2049–69.

Schwartz, D. and Coppersmith, K. 1984. Fault behavior and characteristic earthquakes: Examples from the Wasatch and San Andreas fault zones. *J. Geophys. Res.* 89: 5681–98.

Seeber, L. and Armbruster, J. 1986. A study of earthquake hazard in New York State and adjacent areas. *Report to U.S. Nuclear Regulatory Commission, NUREG / CR-4750*.

Segall, P. and Pollard, D. 1983. Joint formation in granitic rock of the Sierra Nevada. *Geol. Soc. Am. Bull.* 94: 563–75.

Sharp, R., Lienkaemper, J., Bonilla, M., Burke, D., Fox, B., Herd, D., Miller, D., Morton, D., Ponti, D., Rymer, M., Tionsley, J., Yount, J., Kahle, J., Hart, E., and Sieh, K. 1982. Surface faulting in the central Imperial Valley. In *The Imperial Valley, California, Earthquake of October 15, 1979. U.S. Geol. Surv. Prof. Paper 1254*, pp. 119–44.

Shimazaki, K. 1986. Small and large earthquakes: The effects of the thickness of the seismogenic layer and the free surface. In *Earthquake Source Mechanics. AGU Geophys. Mono. 37*, ed. S. Das, J. Boatwright, and C. Scholz. Washington, D.C.: American Geophysical Union, pp. 209–16.

Sibson, R. 1981. Fluid flow accompanying faulting: field evidence and models. In *Earthquake Prediction, an International Review. M. Ewing Ser. 4*, ed. D. Simpson and P. Richards. Washington, D.C.: American Geophysical Union, pp. 593–603.

Singh, S., Rodriguez, M., and Esteva, L. 1983. Statistics of small earthquakes and frequency of occurrence of large earthquakes along the Mexican subduction zone. *Bull. Seismol. Soc. Am.* 73: 1779–96.

Slemmons, D. B. 1957. Geological effects of the Dixie Valley–Fairview Peak, Nevada, earthquakes of December 16, 1954. *Bull. Seismol. Soc. Am.* 47: 353–75.

Smith, S. and Wyss, M. 1968. Displacement on the San Andreas fault subsequent to the 1966 Parkfield earthquake. *Bull. Seismol. Soc. Am.* 58: 1955–73.

412   *References*

Starr, A. 1928. Slip in a crystal and rupture in a solid due to shear. *Proc. Cambridge Philos. Soc.* 24: 489–500.

Stauder, W. 1968. Mechanism of the Rat Island earthquake sequence of February 4, 1965, with relationships to island arcs and sea-floor spreading. *J. Geophys. Res.* 73: 3847–54.

Stein, R. and Lisowski, M. 1983. The 1979 Homestead Valley earthquake sequence, California: Control of aftershocks and postseismic deformations. *J. Geophys. Res.* 88: 6477–90.

Susong, D., Bruhn, R., and Janecke, S. 1990. Structure and mechanics of a rupture barrier in the Lost River fault zone, Idaho: Implications for the 1983 Borah Peak earthquakes. *Bull. Seismol. Soc. Am.* 80: 57–68.

Suyehiro, S., Asada, T., and Ohtake, M. 1964. Foreshocks and aftershocks accompanying a perceptible earthquake in central Japan – on a peculiar nature of foreshocks. *Papers Meteorol. Geophys.* 15: 71–88.

Sykes, L. R. 1970. Earthquake swarms and sea-floor spreading. *J. Geophys. Res.* 75: 6598–611.

Tajima F. and Kanamori, H. 1985. Global survey of aftershock area expansion patterns. *Phys. Earth Planet. Int.* 40: 77–134.

Tse, S. and Rice, J. 1986. Crustal earthquake instability in relation to the depth variation of frictional slip properties. *J. Geophys. Res.* 91: 9452–72.

Utsu, T. 1971. Aftershocks and earthquake statistics (III). *J. Fac. Science, Hokkaido Univ. Ser. VII (Geophysics)* 3: 379–441.

Ward, S. and Barrientos, S. 1986. An inversion for slip distribution and fault shape from geodetic observations of the 1983, Borah Peak, Idaho, earthquake. *J. Geophys. Res.* 91: 4909–19.

Wesnousky, S., Scholz, C., and Shimazaki, K. 1983. Earthquake frequency distribution and the mechanics of faulting. *J. Geophys. Res.* 88: 9331–40.

Wesson, R., Lee, W., and Gibbs, J. 1971. Aftershocks of the earthquake. In *The San Fernando Earthquake of February 9, 1971. U.S. Geol. Surv. Prof. Paper 733*, pp. 24–9.

Whitcomb, J., Allen, C., Garmany, J., and Hileman, J. 1973. The 1971 San Fernando earthquake series: Focal mechanisms and tectonics. *Rev. Geophys. Space Phys.* 11: 693–730.

Wyss, M. and Brune, J. 1967. The Alaska earthquake of 28 March 1964: A complex multiple rupture. *Bull. Seismol. Soc. Am.* 57: 1017–23.

Chapter 5

Ambrayseys, N. 1970. Some characteristic features of the Anatolian fault zone. *Tectonophysics* 9: 143–65.

Anderson, J. G. and Bodin, P. 1987. Earthquake recurrence models and historical seismicity in the Mexicali–Imperial Valley. *Bull. Seismol. Soc. Am.* 77: 562–78.

Ando, M. 1974. Seismotectonics of the 1923 Kanto earthquake. *J. Phys. Earth* 22: 263–77.

Ando, M. 1975. Source mechanisms and tectonic significance of historic earthquakes along the Nankai trough, Japan. *Tectonophysics* 27: 119–40.

Bak, P., Tang, C., and Wiesenfeld, K. 1988. Self-organized criticality. *Phys. Rev. A* 38: 364–74.

Bakun, W. and Lindh, A. 1985. The Parkfield, California, earthquake prediction experiment. *Science* 229: 619–24.

Bakun, W. H. and McEvilly, T. V. 1984. Recurrence models and Parkfield, California, earthquakes. *J. Geophys. Res.* 89: 3051–8.

Burford, R. O. 1988. Retardations in fault creep rates before local moderate earthquakes along the San Andreas fault system, central California. *Pageoph* 126: 499–529.

Burridge, R. and Knopoff, L. 1967. Model and theoretical seismicity. *Bull. Seismol. Soc. Am.* 57: 341–71.

Cao, T. and Aki, K. 1984. Seismicity simulation with a mass–spring model and a displacement hardening–softening friction law. *Pageoph* 122: 10–23.

Chinnery, M. A. 1961. Deformation of the ground around surface faults. *Bull. Seismol. Soc. Am.* 51: 355–72.

Christensen, D.H. and Ruff, L. J. 1983. Outer rise earthquakes and seismic coupling. *Geophys. Res. Lett.* 10: 697–700.

DeMets, C., Gordon, R. G., Stein, S., and Argus, D. F. 1987. A revised estimate of Pacific–North America motion and implications for western North America Plate boundary tectonics. *Geophys. Res. Lett.* 14: 911–14.

Dmowska, R., Rice, J. R., Lovison, L. C., and Josell, D. 1988. Stress transfer and seismic phenomena in coupled subduction zones during the earthquake cycle. *J. Geophys. Res.* 93: 7869–85.

Ellsworth, W. L., Lindh, A. G., Prescott, W. H., and Herd, D. G. 1981. The 1906 San Francisco earthquake and the seismic cycle. In *Earthquake Prediction, an International Review. M. Ewing Ser. 4*, ed. D. Simpson and P. Richards. Washington, D.C.: American Geophysical Union, pp. 126–40.

Elsasser, W. M. 1969. Convection and stress propagation in the upper mantle. In *The Application of Modern Physics to the Earth and Planetary Interiors*, ed. S. K. Runcorn. New York: Wiley Interscience, pp. 223–46.

Fedotov, S. A. 1965. Regularities in the distribution of strong earthquakes in Kamchatka, the Kuriles, and northeastern Japan. *Akad. Nauk USSR Inst. Fiz. Zeml.: Trudy* 36: 66–95.

Fitch, T. J. and Scholz, C. H. 1971. Mechanism of underthrusting in southwest Japan: A model of convergent plate interactions. *J. Geophys. Res.* 76: 7260–92.

Gilbert, G. K. 1909. Earthquake forecasts. *Science* XXIX: 121–38.

Grapes, R. and Wellman, H. The 12 m dextral displacement along the Wairarapa fault, New Zealand, during the 1855 earthquake. *Geology*, in press.

Gu, J.-C., Rice, J. R., Ruina, A. L., and Tse, S. T. 1984. Slip motion and stability of a single degree of freedom elastic system with rate and state dependent friction. *J. Mech. Phys. Solids* 32: 167–96.

Hall, N. T. 1984. Holocene history of the San Andreas fault between Crystal Springs Reservoir and the San Andreas Dam, San Mateo County, California. *Bull. Seismol. Soc. Am.* 74: 281–99.

Harris, R. A. and Segall, P. 1987. Detection of a locked zone at depth on the Parkfield, California, segment of the San Andreas fault. *J. Geophys. Res.* 92: 7945–62.

Hill, D. P., Eaton, J. P., and Jones, L. M. Seismicity of the San Andreas fault system: 1980–1986. In *The San Andreas Fault System. U.S. Geol. Surv. Prof. Paper*, ed. R. E. Wallace, in press.

Huberman, B. A. and Crutchfield, J. P. 1979. Chaotic states of anharmonic systems in periodic fields. *Phys. Rev. Lett.* 43: 1743–6.

Kanamori, H. and McNally, K. C. 1982. Variable rupture mode of the subduction zone along the Equador–Colombia coast. *Bull. Seismol. Soc. Am.* 72: 1241–53.

Kasahara, K. 1981. *Earthquake Mechanics*. Cambridge: Cambridge Univ. Press.

King, N. E. and Savage, J. C. 1984. Regional deformation near Palmdale, California, *J. Geophys. Res.* 89: 2471–7.

Kroger, P. M., Lyzenga, G. A., Wallace, K. S., and Davidson, J. M. 1987. Tectonic motion in the western United States inferred from very long baseline interferometry measurements, 1980–1986. *J. Geophys. Res.* 92: 14151–63.

Li, V. C. and Rice, J. R. 1987. Crustal deformation in great California earthquake cycles. *J. Geophys. Res.* 92: 11533–51.

Lyell, Sir Charles, 1868. *Principles of Geology*. London: John Murray.

Matsuda, T., Ota, Y., Ando, M., and Yonekura, N. 1978. Fault mechanism and recurrence time of major earthquakes in the southern Kanto district. *Geol. Soc. Am. Bull.* 89: 1610–18.

Mavko, G. M. 1981. Mechanics of motion on major faults. *Ann. Rev. Earth Planet. Sci.* 9: 81–111.

McCann, W. R., Nishenko, S. P., Sykes, L. R., and Krause, J. 1979. Seismic gaps and plate tectonics: Seismic potential for major boundaries. *Pageoph* 117: 1082–147.

McNally, K. C. and Gonzalez-Ruis, J. R. 1986. Predictability of the whole earthquake cycle and source mechanics for large $(7.0 < M_w < 8.1)$ earthquakes along the Middle America trench offshore Mexico. *Earthquake Notes* 57: 22.

Minster, J. B. and Jordan, T. H. 1978. Present day plate motions. *J. Geophys. Res.* 83: 5531–54.

Minster, J. B. and Jordan, T. H. 1987. Vector constraints on western U.S. deformation from space geodesy, neotectonics, and plate motions. *J. Geophys. Res.* 92: 4798–809.

Mogi, K. 1979. Two kinds of seismic gaps. *Pageoph* 117: 1172–86.

Mogi, K. 1981. Seismicity in western Japan and long term earthquake forecasting. In *Earthquake Prediction, an International Review. M. Ewing Ser. 4*, ed. D. Simpson and P. Richards. Washington, D.C.: American Geophysical Union, pp. 43–51.

Nakata, T., Takahashi, T., and Koba, M. 1978. Holocene emerged coral reefs and sea level changes in the Ryukyu Islands. *Geograph. Rev. Japan* 51: 87–108.

Nishenko, S. P. and Buland, R. 1987. A generic recurrence interval distribution for earthquake forecasting. *Bull. Seismol. Soc. Am.* 77: 1382–99.

Nishenko, S. P. and McCann, W. R. 1981. Seismic potential of the world's major plate boundaries. In *Earthquake Prediction, an International Review.*

M. Ewing Ser. 4, ed. D. Simpson and P. Richards. Washington, D.C.: American Geophysical Union, pp. 20–8.

Nur, A. and Mavko, G. 1974. Postseismic viscoelastic rebound. *Science* 183: 204–6.

Nussbaum, J. and Ruina, A. 1987. A two degree-of-freedom earthquake model with static/dynamic friction. *Pageoph* 125: 629–56.

Ota, Y., Machida, H., Hori, N., Kunishi, K., and Omur, A. 1978. Holocene raised coral reefs of Kinkaijima (Ryukyu Is.) – an approach to Holocene sea level study. *Geograph. Rev. Japan*. 51: 109–30.

Peltier, W. R. 1981. Ice age geodynamics. *Ann. Rev. Earth and Planet. Sci.* 9: 199–226.

Perez, O. 1983. Spatial–temporal–energy characteristics of seismicity occurring during the seismic cycle. PhD thesis, Columbia University.

Plafker, G. and Rubin, M. 1978. Uplift history and earthquake recurrence as deduced from marine terraces on Middleton Island, Alaska. In *Proc. Conf. VI: Methodology for Identifying Seismic Gaps and Soon-to-break Gaps. U.S. Geol. Surv. Open-file Rept. 78-943*, pp. 857–68.

Prescott, W. H., Lisowski, M., and Savage, J. C. 1981. Geodetic measurements of crustal deformation on the San Andreas, Hayward, and Calaveras faults near San Francisco, California. *J. Geophys. Res.* 86: 10853–69.

Prescott, W. H. and Yu, S. B. 1986. Geodetic measurement of horizontal deformation in the northern San Francisco Bay region, California. *J. Geophys. Res.* 91: 7475–84.

Lord Rayleigh. 1877. *The Theory of Sound*. Dover: New York.

Reid, H. F. 1910. The mechanism of the earthquake. In *The California Earthquake of April 18, 1906, Report of the State Earthquake Investigation Commission, 2*. Washington, D.C.: Carnegie Institution, pp. 1–192.

Rice, J. R. and Tse, S. T. 1986. Dynamic motion of a single degree of freedom system following a rate and state dependent friction law. *J. Geophys. Res.* 91: 521–30.

Ruff, L. and Kanamori, H. 1983. Seismic coupling and uncoupling at subduction zones. *Tectonophysics* 99: 99–117.

Savage, J. C. 1983. A dislocation model of strain accumulation and release at a subduction zone. *J. Geophys. Res.* 88: 4984–96.

Savage, J. C. and Burford, R. O. 1973. Geodetic determination of relative plate motion in central California. *J. Geophys. Res.* 78: 832–45.

Savage, J. C. and Hastie, L. M. 1966. Surface deformation associated with dip-slip faulting. *J. Geophys. Res.* 71: 4897–904.

Savage, J. C. and Lisowski, M. 1984. Deformation in the White Mountain seismic gap, California–Nevada, 1972–1982. *J. Geophys. Res.* 89: 7671–88.

Savage, J. C. and Prescott, W. H. 1978. Asthenosphere readjustment and the earthquake cycle. *J. Geophys. Res.* 83: 3369–76.

Savage, J. C., Prescott, W. H., Lisowski, M., and King, N. 1979. Deformation across the Salton Trough, California, 1973–1977. *J. Geophys. Res.* 84: 3069–80.

Savage, J. C., Prescott, W. H., Lisowski, M., and King, N. 1981. Strain accumulation in southern California, 1973–1980. *J. Geophys. Res.* 86: 6991–7001.

Scholz, C. H. 1988. Mechanisms of seismic quiescences. *Pageoph* 126: 701–18.

Scholz, C. H. and Kato, T. 1978. The behavior of a convergent plate boundary: Crustal deformation in the south Kanto District, Japan. *J. Geophys. Res.* 83: 783–91.

Scholz, C. H., Wyss, M., and Smith, S. W. 1969. Seismic and aseismic slip on the San Andreas fault. *J. Geophys. Res.* 74: 2049–69.

Schwartz, D. P. and Coppersmith, K. J. 1984. Fault behavior and characteristic earthquakes: Examples from the Wasatch and San Andreas faults. *J. Geophys. Res.* 89: 5681–98.

Schwartz, D. P., Hanson, K., and Swan, F. H. 1983. Paleoseismic investigations along the Wasatch fault zone: An update. In *Paleoseismicity along the Wasatch Front and Adjacent Areas, Central Utah*, ed. A. J. Crone. Utah Geological and Mineral Survey Special Studies 62: 45–9.

Segall, P. and Harris, R. 1987. Earthquake deformation cycle on the San Andreas fault near Parkfield, California. *J. Geophys. Res.* 92: 10511–25.

Seno, T. 1979. Pattern of intraplate seismicity in southwest Japan before and after great interplate earthquakes. *Tectonophysics* 57: 267–83.

Sharp, R. V. 1981. Variable rates of late Quaternary strike slip on the San Jacinto fault zone, southern California. *J. Geophys. Res.* 86: 1754–62.

Shimazaki, K. 1976. Intraplate seismicity and interplate earthquakes – historical activity in southwest Japan. *Tectonophysics* 33: 33–42.

Shimazaki, K. and Nakata, T. 1980. Time-predictable recurrence model for large earthquakes. *Geophys. Res. Lett.* 7: 279–82.

Sieh, K. 1978. Slip along the San Andreas fault associated with the great 1857 earthquake. *Bull. Seis. Soc. Am.* 68: 1421–8.

Sieh, K. 1981. A review of geological evidence for recurrence times of large earthquakes. In *Earthquake Prediction, an International Review. M. Ewing Ser. 4*, ed. D. Simpson and P. Richards. Washington, D.C.: American Geophysical Union, pp. 209–16.

Sieh, K. 1984. Lateral offsets and revised dates of large prehistoric earthquakes at Pallett Creek, southern California. *J. Geophys. Res.* 89: 7641–70.

Sieh, K. and Jahns, R. 1984. Holocene activity of the San Andreas fault at Wallace Creek, California. *Geol. Soc. Am. Bull.* 95: 883–896.

Sieh, K., Stuiver, M., and Brillinger, D. 1989. A more precise chronology of earthquakes produced by the San Andreas fault in southern California. *J. Geophys. Res.* 94: 603–24.

Stoker, J. J. 1950. *Nonlinear Vibrations*. New York: Interscience.

Stuart, W. D. 1988. Forecast model for great earthquakes at the Nankai Trough subduction zone. *Pageoph* 126: 619–42.

Swan, F. H., Schwartz, D. P., and Cluff, L. S. 1980. Recurrence of moderate to large magnitude earthquakes produced by surface faulting on the Wasatch fault, Utah. *Bull. Seismol. Soc. Am.* 70: 1431–62.

Sykes, L. R. 1971. Aftershock zones of great earthquakes, seismicity gaps, and earthquake prediction for Alaska and the Aleutians. *J. Geophys. Res.* 76: 8021–41.

Sykes, L. R., Kisslinger, J., House, L., Davies, J., and Jacob, K. H. 1981. Rupture zones and repeat times of great earthquakes along the Alaska–Aleutian arc, 1784–1980. In *Earthquake Prediction, an International Review.*

M. Ewing Ser. 4, ed. D. Simpson and P. Richards. Washington, D.C.: American Geophysical Union, pp. 217–47.

Sykes, L. R. and Nishenko, S. P. 1984. Probabilities of occurrence of large plate rupturing earthquakes for the San Andreas, San Jacinto, and Imperial faults, California. *J. Geophys. Res.* 89: 5905–27.

Sykes, L. R. and Quittmeyer, R. C. 1981. Repeat times of great earthquakes along simple plate boundaries. In *Earthquake Prediction, an International Review. M. Ewing Ser. 4*, ed. D. Simpson and P. Richards. Washington, D.C.: American Geophysical Union, pp. 217–47.

Taylor, F. W., Frohlich, C., Lecolle, J., and Strekler, M. 1987. Analysis of partially emerged corals and reef terraces in the central Vanuatu arc: Comparison of contemporary coseismic and nonseismic with Quarternary vertical movements. *J. Geophys. Res.* 92: 4905–33.

Thatcher, W. 1979. Systematic inversion of geodetic data in central California. *J. Geophys. Res.* 84: 2283–95.

Thatcher, W. 1982. Seismic triggering and earthquake prediction. *Nature* 299: 12–13.

Thatcher, W. 1983. Nonlinear strain buildup and the earthquake cycle on the San Andreas fault. *J. Geophys. Res.* 88: 5893–902.

Thatcher, W. 1984a. The earthquake deformation cycle at the Nankai Trough, southwest Japan. *J. Geophys. Res.* 89: 3087–101.

Thatcher, W. 1984b. The earthquake deformation cycle, recurrence, and the time-predictable model. *J. Geophys. Res.* 89: 5674–80.

Thatcher, W. 1986. The crustal deformation cycle at convergent plate margins. *R. Soc. New Zealand Bull.* 24: 317–32.

Thatcher, W. and Fujita, N. 1984. Deformation of the Mitaka rhombus: Strain buildup following the 1923 Kanto earthquake, central Honshu, Japan. *J. Geophys. Res.* 89: 3102–6.

Thatcher, W. and Rundle, J. B. 1979. A model for the earthquake cycle in subduction zones. *J. Geophys. Res.* 84: 5540–56.

Thatcher, W. and Rundle, J. B. 1984. A viscoelastic coupling model for the cyclic deformation due to periodically repeated earthquakes at subduction zones. *J. Geophys. Res.* 89: 7631–40.

Tocher, D. 1959. Seismic history of the San Francisco region. In *Calif. Div. Mines Spec. Rept. 57*. Sacramento, Calif.: Calif. Div. of Mines, pp. 39–49.

Tse, S. T. and Rice, J. R. 1986. Crustal earthquake instability in relation to the depth variation of frictional slip properties. *J. Geophys. Res.* 91: 9452–72.

Turcotte, D. L., Liu, J. Y., and Kulhawy, F. H. 1984. The role of an intracrustal asthenosphere on the behavior of major strike-slip faults. *J. Geophys. Res.* 89: 5801–16.

Walcott, R. I. 1973. Structure of the earth from glacio-isostatic rebound. *Ann. Rev. Earth Planet. Sci.* 1: 15–37.

Wallace, R. E. 1981. Active faults paleoseismology, and earthquake hazards in the Western United States. In *Earthquake Prediction, an International Review. M. Ewing Ser. 4*, ed. D. Simpson and P. Richards. Washington, D.C.: American Geophysical Union, pp. 209–16.

Wallace, R. E. 1987. Grouping and migration of surface faulting and variations of slip rate on faults in the Great Basin province. *Bull. Seismol. Soc. Am.* 77: 868–76.

Weber, G. E. and Lajoie, K. R. 1977. Late Pleistocene and Holocene tectonics of the San Gregorio fault zone between Moss Beach and Point Ano Nuevo, San Mateo County, California. *Geol. Soc. Am. Abstracts with Programs* 9: 524.

Weldon, R. J. and Humphreys, F. 1986. A kinematic model of southern California. *Tectonics* 5: 33–48.

Weldon, R. J. and Sieh, K. E. 1985. Holocene rate of slip and tentative recurrence interval for large earthquakes on the San Andreas fault, Cajon Pass, southern California. *Geol. Soc. Am. Bull.* 96: 793–812.

Yonekura, N. 1975. Quaternary tectonic movements in the outer arc of southwest Japan with special reference to seismic crustal deformation. *Bull. Dept. Geogr. Univ. Tokyo* 7: 19–71.

Chapter 6

Adams, J. 1980a. Contemporary uplift and erosion of the Southern Alps, New Zealand. Summary. *Geol. Soc. Am. Bull.* 91: 2–4.

Adams, J. 1980b. Paleoseismicity of the Alpine fault seismic gap, New Zealand. *Geology* 8: 72–6.

Allen, C. R. 1968. The tectonic environment of seismically active and inactive areas along the San Andreas fault system. In *Proc. Conf. on Geological Problems of the San Andreas Fault System. Stanford Univ. Publ. Geol. Sci. 11*, pp. 70–82.

Allis, R. G., Henley, R. W., and Carman, R. W. 1979. The thermal regime beneath the Southern Alps. *Bull. R. Soc. N.Z.* 18: 79–85.

Ambrayseys, N. H. 1975. Studies of historical seismicity and tectonics. In *Geodynamics Today, A Review of the Earth's Dynamic Processes.* London: The Royal Society, pp. 7–16.

Atwater, T. 1970. Implications of plate tectonics for the Cenozoic tectonic evolution of western North America. *Geol. Soc. Am. Bull.* 81: 3513–36.

Baker, B. H. and Wohlenberg, J. 1971. Structure and evolution of the Kenya rift valley. *Nature* 229: 538.

Bell, M. L. and Nur, A. 1978. Strength changes due to reservoir induced pore pressure and stresses and application to Lake Oroville. *J. Geophys. Res.* 83: 4469–83.

Bergman, E. A. 1986. Intraplate earthquakes and the state of stress in oceanic lithosphere. *Tectonophysics* 132: 1–35.

Bergman, E. and Solomon, S. 1988. Transform fault earthquakes in the North Atlantic: Source mechanism and depth of faulting. *J. Geophys. Res.* 93: 9027–57.

Berryman, K. R., Cutten, H. N. C., Fellows, D. L., Hull, A. G., and Sewell, R. J. Alpine fault reconnaissance survey from Arawata River to Pyke River, South Westland, New Zealand. *N.Z. J. Geol. Geophys.*, in press.

Bibby, H. M., Haines, A. J., and Walcott, R. I. 1986. Geodetic strain and the present day plate boundary zone through New Zealand. *R. Soc. N.Z. Bull.* 24: 427–38.

Bilham, R. and Williams, P. 1985. Sawtooth segmentation and deformation processes on the southern San Andreas fault, California. *Geophys. Res. Lett.* 12: 557–60.

Bjarnason, I. and Einarsson, P. Source mechanism of the 1987 Vatnafjoll earthquake in south Iceland. *J. Geophys. Res.*, in press.

Bowman, J. Jones, T., Gibson, G., Corke, A., Thompson, R., and Comacho, A. 1988. Tennant Creek earthquakes of 22 January 1988: Reactivation of a fault zone in the Proterozoic Australian shield. *Eos* 69: 400.

Bridgman, P. W. 1945. Polymorphic transitions and geological phenomena. *Am. J. Sci.* 243A: 90–7.

Brune, J. N. 1968. Seismic moment, seismicity, and rate of slip along major fault zones. *J. Geophys. Res.* 73: 777–84.

Buck, W. R. 1988. Flexural rotation of normal faults. *Tectonics* 7: 959–73.

Burford, R. O. 1972. Continued slip on the Coyote Creek fault after the Borrego Mountain earthquake. In *The Borrego Mountain Earthquake of April 9, 1968. U.S. Geol. Surv. Prof. Paper 787*, pp. 105–11.

Burford, R. O. and Harsh, P. W. 1980. Slip on the San Andreas fault in central California from alinement array surveys. *Bull. Seismol. Soc. Am.* 70: 1233–61.

Burr, N. and Solomon, S. 1978. The relationship of source parameters of oceanic transform earthquakes to plate velocity and transform length. *J. Geophys. Res.* 83: 1193–205.

Byrne, D. E., Davis, D. M., and Sykes, L. R. 1988. Loci and maximum size of thrust earthquakes and the mechanics of the shallow region of subduction zones. *Tectonics* 7: 833–57.

Cao, T. and Aki, K. 1986. Effect of slip rate on stress drop. *Pageoph* 124: 515–30.

Carder, D. S. 1945. Seismic investigations in the Boulder Dam area, 1940–1945, and the influence of reservoir loading on earthquake activity. *Bull. Seismol. Soc. Am.* 35: 175–92.

Cessaro, R. K. and Hussong, D. M. 1986. Transform seismicity at the intersection of the oceanographer fracture zone and the Mid-Atlantic ridge. *J. Geophys. Res.* 91: 4839–53.

Chase, C. G. 1978. Plate kinematics: The Americas, East Africa and the rest of the world. *Earth Planet. Sci. Lett.* 37: 353–68.

Chen, A., Frohlich, C., and Latham, G. 1982. Seismicity of the forearc marginal wedge (accretionary prism). *J. Geophys. Res.* 87: 3679–90.

Chen, W.-P. and Molnar, P. 1983. Focal depths of intracontinental and intraplate earthquakes and their implications for the thermal and mechanical properties of the lithosphere. *J. Geophys. Res.* 88: 4183–215.

Chen, Y. 1988. Thermal model of oceanic transform faults. *J. Geophys. Res.* 93: 8839–51.

Christoffel, D. A. 1971. Motion of the New Zealand Alpine fault deduced from the pattern of sea-floor spreading. *Bull. R. Soc. N.Z.* 9: 25–30.

Coward, M. P., Dewey, J. F., and Hancock, P. L. (eds.) 1987. *Continental Extensional Tectonics. Geol. Soc. Spec. Publ. 28*. London: Blackwell.

Crook, C. N. 1984. Geodetic measurement of the horizontal crustal deformation associated with the Oct. 15, 1979 Imperial Valley (California) earthquake. PhD thesis, University of London.

Crone, A. J. and Luza, K. V. 1986. Holocene deformation associated with the Meers fault, southwestern Oklahoma. In *The Slick Hills of Southwestern Oklahoma – Fragments of an Aulachogen? Oklahoma Geol. Surv. Guidebook 24*, ed. R. N. Donovan. Norman, Okla: Univ. of Oklahoma, pp. 68–74.

Cutten, H. N. C. 1979. Rappahannock group: Late Cenozoic sedimentation and tectonics contemporaneous with Alpine fault movement. *N.Z. J. Geol. Geophys.* 22: 535–53.

Davies, G. and Brune, J. N. 1971. Global plate motion rates from seismicity data. *Nature (Phys. Sci.)* 229: 101–7.

Davis, D., Dahlen, F. A., and Suppe, J. 1983. Mechanics of fold-and-thrust belts and accretionary wedges. *J. Geophys. Res.* 88: 1153–72.

DeMets, C., Gordon, R. G., Stein, S., and Argus, D. F. 1987. A revised estimate of Pacific–North America motion and implications for western North America Plate boundary tectonics. *Geophys. Res. Lett.* 14: 911–14.

Dewey, J. F. and Bird, J. M. 1970. Mountain belts and the new global tectonics. *J. Geophys. Res.* 75: 2625–47.

Doser, D. I. 1988. Source parameters of earthquakes in the Nevada seismic zone, 1915–43. *J. Geophys. Res.* 93: 15001–15.

Engdahl, E. R. 1977. Seismicity and plate subduction in the central Aleutians. In *Island Arcs and Deep Sea Trenches and Back-arc Basins. M. Ewing Ser. 1*, ed. M. Talwani and W. Pittman, III. Washington, D.C.: American Geophysical Union, pp. 259–72.

England, P. C. and McKenzie, D. P. 1982. A thin viscous sheet model for continental deformation. *Geophys. J. R.A.S.* 70: 295–321.

Ernst, W. G. 1975. Summary. In *Metamorphism and Plate Tectonic Regimes*, ed. W. G. Ernst. Stroudsburg, Pa.: Dowden, Hutchinson & Ross, pp. 423–6.

Ewing, M. and Heezen, B. 1956. Some problems of Antarctic submarine geology in Antarctica. In *The International Geophysical Year. AGU Geophys. Mono. 1*, ed. A. Crary. Washington, D.C.: American Geophysical Union, p. 75.

Fletcher, R. and Pollard, D. D. 1981. An anticrack mechanism for stylolites. *Geology* 9: 419–24.

Forsyth, D. and Uyeda, S. 1975. On the relative importance of driving forces of plate motion. *Geophys. J. R.A.S.* 43: 163–200.

Frohich, C. 1989. The nature of deep-focus earthquakes. *Ann. Rev. Earth Planet. Phys.* 17: 227–54.

Gordon, F. and Lewis, J. 1980. The Meckering and Caligiri earthquakes of October 1968 and March, 1970. *Geol. Surv. Western Australia, Bull. 126*.

Gough, D. I., Fordjor, C. K., and Bell, J. S. 1983. A stress province boundary and tractions on the North American plate. *Nature* 305: 619–21.

Grapes, R. H. 1987. Some thoughts on the Sanbagawa metamorphism, Shikoku, Japan. *Mem. Ehime Univ. Sci. Ser. D (Earth Sci)* X: 29–30.

Grapes, R. H., Sissons, B. A., and Wellman, H. W. 1987. Widening of the Taupo Volcanic Zone, New Zealand and the Edgecumbe earthquake of March, 1987. *Geology* 15: 1123–5.

Grapes, R. and Wellman, H. 1989. The 12 m dextral displacement along the Wairarapa fault, New Zealand, during the 1855 earthquake. *Geology*, submitted.

Green, H. W. and Burnley, P. C. 1989. The mechanism of failure responsible for deep-focus earthquakes. *Nature, 341,* 733–7.

Grindley, G. W. 1975. New Zealand. *Spec. Publ. Geol. Soc. London.* 4: 387–416.

Gupta, H. K. and Rastogi, B. K. 1976. *Dams and Earthquakes.* Amsterdam: Elsevier.

Harsh, P. W. 1982. Distribution of afterslip along the Imperial fault. In *The Imperial Valley, California, Earthquake of October 15, 1979. U.S. Geol. Surv. Prof. Paper 1254,* pp. 193–204.

Hatherton, T. 1968. Through the looking glass: A comparative study of New Zealand and California. *Nature* 220: 660–3.

Hauksson, E. et al. 1988. The 1987 Whittier Narrows earthquake in the Los Angeles metropolitan area, California. *Science* 239: 1409–12.

Hill, M. L. and Diblee, T. W., Jr. 1953. San Andreas, Garlock and Big Pine faults, California. *Geol. Soc. Am. Bull.* 64: 443–58.

Hudnut, K. W., Seeber, L., and Pacheco, J. 1989. Cross-fault triggering in the November 1987 Superstition Hills earthquake sequence, southern California. *Geophys. Res. Lett.* 16: 199–202.

Hull, A. G. and Berryman, K. R. 1986. Holocene tectonism in the vicinity of the Alpine fault at Lake McKerrow, Fiordland, New Zealand. *R. Soc. N.Z. Bull.* 24: 317–31.

Irwin, W. P. and Barnes, I. 1975. Effects of geological structure and metamorphic fluids on seismic behavior of the San Andreas fault system in central and northern California. *Geology* 3: 713–16.

Isacks, B. L., Oliver, J., and Sykes, L. R. 1968. Seismology and the new global tectonics. *J. Geophys. Res.* 73: 5855–99.

Jackson, J. A. 1987. Active normal faulting and crustal extension. In *Continental Extensional Tectonics,* ed. M. Coward, J. Dewey, and P. Hancock. London: Blackwell, pp. 3–18.

Jackson, J. and McKenzie, D. 1988. The relationship between plate motions and seismic moment tensors, and the rates of active deformation in the Mediterranean and Middle East. *Geophys. J. R.A.S.* 93: 45–73.

Jacob, K. H. 1984. Estimates of long-term probabilities for future great earthquakes in the Aleutians. *Geophys. Res. Lett.* 11: 295–8.

Jacob, K. H., Armbruster, J., Seeber, L., Pennington, W., and Farhatulla, S. 1979. Tarbela reservoir, Pakistan: A region of compressive tectonics and reduced seismicity upon initial reservoir filling. *Bull. Seismol. Soc. Am.* 69: 1175–92.

Jarrard, R. D. 1986a. Relations among subduction parameters. *Rev. Geophys.* 24: 217–84.

Jarrard, R. D. 1986b. Causes of compression and extension behind trenches. *Tectonophysics* 132: 89–102.

Jaume, S. C. and Lillie, R. J. 1988. Mechanics of the Salt Range – Potwar Plateau, Pakistan: A fold and thrust belt underlain by evaporites. *Tectonics* 7: 57–71.

Kanamori, H. 1971. Great earthquakes at island arcs and the lithosphere. *Tectonophysics* 12: 187–98.

Kanamori, H. 1986. Rupture process of subduction zone earthquakes. *Ann. Rev. Earth Planet. Sci.* 14: 293–322.

Kanamori, H. and Allen, C. R. 1986. Earthquake repeat time and average stress drop. In *Earthquake Source Mechanics. AGU Geophys. Mono. 37*, ed. S. Das, J. Boatwright, and C. Scholz. Washington, D.C.: American Geophysical Union, pp. 227–36.

Kanamori, H. and Stewart, G. S. 1976. Mode of strain release along the Gibbs fracture zone, Mid-Atlantic ridge. *Phys. Earth Planet. Int.* 11: 312–32.

Karig, D. E. 1970. Kermadec arc–New Zealand tectonic confluence. *N.Z. J. Geol. Geophys.* 13: 21–9.

Kawasaki, I., Kawahara, Y., Takata, I., and Kosugi, N. 1985. Mode of seismic moment release at transform faults. *Tectonophysics* 118: 313–27.

Kelleher, J. and McCann, W. 1976. Bouyant zones, great earthquakes, and unstable boundaries of subduction. *J. Geophys. Res.* 81: 4885–96.

Kelleher, J., Savino, J., Rowlett, H., and McCann, W. 1974. Why and where great thrust earthquakes occur along island arcs. *J. Geophys. Res.* 79: 4889–99.

Kikuchi, M. and Fukao, Y. 1987. Inversion of long period P waves from great earthquakes along subduction zones. *Tectonophysics* 144: 231–48.

Kirby, S. 1987. Localized polymorphic phase transformations in high pressure faults and applications to the physical mechanism of deep earthquakes. *J. Geophys. Res.* 92: 13789–800.

Kostrov, B. V. 1974. Seismic moment and energy of earthquakes, and seismic flow of rock. *Izv. Acad. Sci. USSR Phys. Solid Earth* 1: 23–44.

Landis, C. A. and Coombs, D. S. 1967. Metamorphic belts and orogenesis in southern New Zealand. *Tectonophysics* 4: 501–18.

Lay, T. and Kanamori, H. 1981. An asperity model of great earthquake sequences. In *Earthquake Prediction, an International Review. M. Ewing Ser. 4*, ed. D. Simpson and P. Richards. Washington, D.C.: American Geophysical Union, pp. 579–92.

Lay, T., Kanamori, H., and Ruff, L. 1982. The asperity model and the nature of large subduction zone earthquakes. *Earthquake Pred. Res.* 1: 1–71.

Lensen, G. 1975. Earth deformation studies in New Zealand. *Tectonophysics* 29: 541–51.

Lisowski, M. and Prescott, W. H. 1981. Short-range distance measurements along the San Andreas fault in central California, 1975 to 1979. *Bull. Seismol. Soc. Am.* 71: 1607–24.

Lisowski, M., Savage, J. C., Prescott, W. H., and Gross, W. K. 1988. Absence of strain accumulation in the Shumagin seismic gap, Alaska, 1980–1987. *J. Geophys. Res.* 93: 7909–22.

Louie, J. N., Allen, C. R., Johnson, D., Haase, P., and Cohn, S. 1985. Fault slip in southern California. *Bull. Seismol. Soc. Am.* 75: 811–34.

Marone, C. and Scholz, C. H. 1988. The depth of seismic faulting and the upper transition from stable to unstable slip regimes. *Geophys. Res. Lett.* 15: 621–4.

McGarr, A. 1976. Seismic moments and volume changes. *J. Geophys. Res.* 81: 1487–94.

McGarr, A. 1977. Seismic moments of earthquakes beneath island arcs, phase changes, and subduction velocities. *J. Geophys. Res.* 82: 256–64.

McGarr, A. 1984. Some applications of seismic source mechanism studies to assessing underground hazard. In *Proc. 1st Int. Cong. Rockbursts and Seismicity in Mines*. ed. N. C. Gay and E. H. Wainwright. Johannesburg: South African Inst. Min. Met. pp. 199–208.

McGarr, A. and Green, R. W. E. 1975. Measurement of tile in a deep-level gold mine and its relationship to mining and seismicity. *Geophys. J. R.A.S.* 43: 327–45.

Minster, J. B. and Jordan, T. 1987. Vector constraints on western U.S. tectonics from space geodesy, neotectonics, and plate motions. *J. Geophys. Res.* 92: 4798–804.

Mogi, K. 1969. Relationship between the occurrence of great earthquakes and tectonic structures. *Bull. Earthquake Res. Inst. Univ. Tokyo* 47: 429–41.

Molnar, P. 1983. Average regional strain due to slip on numerous faults of different orientations. *J. Geophys. Res.* 88: 6430–2.

Molnar, P., Atwater, T., Mammerickx, J., and Smith, S. M. 1975. Magnetic anomalies, bathymetry and tectonic evolution of the South Pacific since the Late Cretaceous. *Geophys. J. R.A.S.* 40: 383–420.

Molnar, P. and Deng, Q. 1984. Faulting associated with large earthquakes and the average rate of deformation in central and eastern Asia. *J. Geophys. Res.* 89: 6203–28.

Mount, V. S. and Suppe, J. 1987. State of stress near the San Andreas fault: Implications for wrench tectonics. *Geology* 15: 1143–6.

Nishenko, S. P. and McCann, W. R. 1981. Seismic potential for the world's major plate boundaries: 1981. In *Earthquake Prediction, an International Review. M. Ewing Ser. 4*, ed. D. Simpson and P. Richards. Washington, D.C.: American Geophysical Union, pp. 20–8.

Norris, R. J., Carter, R. M., and Turnbull, I. M. 1978. Cainozoic sedimentation in basins adjacent to a major continental transform boundary in southern New Zealand. *J. Geol. Soc. London* 135: 191–205.

Nuttli, O. 1973. The Mississippi Valley earthquakes of 1811–12: Intensities, ground motion and magnitudes. *Bull. Seismol. Soc. Am.* 63: 227–48.

O'Reilly, W. and Rastogi, B. K. (eds.) 1986. Induced seismicity. *Phys. Earth Planet. Int.* 44(2): 73–199.

Peterson, E. T. and Seno, T. 1984. Factors affecting seismic moment release rates in subduction zones. *J. Geophys. Res.* 89: 10233–48.

Pomeroy, P. W., Simpson, D. W., and Sbar, M. L. 1976. Earthquakes triggered by surface quarrying – Wappinger Falls, New York sequence of June, 1974. *Bull. Seismol. Soc. Am.* 66: 685–700.

Prothero, W. A., Jr. and Reid, I. D. 1982. Microearthquakes on the East Pacific Rise at 21°N and the Rivera fracture zone. *J. Geophys. Res.* 87: 8509–18.

Rice, J. R. and Cleary, M. P. 1976. Some basic stress diffusion solutions for fluid-saturated elastic porous media with compressible constituents. *Rev. Geophys. Space Phys.* 14: 227–41.

Richter, C. F. 1958. *Elementary Seismology.* San Fancisco: W. H. Freeman.

Richter, F. M. and McKenzie, D. P. 1978. Simple plate models of mantle convection. *J. Geophys.* 44: 441–71.

Roeloffs, E. A. 1988. Fault stability changes induced beneath a reservoir with cyclic variations in water level. *J. Geophys. Res.* 93: 2107–24.

Ruff, L. D. 1989. Do trench sediments affect great earthquake occurrence in subduction zones? *Pageoph* 129: 263–82.

Ruff, L. and Kanamori, H. 1980. Seismicity and the subduction process. *Phys. Earth Planet. Int.* 23: 240–52.

Savage, J. C. and Burford, R. O. 1973. Geodetic determination of relative plate motion in central California. *J. Geophys. Res.* 78: 832–45.

Sbar, M. L. and Sykes, L. R. 1977. Seismicity and lithospheric stress in New York and adjacent areas. *J. Geophys. Res.* 82: 5771–86.

Scholz, C. H. 1972. Crustal movements in tectonic areas. *Tectonophysics* 14: 201–17.

Scholz, C. H. 1977. Transform fault systems of California and New Zealand: Similarities in their tectonic and seismic styles. *J. Geol. Soc. London* 133: 215–29.

Scholz, C. H. 1980. Shear heating and the state of stress on faults. *J. Geophys. Res.* 85: 6174–84.

Scholz, C. H., Aviles, C., and Wesnousky, S. 1986. Scaling differences between large intraplate and interplate earthquakes. *Bull. Seismol. Soc. Am.* 76: 65–70.

Scholz, C. H., Barazangi, M., and Sbar, M. L. 1971. Late Cenozoic evolution of the Great Basin, western United States, as an ensialic interarc basin. *Geol. Soc. Am. Bull.* 82: 2979–90.

Scholz, C. H., Koczynski, T., and Hutchins, J. 1976. Evidence for incipient rifting in southern Africa. *Geophys. J. R.A.S.* 44: 135–44.

Scholz, C. H., Rynn, J. M. W., Weed, R. F., and Frohlich, C. 1973. Detailed seismicity of the Alpine fault zone and Fiordland region, New Zealand. *Geol. Soc. Am. Bull.* 84: 3297–316.

Scholz, C. H., Sykes, L. R., and Aggarwal, Y. P. 1973. Earthquake prediction: A physical basis. *Science* 181: 803–10.

Scholz, C. H., Wyss, M., and Smith, S. W. 1969. Seismic and aseismic slip on the San Andreas fault. *J. Geophys. Res.* 74: 2049–69.

Schulz, S. S., Mavko, G., Burford, R. O., and Stuart, W. D. 1982. Long-term fault creep observations in central California. *J. Geophys. Res.* 87: 6977–82.

Seeber, L. and Armbruster, J. 1981. The 1886 Charleston, South Carolina earthquake and the Appalachian detachment. *J. Geophys. Res.* 86: 7874–94.

Seeber, L., Armbruster, J., and Quittmeyer, R. 1981. Seismicity and continental subduction in the Himalayan arc. In *Zagros, Hindu Kush, Himalaya, Geodynamic Evolution. Geodynam. Ser.*, vol. 3, ed. H. Gupta and F. Delaney, Washington, D.C.: American Geophysical Union, pp. 215–42.

Sharp, R. V. 1981. Variable rates of late Quaternary strike slip on the San Jacinto fault zone, southern California. *J. Geophys. Res.* 86: 1754–62.

Simpson, D. W. 1986. Triggered earthquakes. *Ann. Rev. Earth Planet. Sci.* 14: 21–42.

Simpson, D. W., Leith, W. S., and Scholz, C. H. 1988. Two types of reservoir induced seismicity. *Bull. Seismol. Soc. Am.* 78: 2025–40.

Simpson, D. W. and Negmatullaev, S. Kh. 1981. Induced seismicity at Nurek reservoir, Tadjikistan, USSR. *Bull. Seismol. Soc. Am.* 71: 1561–86.

Smith, R. B. and Bruhn, R. L. 1984. Intraplate extensional tectonics of the eastern Basin–Range: Inferences on structural style from seismic reflection data, regional tectonics, and thermal–mechanical models of brittle–ductile deformation. *J. Geophys. Res.* 89: 5733–62.

Smith, S. W. and Wyss, M. 1968. Displacement on the San Andreas fault initiated by the 1966 Parkfield earthquake. *Bull. Seismol. Soc. Am.* 58: 1955–74.

Snow, D. T. 1972. Geodynamics of seismic reservoirs. In *Proc. Symp. Percolation through Fissured Rocks.* Stuttgart: Ges. Erd- und Grundbau, T2-J: 1–19.

Solomon, S. C., Huang, P. Y., and Meinke, L. 1988. The seismic moment budget of slowly spreading ridges. *Nature* 334: 58–60.

Somerville, P. G., McLaren, J. P., LeFevre, L. V., Burger, R. W., and Helmberger, D. V. 1987. Comparison of source scaling relations of eastern and western North American earthquakes. *Bull. Seismol. Soc. Am.* 77: 322–46.

Spottiswoode, S. M. 1984. Seismic deformation around Blyvooruitzicht Gold Mine. In *Proc. 1st Int. Cong. Rockbursts and Seismicity in Mines*, ed. N. C. Gay and E. H. Wainwright. South African Inst. Min. Met.: Johannesburg, pp. 29–37.

Spottiswoode, S. M. and McGarr, A. 1975. Source parameters of tremors in a deep-level gold mine. *Bull. Seismol. Soc. Am.* 65: 93–112.

Stein, R. S. and King, G. C. 1984. Seismic potential revealed by surface folding: The 1983 Coalinga, California earthquake. *Science* 224: 869–71.

Steinbrugge, K. V., Zacher, E. G., Tocher, D., Whitten, C. A., and Claire, C. N. 1960. Creep on the San Andreas fault. *Bull. Seismol. Soc. Am.* 50: 389–415.

Stock, J. and Molnar, P. 1987. Revised history of early Tertiary plate motion in the southwest Pacific. *Nature* 325: 495–9.

Stuart, W. D. 1988. Forecast model for great earthquakes at the Nankai trough, southwest Japan. *Pageoph* 126: 619–42.

Suggate, R. P. 1963. The Alpine fault. *Trans. Roy. Soc. N.Z., Geology* 2: 105–29.

Sykes, L. R. 1967. Mechanism of earthquakes and nature of faulting on the mid-oceanic ridges. *J. Geophys. Res.* 72: 2131.

Sykes, L. R. 1970a. Earthquake swarms and sea-floor spreading. *J. Geophys. Res.* 75: 6598–611.

Sykes, L. R. 1970b. Seismicity of the Indian Ocean and a possible nascent island arc between Ceylon and Australia. *J. Geophys. Res.* 75: 5041–55.

Sykes, L. R. 1978. Intra-plate seismicity, reactivation of pre-existing zones of weakness, alkaline magmatism, and other tectonics post-dating continental separation. *Rev. Geophys. Space Phys.* 16: 621–88.

Sykes, L. R. and Quittmeyer, R. C. 1981. Repeat times of great earthquakes along simple plate boundaries. In *Earthquake Prediction, an International Review. M. Ewing Ser. 4*, ed. D. Simpson and P. Richards. Washington, D.C.: American Geophysical Union, pp. 217–47.

Sykes, L. R. and Sbar, M. L. 1973. Intraplate earthquakes, lithospheric stresses and the driving mechanism of plate tectonics. *Nature* 245: 298–302.

Tajima, F. and Kanamori, H. 1985. Global survey of aftershock area expansion patterns. *Phys. Earth Planet. Int.* 40: 77–134.

Talwani, P. and Acree, S. 1985. Pore-pressure diffusion and the mechanism of reservoir-induced seismicity. *Pageoph* 122: 947–65.

Thatcher, W. and Lisowski, M. 1987. Long-term seismic potential of the San Andreas fault southeast of San Francisco, California. *J. Geophys. Res.* 92: 4771–84.

Tocher, D. 1960. Creep rate and related measurements at Vineyard, California. *Bull. Seismol. Soc. Am.* 50: 396–404.

Toomey, D. R., Solomon, S. C., Purdy, G. M., and Murray, M. H. 1985. Microearthquakes beneath the median valley of the Mid-Atlantic ridge near 23°N: Hypocenters and focal mechanisms. *J. Geophys. Res.* 90: 5443–58.

Toriumi, M. 1982. Strain, stress, and uplift. *Tectonics* 1: 57–72.

Toyoda, H. and Noma, Y. 1952. Study for underground condition of the Dogo hot spring area, Ehime Prefecture. *Mem. Ehime Univ., Sec. II (Science)* I: 139–46.

Trehu, A. M. and Solomon, S. C. 1983. Earthquakes in the Orozco Transform zone: Seismicity, source mechanisms, and earthquakes. *J. Geophys. Res.* 88: 8203–25.

Tse, S., Dmowska, R., and Rice, J. R. 1985. Stressing of locked patches along a creeping fault. *Bull. Seismol. Soc. Am.* 75: 709–36.

Tse, S. and Rice, J. R. 1986. Crustal earthquake instability in relation to the depth variation of frictionial slip processes. *J. Geophys. Res.* 91: 9452–72.

Uyeda, S. 1982. Subduction zones: An introduction to comparative subductology. *Tectonophysics* 81: 133–59.

Uyeda, S. and Kanamori, H. 1979. Back-arc opening and the mode of subduction. *J. Geophys. Res.* 84: 1049–61.

Vail, J. R. 1967. The southern extension of the East Africa rift system and related igneous activity. *Geol. Runds.* 57: 601–14.

Vanigen, R. and Beanland, S. Rates of motion of the Hope fault. *N.Z. J. Geol. Geophys.*, in press.

Versfelt, J. and Rosendahl, B. R. 1989. Relationships between pre-rift structure and rift architecture in Lakes Tanganyika and Malawi, East Africa. *Nature* 337: 354–7.

Wadati, K. 1928. Shallow and deep earthquakes. *Geophys. Mag.* 1: 161–202.

Walcott, R. I. 1978. Present tectonics and Late Cenozoic evolution of New Zealand. *Geophys. J. R.A.S.* 52: 137–64.

Walcott, R. I. 1984. The kinematics of the plate boundary zone through New Zealand: A comparison of long and short term deformation. *Geophys. J. R.A.S.* 79: 613–33.

Walcott, R. I. 1987. Geodetic strain and the deformational history of the North Island of New Zealand during the late Cainozoic. *Phil. Trans. Roy. Soc. London, Ser. A* 321: 163–81.

Watanabe, T., Langseth, M., and Anderson, R. N. 1977. Heat flow in back-arc basins of the western Pacific. In *Island Arcs, Deep Sea Trenches and Back-arc Basins. M. Ewing Ser. 4*, ed. M. Talwani and W. C. Pittman, III. Washington, D.C.: American Geophysical Union, pp. 137–61.

Weldon, R. J. and Humphreys, G. 1986. A kinematic model of southern California. *Tectonics* 5: 33–48.

Weldon, R. J. and Sieh, K. E. 1985. Holocene rate of slip and tentative recurrence interval for large earthquakes on the San Andreas fault, Cajon Pass, southern California. *Geol. Soc. Am. Bull.* 96: 793–812.

Wellman, H. R. 1955. New Zealand Quaternary tectonics. *Geol. Runds.* 43: 248–57.

Wellman, H. W. and Wilson, A. T. 1964. Notes on the geology and archeology of the Martins Bay District. *N.Z. J. Geol. Geophys.* 7: 702–21.

Wesson, R. L. 1987. Modelling aftershock migration and afterslip of the San Juan Bautista, California, earthquake of October 3, 1972. *Tectonophysics* 144: 215–29.

Wesson, R. L. and Ellsworth, W. L. 1973. Seismicity preceding moderate earthquakes in California. *J. Geophys. Res.* 78: 8527–46.

Wesnousky, S. G. 1986. Earthquakes, Quaternary faults, and seismic hazard in California. *J. Geophys. Res.* 91: 12587–631.

Wesnousky, S. G. and Scholz, C. H. 1980. The craton: Its effect on the distribution of seismicity and stress in North America. *Earth Planet. Sci. Lett.* 48: 348–55.

Wesnousky, S. G., Jones, L., Scholz, C. H., and Deng, Q. 1984. Historical seismicity and rates of crustal deformation along the margin of the Ordos block, north China. *Bull. Seismol. Soc. Am.* 74: 1767–83.

Wesnousky, S. G., Scholz, C. H. and Shimazaki, K. 1982. Deformation of an island arc: Rates of moment release and crustal shortening in intraplate Japan determined from seismicity and Quaternary fault data. *J. Geophys. Res.* 87: 6829–52.

Wiens, D. and Stein, S. 1983. Age dependence of oceanic intraplate seismicity and stresses in young oceanic lithosphere. *J. Geophys. Res.* 88: 6455–68.

Wiens, D., Stein, S., DeMets, C., Gordon, R., and Stein, C. 1986. Plate tectonic models for Indian Ocean "intraplate" deformation. *Tectonophysics* 132: 37–48.

Wilson, J. T. 1965. A new class of faults and their bearing on continental drift. *Nature* 207: 343–7.

Yeats, R. S. 1986. Faults related to folding with examples from New Zealand. *R. Soc. N.Z. Bull.* 24: 273–92.

Yeats, R. S. and Berryman, K. R. 1987. South Island, New Zealand, and Transverse Ranges, California: A seismotectonic comparison. *Tectonics* 6: 363–76.

Yoshii, T. 1979. A detailed cross-section of the deep seismic zone beneath northeastern Honshu, Japan. *Tectonophysics* 55: 349–60.

Zhao, Y. and Wong, T.-F. 1990. Effects of load point velocity on frictional instability behavior. *Tectonophysics*, 175, 177–95.

Zoback, M. L. and Zoback, M. D. 1980. State of stress in the conterminous United States. *J. Geophys. Res.* 85: 6113–56.

Chapter 7

Aggarwal, Y. P., Sykes, L. R., Simpson, D. W., and Richards, P. G. 1973. Spatial and temporal variations of t_s/t_p and in P wave residuals at Blue Mountain Lake, New York: Application to earthquake prediction. *J. Geophys. Res.* 80: 718–32.

Aki, K. 1985. Theory of earthquake prediction with special references to monitoring of the quality factor of lithosphere by coda method. *Earthquake Pred. Res.* 3: 219–30.

Bakun, W. and Lindh, A. 1985. The Parkfield, California, earthquake prediction experiment. *Science* 229: 619–24.

Bilham, R. G. and Beavan, J. 1978. Tilts and strains on crustal blocks. *Tectonophysics* 52: 121–38.

Bowman, J., Jones, T., Gibson, G., Corke, A., Thompson, R., and Camacho, A. 1988. Tennant Creek earthquakes of 22 January 1988: Reactivation of a fault zone in the Proterozoic Australian Shield. *Eos* 69: 400.

Cao, T. and Aki, K. 1985. Seismicity simulation with a mass–spring model and a displacement hardening–softening friction law. *Pageoph* 122: 10–24.

Castle, R., Elliot, M., Church, J., and Wood, S. 1984. *The Evolution of the Southern California Uplift, 1955 through 1976. U.S. Geol. Surv. Prof. Paper 1342.*

Cifuentes, I. L. and Silver, P. G. 1989. Low-frequency source characteristics of the great 1960 Chilean earthquake. *J. Geophys. Res.* 643–65.

Crampin, S. 1987. Geological and industrial applications of extensive-dilatancy anisotropy. *Nature* 328: 491–6.

Crampin, S., Evans, R., and Atkinson, B. K. 1984. Earthquake prediction: A new physical basis. *Geophys. J. R.A.S.* 76: 147–56.

Das, S. and Scholz, C. H. 1981. Theory of time dependent rupture in the earth. *J. Geophys. Res.* 86: 6039–51.

Dieterich, J. H. 1986. A model for the nucleation of earthquake slip. In *Earthquake Source Mechanics. AGU Geophys. Mono. 37*, ed. S. Das, J. Boatwright, and C. Scholz. Washington, D.C.: American Geophysical Union, pp. 37–47.

Eaton, J., Cockerham, R., and Lester, F. 1983. Study of the May 2, 1983 Coalinga earthquake and its aftershocks, based on the U.S.G.S. seismic network in northern California. In *The 1983 Coalinga, California, Earthquakes. Spec. Pub. 66*, ed. J. Bennet and R. Sherburne. Sacramento: California Department of Conservation, Division of Mines.

Evison, F. 1977. Fluctuations of seismicity before major earthquakes. *Nature* 266: 710–12.

Fedotov, S. A. 1965. Regularities of the distribution of strong earthquakes in Kamchatka, the Kurile Islands, and northeast Japan. *Tr. Inst. Fiz. Zemli, Akad. Nauk SSSR* 36: 66–93.

Gilbert, G. K. 1909. Earthquake forecasts. *Science* XXIX: 121–38.

Gu, J. C. 1984. Frictional resistance to accelerating slip. *Pageoph* 122: 662–79.

Habermann, R. E. 1981. Precursory seismicity patterns: Stalking the mature seismic gap. In *Earthquake Prediction, an International Review. M. Ewing Ser. 4*, ed. D. Simpson and P. G. Richards. Washington, D.C.: American Geophysical Union, pp. 29–42.

Habermann, R. E. 1988. Precursory seismic quiescence: Past, present, and future. *Pageoph* 126: 277–318.

Hadley, K. 1973. Laboratory investigation of dilatancy and motion of fault surfaces at low confining pressures. In *Proc. Conf. Tect. Problems San Andreas Fault System. Publ. Geol. Sci. vol. XIII*, ed. R. Kovach and A. Nur. Stanford, Calif.: Stanford Univ., pp. 427–35.

Hagiwara, Y. 1974. Probability of earthquake occurrence as obtained from a Weibull distribution analysis of crustal strain. *Tectonophysics* 23: 313–18.

Hanks, T. C. and McGuire, R. 1981. The character of high-frequency strong ground motion. *Bull. Seismol. Soc. Am.* 71: 2071–95.

Imamura, A. 1937. *Theoretical and Applied Seismology*. Tokyo: Maruzen.

Ishibashi, K. 1988. Two categories of earthquake precursors, physical and tectonic, and their role in intermediate-term earthquake prediction. *Pageoph* 126: 687–700.

Jones, L. M. and Molnar, P. 1979. Some characteristics of foreshocks and their possible relationship to earthquake prediction and premonitory slip on faults. *J. Geophys. Res.* 84: 3596–608.

Kagan, Y. and Knopoff, L. 1978. Statistical study of the occurrence of shallow earthquakes. *Geophys. J. R.A.S.* 55: 67–86.

Kanamori, H. 1973. Mode of strain release associated with major earthquakes in Japan. *Ann. Rev. Earth Planet. Sci.* 5: 129–39.

Kanamori, H. 1981. The nature of seismicity patterns before large earthquakes. In *Earthquake Prediction, an International Review. M. Ewing Ser. 4*, ed. D. Simpson and P. Richards. Washington, D.C.: American Geophysical Union, pp. 1–19.

Kanamori, H. and Cipar, J. 1974. Focal process of the great Chilean earthquake May 22, 1960. *Phys. Earth Planet. Int.* 9: 128–36.

Kanamori, H. and Fuis, G. 1976. Variations of P-wave velocity before and after the Galway Lake earthquake ($M_L = 5.2$) and the Goat Mountain earthquakes ($M_L = 4.7, 4.7$), 1975, in the Mojave Desert, California. *Bull. Seismol. Soc. Am.* 66: 2017–38.

Keilis-Borok, V. I., Knopoff, L., Rotwain, I. M., and Allen, C. R. 1988. Intermediate-term prediction of occurrence times of strong earthquakes. *Nature* 335: 690–4.

Kisslinger, C. 1975. Processes during the Matsushiro earthquake swarm as revealed by leveling, gravity, and spring-flow observations. *Geology* 3: 57–62.

Li, V. C. and Rice, J. R. 1983a. Preseismic rupture progression and great earthquake instabilities at plate boundaries. *J. Geophys. Res.* 88: 4231–46.

Li, V. C. and Rice, J. R. 1983b. Precursory surface deformation in great plate boundary earthquake sequences. *Bull. Seismol. Soc. Am.* 73: 1415–34.

Linde, A., Suyehiro, K., Miura, S., Sacks, I., and Takagi, A. 1988. Episodic aseismic earthquake precursors. *Nature* 334: 513–15.

Lindh, A. G. 1983. Preliminary assessment of long-term probabilities for large earthquakes along selected fault segments of the San Andreas fault system in California. *U.S. Geol. Surv. Open-file Rept. 83-63*.

Matsuda, T. 1977. Estimation of future destructive earthquakes from active faults in Japan. *J. Phys. Earth, Suppl.* 25: 795–855.

McEvilly, T. V. and Johnson, L. R. 1974. Stability of p and s velocities from central California quarry blasts. *Bull. Seismol. Soc. Am.* 64: 343–53.

McNally, K. 1981. Plate subduction and prediction of earthquakes along the Middle America trench. In *Earthquake Prediction, an International Review. M. Ewing Ser. 4*, ed. D. Simpson and P. Richards. Washington, D.C.: American Geophysical Union, pp. 63–71.

Mjachkin, V., Brace, W., Sobolev, G., and Dieterich, J. 1975. Two models of earthquake forerunners. *Pure Appl. Geophys.* 113: 169–81.

Mogi, K. 1977. Seismic activity and earthquake prediction. In *Proc. Earthquake Prediction Symposium*. Tokyo, pp. 203–14.

Mogi, K. 1981. Seismicity in western Japan and long-term earthquake forecasting. In *Earthquake Prediction, an International Review. M. Ewing Ser. 4*, ed. D. Simpson and P. Richards. Washington, D.C.: American Geophysical Union, pp. 43–52.

Mogi, K. 1982. Temporal variation of the precursory crustal deformation just prior to the 1944 Tonankai earthquake. *J. Seismol. Soc. Japan (2)* 35: 145–8.

Mogi, K. 1985. *Earthquake Prediction*. Tokyo: Academic Press.

Musha, K. 1943. *Zotei Dainihon Jisin Shiryo*, vol. 3. Tokyo: Shinsai Yobo Hyogi Kai, pp. 142–9.

Nishenko, S. P. 1985. Seismic potential for large and great interplate earthquakes along the Chilean and southern Peruvian margins of South America: A quantitative reappraisal. *J. Geophys. Res.* 90: 3589–615.

Nishenko, S. P. and Buland, R. 1987. A generic recurrence interval distribution for earthquake forecasting. *Bull. Seismol. Soc. Am.* 77: 1382–99.

Nur, A. 1972. Dilatancy, pore fluids, and premonitory variations in t_s/t_p travel times. *Bull. Seismol. Soc. Am.* 62: 1217–22.

Nur, A. 1974. Matsushiro, Japan, earthquake swarm: Confirmation of the dilatancy–fluid diffusion model. *Geology* 2: 217–22.

O'Connell, R. and Budiansky, B. 1974. Seismic velocities in dry and saturated cracked solids. *J. Geophys. Res.* 79: 5412–26.

Ohtake, M., Matumoto, T., and Latham, G. 1977. Seismicity gap near Oaxaca, southern Mexico as a probable precursor to a large earthquake. *Pageoph* 115: 375–85.

Ohtake, M., Matumoto, T., and Latham, G. 1981. Evaluation of the forecast of the 1978 Oaxaca, southern Mexico earthquake based on a precursory seismic quiescence. In *Earthquake Prediction, an International Review. M. Ewing Ser. 4*, ed. D. Simpson and P. Richards. Washington, D.C.: American Geophysical Union, pp. 53–62.

Ota, Y., Matsuda, T., and Naganuma, K. 1976. Tilted marine terraces of the Ogi Peninsula, Sado Island, related to the Ogi earthquake of 1802. *Zisin II* 29: 55–70.

Papazachos, B. 1975. Foreshocks and earthquake prediction. *Tectonophysics* 28: 213–26.

Perez, O. and Scholz, C. H. 1984. Heterogeneities of the instrumental seismicity catalog (1904–1980) for strong shallow earthquakes. *Bull. Seismol. Soc. Am.* 74: 669–86.

Rabotnov, Y. N. 1969. *Creep Problems in Structural Members.* Amsterdam: North-Holland.

Raleigh, C. B., Bennet, G., Craig, H., Hanks, T., Molnar, P., Nur, A., Savage, J., Scholz, C., Turner, R., and Wu, F. 1977. Prediction of the Haicheng earthquake. *Eos* 58: 236–72.

Ramsay, J. G. 1983. The crack–seal mechanism of rock deformation. *Nature* 284: 135–9.

Research Group for Active Faults in Japan. 1980. *Active Faults in Japan: Sheet Maps and Inventories* Tokyo: University of Tokyo Press.

Rice, J. R. 1983. Constitutive relations for fault slip and earthquake instabilities. *Pageoph* 121: 443–75.

Rice, J. R. and Rudnicki, J. W. 1979. Earthquake precursory effects due to pore fluid stabilization of a weakened fault zone. *J. Geophys. Res.* 84: 2177–84.

Rikitake, T. 1976. *Earthquake Prediction.* Amsterdam: Elsevier.

Rikitake, T. 1982. *Earthquake Forecasting and Warning.* Tokyo: Center for Academic Publications Japan, D. Reidel Publishing Co.

Roeloffs, E. 1988. Hydrological precursors to earthquakes: A review. *Pageoph* 126: 177–209.

Rudnicki, J. W. 1988. Physical models of earthquake instability and precursory processes. *Pageoph* 126: 531–54.

Sato, H. 1988. Temporal change in scattering and attenuation associated with the earthquake occurrence – a review of recent studies on coda waves. *Pageoph* 126: 465–98.

Scholz, C. H. 1968. The frequency–magnitude relation of microfracturing in rock and its relation to earthquakes. *Bull. Seismol. Soc. Am.* 58: 399–415.

Scholz, C. H. 1974. Post-earthquake dilatancy recovery. *Geology* 2: 551–4.

Scholz, C. H. 1978. Velocity anomalies in dilatant rock. *Science* 201: 441–2.

Scholz, C. H. 1988. Mechanisms of seismic quiescences. *Pageoph* 126: 701–18.

Scholz, C. H., Sykes, L. R., and Aggarwal, Y. P. 1973. Earthquake prediction: A physical basis. *Science* 181: 803–9.

Semenov, A. N. 1969. Variations of the travel time of transverse and longitudinal waves before violent earthquakes. *Izv. Acad. Sci. USSR, Phys. Solid Earth* (Eng. trans.) 3: 245–58.

Sibson, R. H. 1987. Earthquake rupturing as a mineralizing agent in hydrothermal systems. *Geology* 15: 701–4.

Stuart, W. D. 1979. Strain softening prior to two-dimensional strike-slip earthquake. *J. Geophys. Res.* 84: 1063–70.

Stuart, W. D. 1988. Forecast model for great earthquakes at the Nankai trough subduction zone. *Pageoph* 126: 619–42.

Stuart, W. D. and Aki, K. (eds.) 1988. Intermediate-term earthquake prediction. *Pageoph* 126(2–4): 175–718.

Stuart, W. D. and Mavko, G. M. 1979. Earthquake instability on a strike-slip fault. *J. Geophys. Res.* 84: 2153–60.

Suyehiro, S., Asada, T., and Ohtake, M. 1964. Foreshocks and aftershocks accompanying a perceptible earthquake in central Japan – on a peculiar nature of foreshocks. *Papers Meteorol. Geophys.* 15: 71–88.

Sykes, L. R. and Nishenko, S. P. 1984. Probabilities of occurrence of large plate rupturing earthquakes for the San Andreas, San Jacinto, and Imperial faults, California. 1983–2003. *J. Geophys. Res.* 89: 5905–27.

Thomas, D. 1988. Geochemical precursors to seismic activity. *Pageoph* 126: 241–67.

Tributsch, H. 1983. *When the Snakes Awake*. Cambridge, Mass.: MIT Press.

Tse, S. T. and Rice, J. R. 1986. Crustal earthquake instability in relation to the depth variation of frictional slip properties. *J. Geophys. Res.* 91: 9452–72.

Tsumura, K., Karakama, I., Ogino, I., and Takahashi, M. 1978. Seismic activities before and after the Izu–Oshima-kinkai earthquake of 1978. *Bull. Earthquake Res. Inst., Univ. Tokyo* 53: 309–15.

Ulomov, V. I. and Mavashev, B. Z. 1971. *The Tashkent Earthquake of 26 April*. Tashkent: Akad. Nauk Uzbek. SSR, FAN.

Usami, T. 1987. *Descriptive Catalogue of Damaging Earthquakes in Japan* (rev. ed.). Tokyo: University of Tokyo Press.

Voight, B. 1989. A relation to describe rate-dependent material failure. *Science* 243: 200–3.

Wakita, H. 1988. Short-term and intermediate-term geochemical precursors. *Pageoph* 126: 267–78.

Wallace, R. E., Davis, J., and McNally, K. 1984. Terms for expressing earthquake potential, prediction, and probability. *Bull. Seismol. Soc. Am.* 74: 1819–25.

Wesnousky, S. 1986. Earthquakes, Quaternary faults, and seismic hazard in California. *J. Geophys. Res.* 91: 12587–631.

Wesnousky, S., Scholz, C., and Shimazaki, K. 1982. Deformation of an island arc: Rates of moment release and crustal shortening in intraplate Japan determined from seismicity and Quaternary fault data. *J. Geophys. Res.* 87: 6829–52.

Wesnousky, S., Scholz, C., Shimazaki, K., and Matsuda, T. 1983. Earthquake frequency distribution and the mechanics of faulting. *J. Geophys. Res.* 88: 9331–40.

Wesnousky, S., Scholz, C., Shimazaki, K., and Matsuda, T. 1984. Integration of geological and seismological data for the analysis of seismic hazard: A case study of Japan. *Bull. Seismol. Soc. Am.* 74: 687–708.

Whitcomb, J. H., Garmony, J. D., and Anderson, D. L. 1973. Earthquake prediction: Variation of seismic velocities before the San Fernando earthquake. *Science* 180: 632–41.

Working Group on California Earthquake Probabilities. 1988. Probabilities of large earthquake occurring in California on the San Andreas fault. *U.S. Geol. Surv. Open-file Rept. 88-398.*

Wyss, M. and Habermann, R. E. 1988. Precursory seismic quiescence. *Pageoph* 126: 319–32.

Wyss, M., Klein, F., and Johnston, A. 1981. Precursors of the Kalapana $M = 7.2$ earthquake. *J. Geophys. Res.* 86: 3881–900.

Index